LEAN SYSTEMS

Applications and Case Studies in Manufacturing, Service, and Healthcare

LEAN SYSTEMS

Applications and Case Studies in Manufacturing, Service, and Healthcare

Elizabeth A. Cudney
Sandra L. Furterer
David M. Dietrich

CRC Press
Taylor & Francis Group
Boca Raton London New York

CRC Press is an imprint of the
Taylor & Francis Group, an **informa** business

CRC Press
Taylor & Francis Group
6000 Broken Sound Parkway NW, Suite 300
Boca Raton, FL 33487-2742

Version Date: 20130516

International Standard Book Number-13: 978-1-4665-5680-5 (Hardback)

Library of Congress Cataloging-in-Publication Data

Lean systems : applications and case studies in manufacturing, service, and healthcare / editors, Elizabeth A. Cudney, Sandra L. Furterer, David M. Dietrich.
pages cm
Includes bibliographical references and index.
ISBN 978-1-4665-5680-5 (hardback)
1. Industrial efficiency--Case studies. 2. Lean manufacturing--Case studies. I. Cudney, Elizabeth A.

T58.8.L35 2013
658.4'013--dc23 2013017595

Visit the Taylor & Francis Web site at
http://www.taylorandfrancis.com

and the CRC Press Web site at
http://www.crcpress.com

This book is dedicated to my husband, Brian Cudney. I am blessed to be married to an amazing, supportive, and loving husband and father. Every day, I am reminded by his words and actions that I am truly lucky to spend my life with him.

Elizabeth A. Cudney

I would like to dedicate this book to my husband Dan and three wonderful children, Kelly, Erik, and Zach.

Sandra L. Furterer

I would like to dedicate this effort to my lovely doctor wife and a specific Japanese video game manufacturer who provided hours of entertainment for my children, thereby allowing me to write the content herein.

David M. Dietrich

Contents

Section III Manufacturing, Service, and Healthcare Case Studies

Preface

This book grew out of the need for our students to better understand how to apply and integrate the Lean philosophy and its associated tools and techniques. Real-world examples and hands-on experience are invaluable resources when instructing the use of methods and tools in training or in a classroom. The instructor may not have access to these resources; thus, he or she can teach only theory and basic examples. Another solution is the use of case studies. Case studies can help enhance the learning experience by allowing the learner a role in a real scenario. The story of the case study adds life to a seemingly lifeless group of tools. With this understanding, Lean methods taught with the aid of case studies may help some novices and even experienced participants to better assimilate the tools, since they are presented as a whole.

The primary objective of this book is to provide in-depth descriptions of the various Lean tools and case studies from real-world Lean projects performed for a wide variety of manufacturing, service, and healthcare processes. The purpose is to facilitate Lean instruction by providing interactive case studies that will enable the learners to apply Lean techniques.

This book would not be possible without the enthusiasm, dedication, and energy that our Lean students and team members exhibit. Our goal as authors and editors is to provide the learners with an understanding of how others applied Lean and a guide for how they might solve their organizations' problems by applying Lean.

Additional material is available from the CRC web site: http://www.crcpress.com/product/isbn/9781466556805

Acknowledgments

Our thanks and appreciation goes to all of the Lean team members, project champions, and mentors who worked so diligently and courageously on these projects.

Special thanks go to several people for their contributions to the development and production of the book at the CRC Press/Taylor & Francis, including Cindy Renee Carelli (senior editor), Amy Blalock (supervisor, editorial project development), and Joette Lynch (project editor).

Editors

Elizabeth A. Cudney, PhD, is an a associate professor in the Engineering Management and Systems Engineering Department at the Missouri University of Science and Technology, Missouri. She earned her bachelor of science in industrial engineering from North Carolina State University, Raleigh, North Carolina. She earned her master of engineering in mechanical engineering with a manufacturing specialization and master of business administration from the University of Hartford, West Hartford, Connecticut, and her doctorate in engineering management from the University of Missouri—Rolla. Her doctoral research focused on pattern recognition and developed a methodology for prediction in multivariate analysis. Dr. Cudney's research was recognized with the 2007 American Society of Engineering Management (ASEM) Outstanding Dissertation Award.

Prior to returning to school for her doctorate, she worked in the automotive industry in various roles including Six Sigma Black Belt, quality/process engineer, quality auditor, senior manufacturing engineer, and manufacturing manager. She is an American Society for Quality (ASQ)-certified Six Sigma Black Belt, certified quality engineer, manager of quality/operational excellence, certified quality inspector, certified quality improvement associate, certified quality technician, and certified quality process analyst. She is a past president of the Rotary Club of Rolla, Missouri. She is also a member of the Japan Quality Engineering Society, American Society for Engineering Education, ASEM, American Society of Mechanical Engineers, ASQ, Institute of Industrial Engineers, and Society of Automotive Engineers.

In 2010, Dr. Cudney was inducted as an associate member into the ASQ's International Academy for Quality. In addition, she received the 2007 ASQ Armand V. Feigenbaum Medal. This international award is given annually to one individual "who has displayed outstanding characteristics of leadership, professionalism, and potential in the field of quality and also whose work has been or, will become, of distinct benefit to mankind." In addition, she received the 2006 Society of Manufacturing Engineers Outstanding Young Manufacturing Engineer Award. This international award is given annually to engineers "who have made exceptional contributions and accomplishments in the manufacturing industry."

Dr. Cudney has published more than 50 conference papers and 30 journal papers. Her first book entitled *Using Hoshin Kanri to Improve the Value Stream* was released in March 2009 through Productivity Press, a division of Taylor & Francis. Her second book entitled *Implementing Lean Six Sigma throughout the Supply Chain: The Comprehensive and Transparent Case Study* was released in November 2010 through Productivity Press, a division of Taylor & Francis.

Her third book entitled *Design for Six Sigma in Product and Service Development: Applications and Case Studies* was released in June 2012 through CRC Press.

Sandy L. Furterer, PhD, is the vice president of process engineering at a financial services firm in the mortgage banking area. She is also currently an adjunct faculty member in the master of science in quality assurance program at Southern Polytechnic State University, Marietta, Georgia. She formerly developed and deployed the enterprise performance excellence program at a hospital system in Florida. She was a business architect in the Information Systems Division, Business and Process Architecture team, with Walmart Stores, Inc., in Bentonville, Arkansas. She led the Business and Process Architecture team in the retail systems application development area, implementing best practices to achieve operational excellence in information systems application development processes and helping to enable the business through application of business architecture and process improvement. She was also a master Black Belt in Walmart's global continuous improvement practice.

Dr. Furterer earned her PhD in industrial engineering with a specialization in quality engineering from the University of Central Florida, in Orlando, Florida, in 2004. She developed a state-of-the-art framework and road map for integrating Lean and Six Sigma for service industries and implemented the framework in a local government's financial services department. She earned a master of business administration from the Xavier University in Cincinnati, Ohio, and a bachelor and master of science in industrial and systems engineering from The Ohio State University in Columbus, Ohio.

Dr. Furterer was an assistant professor in the industrial distribution and logistics program at East Carolina University, Greenville, North Carolina, from 2006 to 2007. She was a visiting assistant professor and assistant department chair in the Industrial Engineering and Management Systems Department at the University of Central Florida from 2004 to 2006. She is also a graduate scholar at the University of Central Florida.

Dr. Furterer has more than 20 years of experience in business process and quality improvements. She is an American Society for Quality (ASQ)-certified Six Sigma Black Belt, certified quality engineer, and a certified Harrington Institute master Black Belt. She is also a senior member of the Institute of Industrial Engineers and a senior member of ASQ. She is an experienced consultant, who has facilitated and implemented quality, statistics, and process improvement projects, using Six Sigma and Lean principles and tools. She has helped Fortune 100 companies, local governments, and nonprofit organizations to streamline their processes and implement information systems.

Dr. Furterer has published and/or presented more than 20 conference papers/proceedings in the areas of Lean Six Sigma, quality, operational excellence, business architecture, and engineering education. She is the editor of *Lean Six Sigma in Service: Applications and Case Studies*, CRC Press, 2009, and coeditor with Dr. Cudney of *Design for Six Sigma in Product and Service Development: Applications and Case Studies*, CRC Press, 2012.

David M. Dietrich, PhD, is a materials process and physics engineer/project manager for a major airframe aerospace company. In this role, he researches emerging technologies for use in aerospace production. He is also currently an adjunct faculty member in the Engineering Management and Systems Engineering department at Missouri University of Science and Technology. He has more than 12 years of manufacturing industry experience that includes both production and research with expertise in the areas of additive manufacturing, quality engineering, Six Sigma, Lean production, and engineering graphics.

Dr. Dietrich earned his PhD in engineering management from Missouri University of Science and Technology in 2010. His dissertation research aligned with the development of emerging technology supply chains for aerospace deployment. He also holds a master of manufacturing engineering from Missouri University of Science and Technology and a master of business administration from Maryville University of St. Louis, Missouri. In addition, he holds a bachelor of science from Murray State University. In 2011, he was awarded the Outstanding Young Manufacturing Engineer distinction from the Society of Manufacturing Engineers. In addition, he has presented at 11 conferences in the field of additive manufacturing and has published several papers on the subject.

Contributors

Mujahid Abjul
Missouri University of Science and Technology
Rolla, Missouri

Shrey Arora
Missouri University of Science and Technology
Rolla, Missouri

Charlie Barclay
Missouri University of Science and Technology
Rolla, Missouri

Marcela Bernardinez
Industrial Engineering and Management Systems
University of Central Florida
Orlando, Florida

Khalid Buradha
Industrial Engineering and Management Systems
University of Central Florida
Orlando, Florida

Kevin S. Cochie
U.S. Army
Washington, D.C.

Elizabeth A. Cudney
Engineering Management and Systems Engineering
Missouri University of Science and Technology
Rolla, Missouri

Kelly M. Davis
Medical Supplies Department
Covidien
Chesterfield, Missouri

Nanday K. Dey
Missouri University of Science and Technology
Rolla, Missouri

David M. Dietrich
Engineering Management and Systems Engineering
Missouri University of Science and Technology
Rolla, Missouri

Snehal Digraskar
Missouri University of Science and Technology
Rolla, Missouri

Rodney Ewing
Missouri University of Science and Technology
Rolla, Missouri

Sandra L. Furterer
Department of Industrial Engineering Technology
Southern Polytechnic State University
New Albany, Ohio

Amita Ghanekar
Missouri University of Science and Technology
Rolla, Missouri

Scott E. Grasman
Industrial and Systems Engineering Department
Rochester Institute of Technology
Rochester, New York

Seth Langston
Missouri University of Science and Technology
Rolla, Missouri

Corbin LeGrand
Missouri University of Science and Technology
Rolla, Missouri

Frank Liou
Mechanical Engineering
Missouri University of Science and Technology
Rolla, Missouri

Sneha Mahajan
Missouri University of Science and Technology
Rolla, Missouri

Lynda Melgarejo
Missouri University of Science and Technology
Rolla, Missouri

Sukhada Mishra
Missouri University of Science and Technology
Rolla, Missouri

Jason Park
Missouri University of Science and Technology
Rolla, Missouri

Neha Pawar
Missouri University of Science and Technology
Rolla, Missouri

Susan Polson
Missouri University of Science and Technology
Rolla, Missouri

Jose Saenz
CRM Program,
Dell, Inc.
Arraijan, Panama

Shirish Sreedharan
Manufacturing Engineering Department
Modine Thermal Systems Shanghai Ltd.
Songjiang, Shanghai, People's Republic of China

Raj Vemulapally
Missouri University of Science and Technology
Rolla, Missouri

Section I

Overview and Introduction to Lean

1

Instructional Strategies for Using This Book

Elizabeth A. Cudney, Sandra L. Furterer, and David M. Dietrich

CONTENTS

Introduction

The purpose of this book is to provide a guide for learners and practitioners of Lean methodologies and tools. The word *Lean* is a broad term that encompasses many different techniques. The term refers to more than just a set of techniques; it is a philosophy or attitude that promotes continual efforts to reduce and ideally eliminate waste in an organization. Lean emphasizes the elimination of waste and creation of flow within an enterprise. The primary focus of Lean is on the customer, to address value-added and non-value-added tasks. Value-added tasks are the only operations for which the customer is ready to pay. The idea in creating flow in Lean manufacturing is to deliver products and services just in time, in the right amounts, and at the right quality levels at the right place. This necessitates that products and services are produced and delivered only when a pull is exerted by the customer through a signal in the form of a purchase. A well-designed Lean system allows for an immediate and effective response to fluctuating customer demands and requirements. Lean tools that are most commonly used to eliminate waste and achieve flow are value stream mapping, standard work, 5-S housekeeping, single-minute exchange of dies (SMED), and visual management. These tools are reinforced with thought-provoking homework problems at the end of chapters to provide stimulating work for the student readers. In addition, the book provides several real-world case studies and applications of Lean tools that have shown significant improvement in meeting customer requirements and streamlining processes. The case studies include various settings in manufacturing, service, and healthcare. These cases can be used by both industry professionals and academics to learn how to apply Lean. The case studies will benefit the reader by showing them the tools and how to integrate them for process improvement. The case studies provide a detailed, step-by-step approach to Lean with clear direction from project infancy to completion.

The book is designed to engage the reader by enabling hands-on experience with real Lean project cases in a safe environment, where experienced Lean managers can help mentor students in the Lean methodologies. Case studies are designed to enable the student to work through the exercises and to provide sufficient background information so that they can apply the tools as if they collected data themselves. This will help prepare them to see actual data and make decisions when they embark on real-world projects.

Lean Project Backgrounds

The Lean case studies consist of manufacturing, service, and healthcare–oriented processes. An overview of each project is provided for the students so that they understand the background of the project as well as sufficient information regarding the processes that need to be improved. Data that were actually collected in the Lean projects are provided for application of Lean tools and analysis.

Lean Case Study Goals

To successfully complete the Lean case studies, participants must apply appropriate problem-solving methods and tools from the Lean toolkit to understand the problem, identify key customers and stakeholders, understand Critical to Satisfaction (CTS) characteristics, and develop potential process improvements and a plan to implement change.

Lean Tools

During the case study, the class will use the Lean tools that are most commonly used in real-world projects including value stream mapping, 5S and visual management, SMED, flow, pull, kanban, mistake proofing, standard work, systems thinking, theory of constraints, and hoshin kanri. These tools will be reinforced with stimulating homework suggestions for student practitioners.

Learning Design

Each case study is designed so that the teams of students experience the following:

- Team interaction, definition of team ground rules, brainstorming, and consensus building, as well as the stages of team growth
- Choosing how to apply Lean tools and problem-solving methods
- Supporting their decisions and application of the tools with data
- Reviewing information for relevant and irrelevant information and data and reframing into what is important to solve the problem
- Greater understanding of the application of specific tools and problem-solving methods
- Development of written and oral communication through customer interaction and written reports and presentations as well as the ability to present technical information
- Application of project management tools to manage activities and complete tasks in a timely manner
- Experience in solving an unstructured problem in a safe learning environment where mentoring is available

The instructor's role is to facilitate the learning process. It is critical for the instructor to act as a coach or mentor to the student teams. It can also be helpful to have Lean mentors experienced in applying Lean tools and methods assigned to each student team to mentor them in the application of Lean projects.

Most Lean programs involve working on projects in teams. Therefore, the instructor can organize the students into teams of four to six students depending on the class size. There is a great deal of value in having students work together as a team to work on a Lean project. They can learn how to work more effectively as a team, and team members can transfer learning across the team members because students grasp the difficult concepts of Lean at different paces. An effective way to organize the teams is to determine the students' experience and balance the team with a group ranging from no experience to extensive experience.

Required Knowledge Levels for Lean Projects

The Lean projects included in this book stem from different knowledge levels and depth of understanding for applying Lean tools and techniques. There are three different student levels defined as follows:

- Beginner—Undergraduate student (usually senior) student with no exposure to Lean and little process improvement background
- Intermediate—Master's student with some exposure (theoretical knowledge) to Lean tools and some process improvement background
- Advanced—Master's or PhD graduate students with theoretical learning of Lean tools and some process improvement background, as well as having worked on a Lean project

The book is divided into four parts to provide an introduction to Lean, case studies, and implementation strategies. Section I (Chapters 1 and 2) provides an overview and introduction to Lean. Section II (Chapters 3 through 10) provides detailed information on each of the Lean tools including step-by-step implementation. Section III (Chapters 11 through 21) provides in-depth manufacturing, service, and healthcare case studies. Section IV (Chapters 22 through 25) provides planning and implementation strategies for Lean initiatives.

Lean Tool Pyramid

The Lean tools pyramid (Figure 1.1) categorizes the Lean tools based on the knowledge level required to implement the tools: basic, intermediate, and advanced.

Figure 1.2 maps the Lean tools applied in each case study. This table can be used to select the case studies that match the desired Lean tool learning objectives.

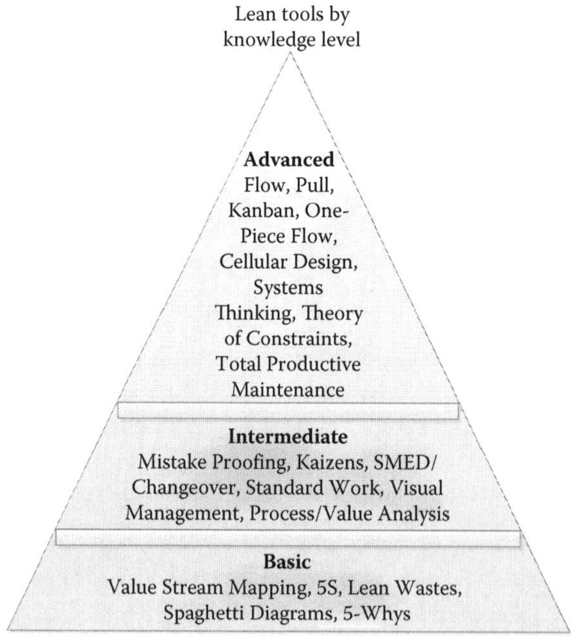

FIGURE 1.1
Lean pyramid.

	Chapter	11	12	13	14	15	16	17	18	19	20	21
Knowledge Level	Lean Tools	Lean Kitchen	Prototyping Lab	Tube Line	Fuel Pump Teardown	Women's Center	Lean Artwork	Motor Grader Assembly	High School Discipline	Financial Management	Pharmacy	Automotive Hoshin Kanri
Basic	Value stream mapping	X	X		X	X	X	X	X		X	X
	5S	X	X		X		X	X			X	X
	Lean wastes	X	X		X	X	X	X	X	X	X	X
	Spaghetti diagrams	X										
	5 Whys				X	X			X	X		
Intermediate	Mistake proofing			X			X	X		X	X	
	Kaizen					X	X		X	X	X	X
	SMED/changeover			X							X	
	Standard work	X		X			X	X		X		
	Visual management/ scorecards		X				X	X	X	X		X
	Process/value analysis					X		X	X	X	X	X
Advanced	Flow	X		X		X	X	X				X
	Pull			X				X				X
	One-piece flow			x				X		X	X	
	Cellular design			X						X		
	Systems thinking			X							X	X
	Theory of constraints											X
	Total productive maintenance			X							X	
	Hoshin kanri	X										

FIGURE 1.2
Lean tools and case study mapping.

2

Lean Six Sigma Roadmap Overview[*]

Sandra L. Furterer

CONTENTS

* Adapted from Furterer, S.L., *Lean Six Sigma in Service: Applications and Case Studies*, Boca Raton, FL, CRC Press, 2009.

Introduction

The purpose of this chapter is to provide a roadmap to applying the Lean tools and to provide a problem-solving methodology that incorporates the elements of the Six Sigma DMAIC (Define–Measure–Analyze–Improve–Control) approach. We will first provide an overview of Lean, Six Sigma, and the combined program of Lean Six Sigma, with a short history of its progression in applications extracted from the literature. We will then describe the steps for applying the Lean tools within the Six Sigma problem-solving framework.

Lean Six Sigma Overview

Lean Six Sigma is an approach focused on improving quality, reducing variation, and eliminating waste in an organization. It is the combination of two improvement programs, Six Sigma and Lean Enterprise. Six Sigma is a quality management philosophy and methodology that focuses on reducing variation, measuring defects (per million output/opportunities), and improving the quality of products, processes, and services. The concept of Six Sigma was developed in the early 1980s at Motorola Corporation. Six Sigma became popularized in the late 1990s by General Electric Corporation and their former CEO Jack Welch. Lean Enterprise is a methodology that focuses on reducing cycle time and waste in processes. Lean Enterprise originated from the Toyota Motor Corporation as the Toyota Production System (TPS) and increased in popularity after the 1973 energy crisis. The term "Lean Thinking" was coined by James P. Womack and Daniel T. Jones (1996) in their book *Lean Thinking*. The term "Lean Enterprise" is used to broaden the scope of a Lean program from manufacturing to embrace the enterprise or entire organization (Alukal, 2003).

Figure 2.1 shows the evolution to the combined methods of Lean and Six Sigma.

In the 1920s, at Western Electric, the concepts of control charts and statistical process control (SPC) were developed by Walter Shewhart. Dr. W. Edwards Deming installed SPC in Japanese manufacturing as he assisted Japan in their rebuilding efforts after World War II. Japan's successes in the 1970 have repopularized SPC in U.S. businesses. Total quality management (TQM) was a natural outgrowth of SPC adding a process improvement methodology. In the 1980s, business process reengineering (BPR) and TQM became popular. BPR encouraged completely throwing out the old process and starting over, many times within the context of implementing major information systems changes. TQM focused on a less structured approach with the principles of quality and process improvement. These methodologies evolved to Six Sigma.

On the productivity side, the Ford Production System was used to assemble cars, which was the basis for the TPS. Just-in-time production philosophies joined with TPS, which evolved into Lean. Now Lean and Six Sigma are merging to capitalize on the best of both improvement philosophies and methodologies.

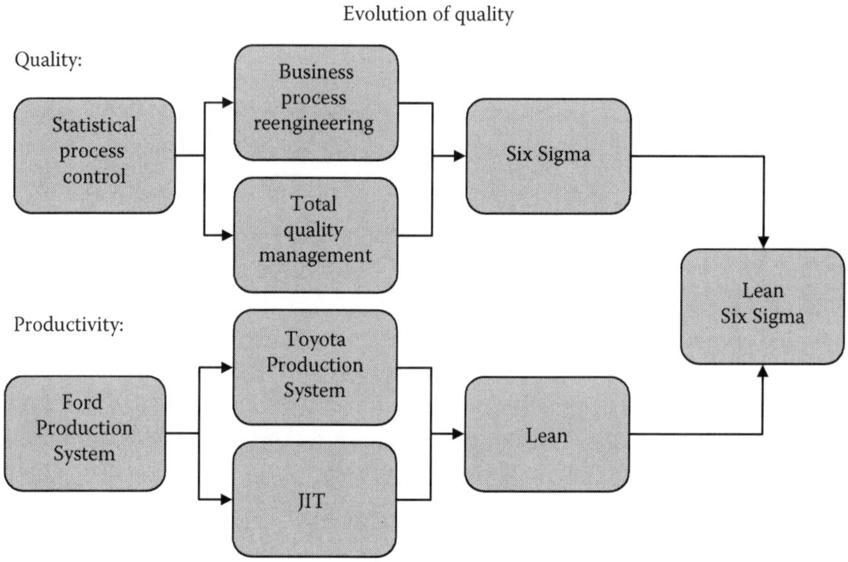

FIGURE 2.1
Evolution of quality and productivity to Lean Six Sigma. (Adapted from Furterer, S.L., ASQ Conference on Quality in the Space and Defense Industries, *Critical Quality Skills of Our Future Engineers,* Cape Canaveral, FL, 2006.)

Six Sigma uses the Define–Measure–Analyze–Improve–Control (DMAIC) problem-solving approach and a wide array of quality problem-solving tools. There are many powerful tools in the Lean tool set that help to eliminate waste and organize and simplify work processes.

Lean Six Sigma Applications in Private Industry

The concept of combining Lean Manufacturing and Six Sigma principles began in the middle to late 1990s and has quickly taken hold as companies recognized the synergies. There are many examples of manufacturing companies implementing a combined effort of Lean and Six Sigma. An early example, starting in 1997, was by an aircraft engine controls firm, BAE Systems Controls, in Fort Wayne, Indiana. They blended Lean Manufacturing principles with Six Sigma quality tools. Their "Lean Sigma" strategy was "designed to increase velocity, eliminate waste, minimize process variation, and secure its future in the evolving aerospace market" (Sheridan, 2000). They started with implementing Lean initiatives and then identified a synergy between Lean and the Six Sigma quality program that had been launched while the company was a part of General Electric. BAE Systems Controls implemented the following Lean initiatives: (1) Kaizen events, (2) takt-time-driven one-piece-flow product cells, (3) kanban pull system and point-of-use storage bins on the plant floor, (4) Lean production cells, (5) mistake proofing, and (6) use of a multiskilled workforce. As part of the Six Sigma program, they implemented statistical methods and team leadership with the use of Black Belts. In BAE Systems Control's implementation of Lean Six Sigma, they improved productivity by 97% and customer lead time by 90%. Their value-added productivity increased 112% in 5 years, work in process was reduced by 70%, product reliability improved by 300%, and there were zero lost workdays in 1999 (Sheridan, 2000).

Another early innovator combining Lean and Six Sigma was Maytag Corporation. Maytag implemented Lean Sigma in 1999. They designed a new production line using the concepts of Lean and Six Sigma. Maytag reduced utilized floor space to one-third of that used by Maytag's other product lines. Maytag also cut production costs by 55%. Their Lean Sigma effort helped Maytag achieve savings worth millions of dollars (Dubai Quality Group, 2003).

Lean Six Sigma has been implemented at Northrop Grumman, an aerospace company. They had already begun to implement Lean Thinking when they embarked upon their Six Sigma program. Northrop integrated WorkOut™ events (a problem-solving process developed at General Electric) with Lean Thinking methods and Kaizen events. They used Six Sigma's strategy and methods within their Product Teams, not as a stand-alone program. Their formal process integrated WorkOut, Kaizen, and DMAIC into the Six Sigma Breakthrough WorkOut. Subject matter experts and a Black Belt were used on their project team. They performed a 4–5-day Define/Measure phase. They then performed the Measure, Analyze, and Improve phases for roughly 30 days each. The final activities included a post-WorkOut phase that included the Control, Integrate, and Realize phases (McIlroy and Silverstein, 2002).

Lockheed Martin Aeronautical Systems reduced costs and improved competitiveness, customer satisfaction, and the first-time quality of all its manufactured goods. They had separate Lean and Six Sigma projects, depending on the objective of the project and the problem that needed to be solved (Kandebo, 1999).

The Six Sigma DMAIC problem-solving methodology is used to improve processes. The phases of the DMAIC are well-defined and standardized, but the steps performed in each phase can vary based on the reference used. The Define phase is where the scope of the project charter is developed. The goal of the Measure phase is to understand and baseline the current process. In the Analyze phase, we analyze the data collected in the Measure phase to identify the root causes of the problems identified. In the Improve phase, the improvement recommendations are developed and implemented. The goal of the Control phase is to ensure that the improvements had a positive impact and that they will be sustained and controlled. In Figure 2.2 is a description of the activities that can be performed within each phase of the DMAIC problem-solving methodology (Brassard and Ritter, 2001).

Following is a description of the DMAIC approach; the detailed steps (Figure 2.2) and most frequently used Lean-oriented tools applied within each phase are shown in Figure 2.3.

DEFINE	MEASURE	ANALYZE	IMPROVE	CONTROL
1. Develop project charter 2. Perform stakeholder analysis 3. Perform initial VOC and identify CTS 4. Select team and launch the project 5. Create project plan	6. Define the current process 7. Define detailed VOC 8. Define the VOP and current performance 9. Validate measurement system 10. Define COPQ and cost/benefit	11. Develop cause and effect relationships 12. Determine and validate root causes 13. Develop process capability	14. Identify breakthrough and select solutions 15. Perform cost/benefit analysis 16. Design future state 17. Establish performance targets, project scorecard 18. Gain approval to implement and implement 19. Train and execute	20. Measure results and manage change 21. Report scorecard data and create process control plan 22. Apply PDCA process 23. Identify replication opportunities 24. Develop future plans

FIGURE 2.2
DMAIC activities.

DEFINE	MEASURE	ANALYZE	IMPROVE	CONTROL
• Project charter • Stakeholder analysis • Suppliers, inputs, process, outputs, and customers (SIPOC) • Communication plan • Project plan • Responsibilities matrix • Ground rules	• Process map • Data collection plan • Value stream map • Spaghetti diagram • VOP matrix • Pareto chart • Benchmarking • Check sheet • Histogram • Measurement systems analysis • Cost of Poor Quality	• Why–why diagram • Basic statistics • Sampling • Process analysis • Failure mode and effects analysis • Waste elimination • 5S • Kaizen • Process capability	• Recommendations improvement plan • Action plan • Cost/benefit analysis • Future state process • Future state value stream map • Dashboards • One-piece flow • Cellular design • SMED • Flow • Pull • Kanban • Revised VOP matrix • Training plans • Procedures	• Mistake proofing/poka-yoke • Standard work • Control plan • Process capability • Kaizen • Dashboards • Visual management • Hypothesis testing

FIGURE 2.3
Most common Lean tools by DMAIC methodology.

The Lean tools that are most commonly used to eliminate waste and achieve flow are discussed at a high level in this chapter and in more detail in the following chapters:

- Value stream mapping (discussed in detail in Chapter 3)
- 5S and visual management (discussed in detail in Chapter 4)
- Single-minute exchange of dies (SMED) (discussed in detail in Chapter 5)
- Flow, pull, and kanban (discussed in detail in Chapter 6)
- Mistake proofing (discussed in detail in Chapter 7)
- Standard work (discussed in detail in Chapter 8)
- Systems thinking and theory of constraints (discussed in detail in Chapter 9)
- Hoshin Kanri (discussed in detail in Chapter 10)

Define Phase

The purpose of the Define phase is to delineate the business problem and scope the project and process to be improved. Whether it is a Lean or Six Sigma project, the steps will be the same to scope and define the business problem that the team is trying to solve. The following steps can be applied to meet the objectives of the Define phase:

1. Develop project charter.
2. Identify customers and stakeholders.
3. Define initial voice of the customer (VOC) and critical to satisfaction (CTS) criteria.
4. Form the team and launch the project.
5. Create project plan.

Figure 2.4 shows the main activities mapped to the tools or deliverables most typically used during each step of the Define phase. A project charter template is provided in Figure 2.5.

Develop Project Charter

The first step in the Define phase is to identify and delineate the problem. The project charter can help to identify the elements that help to scope the project and identify the project goals.

Following is a description of the elements of the project charter that helps to scope and define the business problem.

	Define activities	Tools/deliverables
1	Develop project charter	• Project charter • Supplier–Inputs–Process–Outputs–Customers (SIPOC) • High-level process map
2	Perform stakeholder analysis	• Stakeholder analysis definition • Stakeholder commitment scale • Communication planning worksheet
3	Perform initial voice of the customer (VOC) and identify critical to satisfaction (CTS)	• Critical to satisfaction (CTS) summary
4	Select team and launch the project	• Responsibilities matrix • Ground rules • Items for resolution (IFR)
5	Create project plan	• Project work plan

FIGURE 2.4
Define activities and tools/deliverables.

Project Name: Name of the Lean Six Sigma project.

Problem Statement: Business problem; describe what, when, impact, consequences.

Project Objectives: Describe the specific objectives of the project in measurable terms.

Customer/Stakeholders: (Internal/External) Key groups impacted by the project.

Critical to Satisfaction—CTS: What is important to the customers, the key business drivers.

Scope Statement: The scope of the project, what is in the scope, and what is out of scope.

Financial and Other Benefit(s): Estimated benefits to the business, tangible and intangible.

Potential Risks: Risks that could impact the success of the project. Can assess risk by probability of occurrence and potential impact to the project.

Project Resources: Champion, Black Belt mentor, process owner, team members.

Milestones: DMAIC phase and estimated completion dates.

FIGURE 2.5
Project charter template.

Project Name: The name of the project. Describe the process to be improved along with the project goal.

Problem Statement: A clear description of the business problem. What is the challenge or the problem that the business is facing? The problem statement should consider the process that is impacted. This defines the measurable impact of the problem. The team should be specific as to what is happening, when it is occurring, and what the impact or consequences are to the business of the existing problem.

Project Objectives: What are the quantifiable objectives of the project? It may be too early in the problem-solving method to identify a clear target, but at least a placeholder should be identified relating to what should be measured and improved.

Customers/Stakeholders: Define the customers, both internal and external, and the stakeholders that are being impacted by the problem or process to be improved.

Critical to Satisfaction: Identify what is important to each customer/stakeholder group. They can be identified by what is critical to quality (CTQ) (defects), delivery (time), and cost.

Scope Statement: The scope should clearly identify the process to be improved and what is included or excluded from the scope for the Lean Six Sigma project. The scope can also address the organizational boundaries to be included and possibly more importantly which should be excluded. It can also include a temporal scope of the timing of the process and data collection activities. The deliverable scope includes what specifics should be delivered from the project, such as improvement recommendations, implementation plan, and so forth.

Projected Financial and Other Benefits: Describes potential savings, revenue growth, cost avoidance, cost reduction, cost of poor quality (COPQ), as well as less tangible benefits such as impact to morale and elimination of waste and inefficiencies.

Potential Risks: Brainstorm the potential risks that could impact the success of the project. Identify the probability that the risk could occur, on a high, medium, or low scale. Identify the potential impact to the project if the risk does occur, on a high, medium, or low scale. The risk mitigation strategy identifies how you would potentially mitigate the impact of the potential risk if it does occur.

Project Resources: Identify the project leader who is in charge of the overall project. Identify the division and department of the project leader or project team. Identify the process owner, the person who is ultimately responsible for implementing the improvement recommendations. The project champion is the person at the director (or above) level who can remove the barriers to successful project implementation. The project sponsor is the person at the executive level who sponsors the project initiative and is the visible representative of the project and the improvements. The Continuous Improvement Mentor or the Master Black Belt is the team's coach who helps mentor the team members in application of the tools and the DMAIC methodology. Finance is the financial representative who approves the financial benefits or savings established during the project. Team members or support resources are those people who are part of the project team or who provide support, information, or data to the project team.

Milestones: The milestones are the estimated key dates when each phase will be completed and when the project improvements will be approved.

SIPOC

The suppliers–inputs–process–outputs–customers (SIPOC) diagram is a useful tool in the Define phase to help scope the project and understand the process. SIPOC shows the interrelationships between the customers and suppliers and how they interact with the process. It also identifies the inputs used in the process steps and the outputs of the process. The process steps transform the inputs into the outputs. The best way to construct the SIPOC diagram is to identify the five to seven high-level process steps that bound the process. For each process step, identify the inputs to the process and who supplies the inputs. Next identify the outputs of each process step and the customer of the output. Figure 2.6 is an example of a SIPOC diagram for processing linen in a hospital.

Suppliers–Inputs–Process–Outputs–Customers (SIPOC)				
Suppliers	**Inputs**	**Process**	**Outputs**	**Customers**
• Materials management • Hospital departments	• Standard order quantities • Order form • Departmental linen usage	**Order linen**	• Linen order	• Linen servicer
• Linen servicer	• Clean linen • Seal for security • Shipping documents	**Receive linen**	• Shipping documents • Seal for security	• Materials management
• Materials management departments	• Unused linen on carts • Clean linen • Par values	**Replenish linen**	• Replenished linen	• Hospital departments
• Hospital departments	• Dirty (used linen)	**Collect linen**	• Collected dirty linen • Carts	• Environmental services
• Environmental services	• Collected dirty linen • Carts	**Send linen back**	• Seal for security • Shipping documents • Dirty (used) linen	• Linen servicer

FIGURE 2.6
SIPOC diagram example.

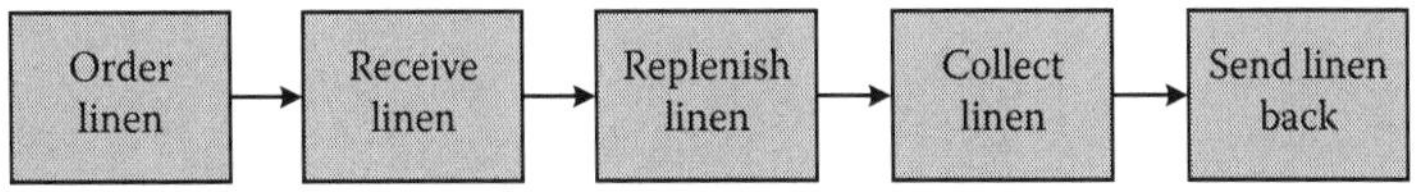

FIGURE 2.7
Level 1 process map.

High-Level Process Map

A process is a description of activities that transforms inputs to outputs. A process map is a graphical representation of the process, interrelationships, and sequence of steps. The high level or level 1 process map utilized in the Define phase can derived from the process steps identified in the SIPOC diagram. The process steps can simply be turned 90° and displayed horizontally instead of vertically. Process maps are a valuable tool in helping to understand the current process, identify the inefficiencies and non-value-added activities, and then create the future state process during the Improve phase. If there is sufficient knowledge of the process, a more detailed, level 2 process can be created in the Define phase, but typically, additional interviews must be held to collect the information, so a level 1 process map is usually sufficient. Figure 2.7 is a process map for the hospital linen process.

Perform Stakeholder Analysis

It is critical to clearly identify the customers and stakeholders that are impacted by the process because the quality of the process is defined by the customers. Quality is measured by first understanding and then exceeding the customers' requirements and expectations. An unhappy customer carriers a high cost: 96% of unhappy customers never complain, 90% of those who are dissatisfied will not buy again, and each unhappy customer will tell his or her story to as many as 14 other people (Pyzdek, 2003).

Customers and stakeholders can be my peers, people who report to me, my boss, other groups within the organization, suppliers, and external customers. The customers can include both internal and external customers of the process. Each process does not always interface directly with an external customer of the company but will have internal customers. The internal customers are people who receive some output from the process, such as information, materials, product, or a service step. It is ultimately the boundary of the process that is being improved that determines who the customer is.

The stakeholder analysis definition identifies the stakeholder groups, their role, and how they are impacted as well as their concerns related to the process. There is an additional column that provides a quick view of whether the impact is positive (+), such as reducing variation, or negative (–), such as resistant to change. This is a high-level view that will be further detailed in the Measure phase. Figure 2.8 is an example of a stakeholder definition.

The next step in the stakeholder analysis is to understand the stakeholders' attitudes toward change as well as potential reasons for resistance. Additionally, the team should understand the barriers to change as a result of the resistance. Next, activities, plans, and actions should be developed that can help the team overcome the resistance and barriers to change. A definition of how and when each stakeholder group should participate in the change effort should be developed in the Define phase and then updated throughout the DMAIC project. Figure 2.9 shows a stakeholder commitment.

The stakeholder commitment scale can be used to summarize where the stakeholders are regarding their acceptance or resistance to change. The team should determine, based on initial interviews and prior knowledge of the stakeholder groups, the current level of support or resistance to the project.

Stakeholder Analysis Definition			
Stakeholders	**Role Description**	**Impact/Concern**	**+/−**
External customer	Customers who receive our marketing efforts related to marketing programs, including advertising circulars and commercials	• Timely information • Accurate information • Coupons	+ + +
Marketing	Internal marketing department that plans, develops, and deploys marketing programs	• Timely deployment • Ability to reach and impact customers	+ +
Information technology	Information technology department that provides technology	• Clear requirements • Accurate data	+ +

FIGURE 2.8
Stakeholder analysis definition.

Stakeholder Commitment Scale							
Stakeholders	**Strongly Against**	**Moderate Against**	**Neutral**	**Moderate Support**	**Strongly Support**	**Communication Plan**	**Action Plan**
External customer		X ———	———	———	→ O	• Surveys • Market research	Test pilot
Marketing			X ———	———	→ O	• E-mail • Meetings	Engage on project
Information technology	X ———	———	———	→ O		• Intranet • E-mail • Meetings	Engage on project Communicate process and requirements
X = At Beginning of Project O = At End of Project							

FIGURE 2.9
Stakeholder commitment scale.

Strongly supportive indicates that these stakeholders are supportive of advocating for and making change happen. Moderately supportive indicates that the stakeholders will help, but not strongly. The neutral stakeholders will allow the change and not stand in the way, but will not go out of their way to advocate for the change. Moderately against stakeholders do not comply with the change and have some resistance to the project. Strongly against stakeholders will not comply with the change and will actively and vocally lobby against the change. A strategy to move the stakeholders from their current state to where the team needs them to be by the end of the project should be developed. This change strategy should include how the team will communicate with the stakeholders and any activities in their action plan to gain support and implement change.

Perform Initial Voice of the Customer and Identify Critical to Satisfaction

In the Define phase, the team can perform an initial VOC data collection to understand the CTS criteria. CTS criteria are the elements of a process that significantly affect the output of the process. It is critical to focus throughout the phases of the DMAIC problem-solving process and the Lean Six Sigma project on the CTS.

In the Define and Measure phases, the focus is on collecting information from the customer to understand what is important to them regarding the process, product, or service. Identify in the Measure phase the measurement of your process directly related to the CTS criteria. Identify the root causes in the Analyze phase related to the CTS. The improvement recommendations implemented are aligned with eliminating the root causes related to the CTS. The variability to be controlled by implementing control mechanisms in the Control phase should reduce the variability related to the CTS.

Some references refer to identifying the CTQ, but the CTS broadens the elements of the CTS by including CTQ, Critical to Delivery, and Critical to Cost. There may also be critical elements in the process to measure related not only to quality, delivery, and cost, but to time as well. For a Lean Six Sigma project, not everything should be a CTS. The CTS should be specific to the scope of the project and the process to be improved. If there are more than a few CTS identified for the project, the scope is probably too large for a reasonable Lean Six Sigma project to be completed in 3–6 months. The CTS should describe the customer need or requirement, not how to solve the problem.

The steps to identify the CTS are as follows (George, Rowlands, Price, and Maxey, 2005):

1. Gather appropriate VOC data from market research, surveys, focus groups, interviews, and so on.
2. Extract key verbatim from the VOC data collections, identifying why a customer would do business with your organization.

3. Sort ideas, find themes, and develop an affinity or tree diagram.
4. Be specific and follow up with customers where needed.
5. Extract CTS and specifications from customer information.
6. Identify where you are missing data and fill in the gaps.

The VOC is a term used to "talk to the customer" to hear their needs and requirements or their "voice." Many different mechanisms can be used to collect VOC, including interviews, focus groups, surveys, customer complaint and warranty data, market research, competitive information, and customer buying patterns. We will further discuss VOC during the Measure phase, where more detailed and extensive VOC can best be performed. The initial VOC is used to identify the CTS. In the Define phase, the CTS summary is a listing of the CTS based on knowledge of the process and the customer to this point.

Select Team and Launch the Project

The Lean Six Sigma project team should be selected based on those team members who have the knowledge of the process and have the commitment to work on the project. The roles and responsibilities of the project team members should be clearly defined.

A team is a group of people working together to achieve a common purpose. Teams need clearly defined purposes and goals, which are provided through the project charter. They also need well-defined roles and responsibilities, which can be provided through developing a responsibilities matrix. The responsibility matrix (Figure 2.10) identifies the team members, along with their roles and high-level responsibilities on the project. Another important

Role Responsibility	Team Leader	Black Belt	Champion	Process Owner	Team Members
Facilitate meetings	X				
Manage project	X				
Mentor team members	X	X			
Transfer knowledge of Lean Six Sigma tools		X			
Remove roadblocks			X		
Monitor project progress			X		
Approve project			X		
Implement improvements				X	
Subject matter expertise				X	
Apply Lean Six Sigma tools					X
Statistical analysis					X
Data collection					X

FIGURE 2.10
Responsibility matrix.

component of forming the team is to brainstorm and identify ground rules. The ground rules identify how the team will interact with each other and ensure that the behavioral expectations are clearly defined at the start of the project. The team's common set of values and ethics can be established during the development of the team ground rules.

Some sample team ground rules follow:

- Treat everyone with respect.
- Listen to everyone's ideas.
- When brainstorming, do not evaluate ideas.
- Contribute fully and actively participate.
- Come to team meetings prepared.
- Make decisions by consensus.
- Identify a backup resource to complete tasks when not available.

Create Project Plan

The project plan is developed in this last step of the Define phase. The resources, time, and effort for the project are planned. A project plan template is provided in Figure 2.11. Additional tasks can be identified within each phase and major activity.

Excel or project planning software such as Microsoft Project can be used to track tasks completed against the project plan. An important part of the project planning is to perform a risk analysis to identify potential risks that could impact the successful completion of the project.

The team can brainstorm potential risks to the project. They can also assess the probability that each risk would occur on a scale of High, Medium, or Low occurrence. The impact of the risk should also be assessed, that is, if the risks were to occur, what level of impact would it have on the successful completion of the project (High, Medium, or Low). It is also important to develop a risk mitigation strategy that identifies how the team will mitigate the impact of the risk to reduce or eliminate the impact of the risk if it occurs. Figure 2.12 shows a simple risk matrix.

Another tool that is useful while planning and managing the project is an Item for Resolution (IFR) form. This helps the team to document and track items that need to be resolved. It enables the team to complete the planned agendas in meetings, by allowing a place to "park" items that arise that cannot be resolved in the meeting, due to either time constraints or lack of data or access to appropriate decision makers. Figure 2.13 shows an IFR form. It includes a description of the item to be resolved. A priority (High, Medium, or Low) should be assigned to each item. The status of the item, whether open (newly opened), closed (resolved), or hold (on hold—not being actively worked), should be identified. The owner who is responsible for resolving

Task Name	Duration	Start Date	End Date	Resources	Predecessor
Define					
1. Develop project charter					
2. Identify stakeholders					1
3. Perform initial VOC and identify CTS					2
4. Select team and launch the project					3
5. Create project plan					4
Measure					
6. Define the current process					5
7. Define the detailed VOC					6
8. Define the VOP and current performance					7
9. Validate measurement system					8
10. Define COPQ and cost/benefit					9
Analyze					
11. Develop cause and effect relationships					10
12. Determine and validate root causes					11
13. Develop process capability					12
Improve					
14. Identify breakthrough and select practical approaches					13
15. Perform cost/benefit analysis					14
16. Design future state					15
17. Establish performance targets, project scorecard					16
18. Gain approval to implement, and implement					17
19. Train and execute					18
Control					
20. Measure results and manage change					19
21. Report scorecard data and create process control plan					20
22. Apply PDCA process					21
23. Identify replication opportunities					22
24. Develop future plans					23

FIGURE 2.11
DMAIC project plan.

Potential Risks	Probability of Risk Occurring (High/Medium/Low)	Impact of Risk (High/Medium/Low)	Risk Mitigation Strategy

FIGURE 2.12
Risk matrix.

Items for Resolution							
#	Issue	Priority	Status	Owner	Open Date	Resolved Date	Resolution

FIGURE 2.13
Items for Resolution form.

the issue, as well as the dates that the item were opened and resolved, should be completed on the IFR form. A description of the resolution should also be included. This helps the team keep track of key decisions and ensures that the items are resolved to the satisfaction of all team members. The log of IFRs can also be used during the lessons learned activity after the project is complete to identify where the problems arose and how they were resolved, so that these items can be incorporated into the risk mitigation strategies for follow-on projects.

Another helpful tool that should be developed in the Define phase, but be used throughout the project, is a communication plan. The communication plan can be used to identify strategies for how the team will communicate with all key stakeholders. It can be useful to help overcome resistance to change by planning how frequently and in what manner the team will communicate with the stakeholders. Each key stakeholder or audience of a communicated message should be identified. The objective or message that will be communicated is then developed. The media or mechanism of how to communicate with the audience is then identified. This can be face-to-face, e-mail, web sites, and so on. The frequency of the communication is important, with higher frequency important for those more resistant to change. The last element of the communication plan is to clearly identify who is responsible for developing and delivering the communication to the audience. The communication plan is shown in Figure 2.14.

Audience	Objectives/Message	Media/Mechanism	Frequency	Responsible

FIGURE 2.14
Communication plan.

Team Meeting Management

Some best practices for team meeting management are as follows:

- Respect people and their time.
- Determine critical/required participants for e-mails, meetings, and decisions.
- Cancel or schedule meetings ahead of time.
- Always create a meeting agenda and send it out in advance of the meeting. The agenda should include required and optional participants.
- Recap action items and meeting minutes.
- Use voting in e-mails to make easy decisions or agree upon a meeting time.
- Track meeting attendance and resolve habitual lack of attendance.

The planned meeting agenda should include the following:

1. Date, time, and proposed length of the meeting
2. Meeting facilitator's name
3. Meeting location
4. Required and optional attendees
5. Purpose of the meeting
6. Desired outcomes
7. Topic with time and proposed outcome for each topic

Some tips that the meeting facilitator can use to keep the meeting productive are as follows:

- Listen, restate what you think you heard.
- Ask for clarification and examples.
- Encourage equal participation, circle the group.
- Summarize ideas and discussion.
- Corral digressions, get back to the agenda.
- Close the discussion.

Summary

Some organizations focus the Lean project on applying major specific Lean tools such as 5S, Lean wastes, changeover or SMED, pull techniques, or other singular applications of Lean tools. We encourage a more robust and integrated approach where the team applies several of the Lean tools with the focus being to improve a specific process. However, defining a project to apply 5S to try to organize a work area and/or applying SMED to reduce changeover time for product changes can also be the goal of a Lean project. We still encourage the use of the DMAIC phases, no matter the number of Lean tools to be applied and the objective of the project. The Define phase is a critical phase of the project. It is important to spend ample time in the Define phase developing the project charter and getting the buy-in of the project champion, the team members, and all stakeholders. The time spent clearly defining the scope of the project will reap dividends by reducing issues during the remaining phases of the project. A process or a problem poorly defined will require the team to revisit the Define phase when improvement efforts bog down or lose focus in subsequent phases.

Measure Phase

The purpose of the Measure phase is to understand and document the current state of the processes to be improved, collect the detailed VOC information, baseline the current state, and validate the measurement system. The activities performed and tools applied during the Measure phase are discussed in the following:

1. Define the current process.
2. Define the detailed VOC.
3. Define the voice of the process (VOP) and current performance.
4. Validate the measurement system.
5. Define the COPQ.

Figure 2.15 shows the main activities mapped to the tools or deliverables most typically used during that step.

Define the Current Process

The first step of the Measure phase is to profile the current state. SIPOC, process mapping, and value stream mapping are excellent tools to document the current process steps, the information that is used, the people who perform the work, and the internal and external customers of the services.

In a process improvement effort, there are typically three levels of process maps that are used to help with documenting the current process. Figure 2.16 shows the three levels and where they should be applied.

An example of a level 2 process map for making a peanut butter and jelly Sandwich is shown in Figure 2.17.

It is also important to identify any process measures and related metrics that are used to measure the quality and productivity of the processes. The current profile of the people and cultural state should be understood, including the level of skills and training of the employees, and their resistance or acceptance levels to change.

	Measure Activities	Tools/Deliverables
1	Define the current process	• Process map • Operational definitions • Metrics • Baseline • Data collection plan • Value stream map • Spaghetti diagrams
2	Define the detailed Voice of the Customer (VOC)	• Surveys, interviews, focus groups • Affinity diagram
3	Define the Voice of the Process (VOP) and current performance	• Pareto charts • VOP matrix • Benchmarking, check sheets, histograms • Statistics
4	Validate the measurement system	• Measurement system validation • Gage R and R (repeatability and reproducibility)
5	Define the cost of poor quality (COPQ)	• Cost of poor quality

FIGURE 2.15
Measure phase activities and tools/deliverables.

Level	Type/Name	Purpose
Level 1	Macro or high level	Scope the improvement project Provide project and process boundaries Provide a high-level view of the process
Level 2	Process map	Identify process improvement areas Identify process inefficiencies Identify waste
Level 3	Process map or process flow chart	Identify improvement areas Identify value-added versus non-value-added activities Helps to identify Lean wastes Provide detailed how-to (almost procedural level)

FIGURE 2.16
Process map level and purpose.

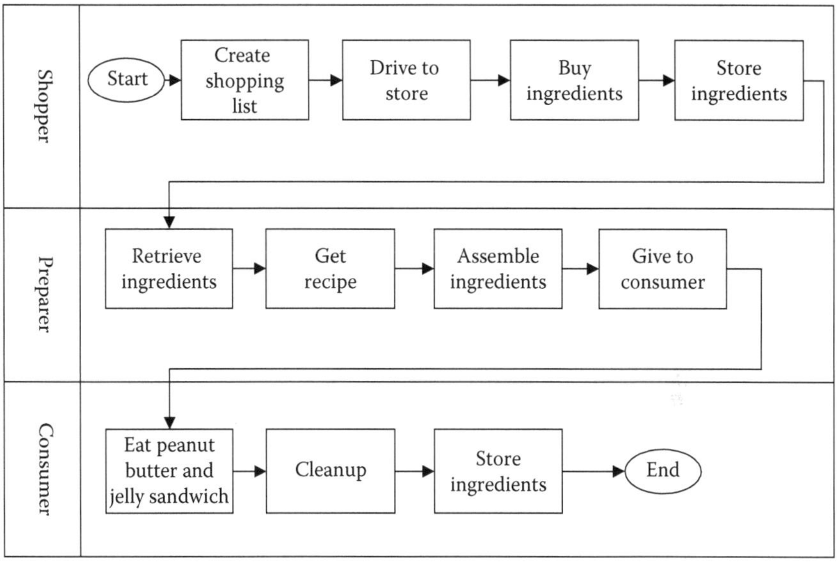

FIGURE 2.17
Process map for making a peanut butter and jelly sandwich.

The steps to completing a process map are as follows:

1. Identify the level (one, two, or three) to map and document.
2. Define the process boundaries.
3. Identify the major activities within the process.
4. Identify the process steps and uncover complexities using the brainstorming and storyboarding techniques.
5. Arrange the steps in time sequence and differentiate operations with symbols.
6. Validate the process map by a "walk through" of the actual process and by having other process experts review it for consistency.

A process map tends to document a more detailed level of the process. The goal of the project may be to understand the handoffs between steps in a process and to understand the material and/or information flow at a higher level of the process. For this purpose, a value stream map may be used to document the information and material flow. Value stream mapping will be discussed in more detail in Chapter 3.

The spaghetti diagram is another tool that can be used in the Measure phase to understand the physical flow of the activities in a process. The spaghetti diagram is an invaluable tool to help understand the path that a product or customer (or customer order) may take to be completed. It helps to understand the wasteful back and forth travel that a product may experience before being completed or the steps that a customer or customer order may travel before being finished.

Define Detailed Voice of the Customer

In the Measure phase, the VOC information should be collected to define the customers' expectations and requirements with respect to the service delivery process. VOC is an expression for listening to the external customer and understanding their requirements for your product or service. Examples of requirements are their expectations for responsiveness, such as turnaround time on vendor (customer) invoices, or error rates, such as employee (customer) expectations of no errors on their paycheck. The VOC can be captured through interviewing, surveys, focus groups with customers, complaint cards, warranty information, and competitive shopping.

Personal interviews are an effective way to gain the VOC; however, it can be expensive and training of interviewers is important to avoid interviewer bias. However, additional questioning can occur to eliminate misunderstanding. The objectives of the interview should be clearly defined before the interviews are held.

Customer Surveys

Customer surveys are a typical way to collect VOC data. The response rate on surveys tends to be low; 20% is a "good" response rate. It can also be extremely difficult to develop a survey that avoids and asks the questions that are desired. Customer survey collection can be quite expensive. The steps to create a customer survey are as follows (Malone, 2005):

1. Conceptualization: Identify the survey objective and develop the concept of the survey, and what questions you are trying to answer from the survey.
2. Construction: Develop the survey questions. A focus group can be used to develop and/or test the questions to see if they are easily understood.
3. Pilot (try out): Pilot the questions by having a focus group of representative people from your population. You would have them review the questions, identify any unclear or confusing questions, and tell you what they think that the questions are asking. You would not use the data collected during the pilot in the actual results of the surveys.
4. Item Analysis: Item analysis provides a statistical analysis to determine which questions answer the same objectives, as a way to reduce

the number of questions. It is important to minimize the number of questions and the total time required to take the survey. Typically, the survey time should be 10 minutes or less.

5. Revision: Revise the survey questions and roll out the customer survey, or pilot again if necessary.

Focus Groups

Focus groups are an effective way to collect VOC data. A small representative group, typically 7 to 10 people, are brought together and asked to respond to predetermined questions. The focus group objective should be developed and the questions should support the objective. The participants should be selected by a common set of characteristics. The goal of a focus group is to gather a common set of themes related to the focus group objective. There is no set sample size for focus groups. Multiple focus groups are typically run until no additional themes are derived. Advantages of focus groups are as follows (University of Wisconsin, 2005):

- They tend to have good face validity, meaning that the responses are in the words of the focus group participants.
- Typically more comments are derived than in an interview with one person at a time.
- The facilitator can probe for additional information and clarification.
- Information is obtained relatively inexpensively.
- They generate speedy results.
- They provide real-life data in a social environment.

Some of the disadvantages of focus groups are

- The facilitator's skills dictates the quality of the responses.
- They can be difficult to schedule.
- It can be difficult to analyze the dialogue due to participant interactions.

Affinity diagrams organize interview, survey, and focus group data after collection. The affinity diagram organizes the data into themes or categories. The themes can first be generated and then the data can be organized into the themes, or the detailed data can be grouped into the themes. An example of a simple affinity diagram for ways to study for a Six Sigma Black Belt exam is shown in Figure 2.18.

Data Collection Plan

A data collection plan should be developed to identify the data to be collected that relate to the CTS criteria.

Resources

- Tab training materials
- Get other references
- Review LSS pocket toolbook
- Discuss with mentor

Preparation

- Study material
- Apply tools on projects
- Study in groups
- Attend refresher

Motivation

- Schedule exam
- Motivate self
- Talk to other candidates

FIGURE 2.18
Affinity diagram for Six Sigma Black Belt exam preparation.

Critical to Satisfaction (CTS)	Metric	Data Collection Mechanism (Survey, Interview, Focus Group, etc.)	Analysis Mechanism (Statistics, Statistical Tests, etc.)	Sampling plan (Sample Size, Sample Frequency)	Sampling Instructions (Who, Where, When, How)
Speed to market	Cycle time	Project management tool	Statistics (mean, variance); t-test	One year of projects	Collect data from project management system for last year
	Functionality delivered	Requirements traceability tool	Count	50 projects (30 development, 20 support)	Extract data based on sampling plan

FIGURE 2.19
Data collection plan for software application development project.

The data collection plan ensures

- Measurement of CTS metrics
- Identification of the right mechanisms to perform the data collection
- Collection and analysis of data
- Definition of how and who is responsible to collect the data

Figure 2.19 shows a data collection plan.

The steps for creating a data collection plan in the Measure phase are as follows:

1. Define the CTS.
2. Develop metrics.

3. Identify data collection mechanism(s).
4. Identify analysis mechanism(s).
5. Develop sampling plans.
6. Develop sampling instructions.

A description of each step in the data collection plan development follows.

Define the Critical to Satisfaction Criteria

CTS is a characteristic of a product or service that fulfills a critical customer requirement or a customer process requirement. CTS criteria are the basic elements to be used in driving process measurement, improvement, and control (George, Rowlands, Price, and Maxey, 2005).

Develop Metrics

In this step, metrics are identified that help to measure and assess improvement related to the identified CTS. Some rules of thumb for selecting metrics are the following (Evans and Lindsey, 2002):

- Consider the vital few versus the trivial many.
- Metrics should focus on the past, the present, and future.
- Metrics should be linked to meet the needs of shareholders, customers, and employees.

It is vital to develop an operational definition for each metric, so it is clearly understood how the data will be collected by anyone that collects it. The operational definition should include a clear description of a measurement, including the process of collection. Include the purpose and metric measurement. It should identify what to measure, how to measure it, and how the consistency of the measure will be ensured. A summary of an operational definition follows.

Operational Definition

Defining the Measure, Definition A clear, concise description of a measurement and the process by which it is to be collected.

Purpose Provides the meaning of the operational definition, to provide a common understanding of how it will be measured.

Clear Way to Measure the Process

- Identifies what to measure
- Identifies how to measure
- Makes sure the measuring is consistent

Identify Data Collection Mechanism(s)

Next you can identify how you will collect the data for the metrics. Some data collection mechanisms include customer surveys, observation, work sampling, time studies, customer complaint data, e-mails, web sites, and focus groups.

Identify Analysis Mechanism(s)

Before collecting data, consider how you will analyze the data to ensure that you collect the data in a manner that enables the analysis. Analysis mechanisms can include the type of statistical tests or graphical analysis that will be performed. The analysis mechanisms can dictate the factors and levels for which you may collect the data.

Develop Sampling Plans

In this step, you should determine how you will sample the data and the sample size for your samples. Several types of sampling are as follows (Pyzdek, 2003):

- Simple random sample: Each unit has an equal chance of being sampled.
- Stratified sample: *N* (population size) items are divided into subpopulations or strata and then a simple random sample is taken from each stratum. This is used to decrease the sample size and cost of sampling.
- Systematic sample: *N* (population size) items are placed into *k* groups. The first item is chosen at random, the rest of the sample selecting every *k*th item.
- Cluster sample: *N* items are divided into clusters. This is used for wide geographic regions.

Develop Sampling Instructions

Clearly identify who will be sampled, where you will sample, and when and how you will take your sample data.

Define the Voice of the Process and Current Performance

There are many tools that can be used to assess the VOP and current performance. We discuss the VOP matrix, Pareto charts, benchmarking, check sheets, and histograms.

VOP Matrix

A VOP matrix, developed by Furterer, can be used to achieve integration and synergy between the DMAIC phases and the critical components of the process to enhance problem solving. The VOP matrix includes the CTS, the

CTS	Process Factors	Operational Definition	Metric	Target
Faculty/staff awareness of process	Procedures exist	Procedures exist and are auditable	Number of departments with procedures	100% of departments have procedures by Jan. 1
	Training in procedures	All faculty will take 1 hour training session within 3 months of hire	Number of faculty trained	100% of faculty are trained within 3 months of hire or Jan. 1
Documented location of assets	Procedures exist	Procedures exist and are auditable	Number of departments with procedures	100% of departments have procedures by Jan. 1
	Training in procedures	All faculty will take 1 hour training session within 3 months of hire	Number of faculty trained	100% of faculty are trained within 3 months of hire or Jan. 1
Identification of assets	Description on PO	All purchasers will input detailed description of asset on PO	Number of POs with detailed description	95% of POs sampled have detailed descriptions
	Description in system	PO description will transfer to asset management system	Number of asset descriptions in asset mgt.	95% of POs sampled have detailed descriptions
Efficiency of yearly scanning	Training	All property managers will be trained in process	Number of property managers trained	100% of property managers trained within 3 months
	Process	Quality of process	Proportion of items found on first try	95% of items found on first scan

FIGURE 2.20
VOP matrix for inventory asset management process. (Adapted from Furterer, S., Total Quality Improvement Course Lecture Notes, ESI 5227, University of Central Florida, Department of Industrial Engineering and Management Systems, Orlando, FL, 2004.)

related process factors that impact the CTS, the operational definition that describes how the CTS will be measured, the metric, and the target for the metric. An example of a VOP matrix for the inventory asset management process for a college in a university is shown in Figure 2.20.

Pareto Chart

A Pareto chart helps to identify critical areas causing the majority of the problems. It provides a summary of the vital few rather than the trivial many. It demonstrates the Pareto principle that 80% of the problems are created by

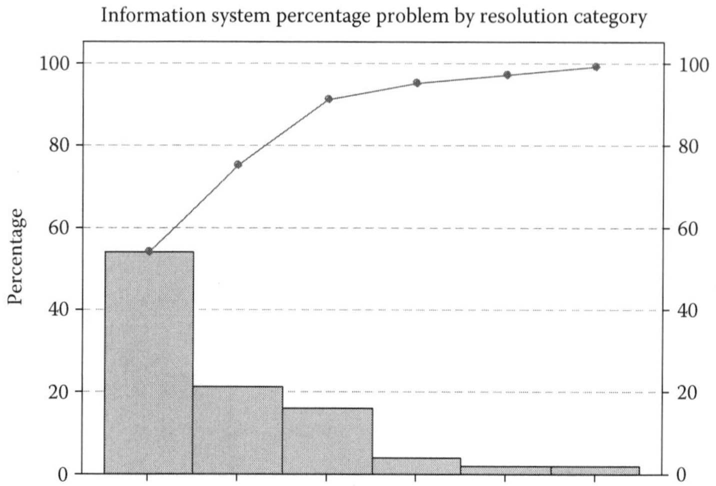

Resolution category	Training	Conversion	Software	Printer	Setup	User security
Percentage	54	21	16	4	2	2
Percent	54.5	21.2	16.2	4.0	2.0	2.0
Cum %	54.5	75.8	91.9	96.0	98.0	100.0

FIGURE 2.21
Pareto chart of resolution categories reported to an information systems help desk created in Minitab.

20% of the causes, so that these root causes can be investigated in the Analyze phase. It helps us to arrange the problems in order of importance and focus on eliminating the problems in the order of highest frequency of occurrence.

Following are the steps for creating a Pareto chart.

Step 1: Define the data categories, defect, or problem types.

Step 2: Determine how the relative importance is defined (dollars, number of occurrences).

Step 3: Collect the data, and compute the cumulative frequency of the data categories.

Step 4: Plot a bar graph, showing the relative importance of each problem area in descending order. Identify the vital few to focus on.

An example of a Pareto chart that identifies the resolution categories for problems reported to a financial information systems help desk is given in Figure 2.21.

Benchmarking

Benchmarking is a tool that provides a review of best practices to be potentially applied to improve your processes. In a Lean Six Sigma project, process benchmarking is typically performed. The organization should document the process that they will benchmark and then select who they will benchmark. It is not necessary to benchmark a company in the same industry, but to focus on the process to be benchmarked, and select an organization that is known for having world class or best practice processes. The next step is to work with the organization to collect the data and understand how the data can be used to identify ways to improve your processes and identify potential improvement opportunities to be implemented in the Improve phase. This is similar to Motorola's benchmarking process (Evans and Lindsey, 2007). It is important to be careful when process benchmarking to ensure that you are comparing apples to apples, meaning that the organization's characteristics are similar to your own, so that the benchmarked process applies to your process.

Check Sheet

A check sheet is a graphical tool that can be used to collect data on the process and types of defects so that root causes can be analyzed in the Analyze phase. The steps to create a check sheet are the following:

Step 1: Choose a characteristic to track, that is, defect types.

Step 2: Set up the data collection check sheet.

Step 3: Collect data using the check sheet.

An example of a check sheet for potential errors when loading data for an online research system is shown in Figure 2.22.

A Pareto chart can then be created from the data collected on a check sheet.

Histogram

A histogram is a graphical tool that provides a picture of the centering, shape, and variance of a data distribution. Minitab® or Excel is commonly used to create a histogram. It is always important to graph the data in a histogram as the first step to understanding the data.

Defect Type	**Tally**	**Total**
Incorrect case name		5
Incorrect docket number		2
Incorrect court name		6
Incorrect cite segment		2
Incorrect decided date		7
Incorrect segment coding		2
Missing text		10
Incorrect primary embedded citations		2
Copyright material included		3

FIGURE 2.22
Check sheet for errors loading data.

Statistics

Statistics can also be used to assess the VOP related to the metrics that are measured. Once the data is collected, it can be tested to see if the data follows a Normal Distribution, using a test for normality. The null hypothesis is that the data is normal. If the null hypothesis is not rejected, then the statistics that would describe the Normal Distribution are the mean and the standard deviation. The mean is the average of the sample data. The mean describes the central location of a Normal Distribution. The sample standard deviation is the square root of the sum of the differences between each data value and the mean divided by the sample size less one. Standard deviation (sigma) is a measure of variation of the data. A value of 99.997% of all data points within the Normal Distribution is within Six Sigma.

Validate the Measurement System

It is important to validate the measurement system to ensure that you are capturing the correct data and that the data reflects what is happening. It is also important to be able to assess a change in the process with our measuring system as well as the measurement system error. We must ensure that the measurement system is stable over time and collecting the data that allows us to make appropriate decisions.

Measurement Systems Analysis

A measurement systems analysis includes the following steps (Gitlow and Levine, 2005):

1. Prepare a flow chart of the ideal measurement system.
2. Prepare a flow chart of the current measurement system.
3. Identify the gaps between the ideal and current measurement systems.
4. Perform a Gage Repeatability & Reproducibility (R&R) study.

Measurement process variation is due to two main types of variation:

- Repeatability related to the gage
- Reproducibility related to the operator

The Gage R&R study assesses both repeatability and reproducibility. Minitab or other statistical software can be used to assess the measurement system error and improve the measurement system if necessary.

Define the Cost of Poor Quality

Cost of Poor Quality

The last step in the Measure phase can be to assess the COPQ related to your Lean Six Sigma project. The COPQ identifies the cost related to poor quality or not doing things right the first time. The COPQ translates defects, errors, and waste into the language of management, which is cost or dollars. There are four categories of COPQ: (1) prevention, (2) appraisal, (3) internal failures, and (4) external failures.

Prevention costs are all the costs expended to prevent errors from being made or the costs involved in helping the employee do the job right every time. Appraisal costs are the results of evaluating already completed output and auditing the process to measure conformance to established criteria and procedures. Internal failure cost is defined as the cost incurred by the company as a result of errors detected before the output is accepted by the company's customer. External failure cost is incurred by the producer because the external customer is supplied with an unacceptable product or service (Harrington, 2005).

Examples of prevention costs are

- Methods of improvement
- Training
- Planning for improvement
- Procedures
- Quality improvement projects
- Quality reporting
- Data gathering and analysis
- Preventive maintenance
- SPC training costs
- ISO 9000 training costs

Examples of appraisal costs are

- Inspections
- Process audits (SPC, ISO)
- Testing activity and equipment depreciation allowances
- Product audits and reviews
- Receiving inspections and testing
- Reviews (meeting time)
- Data collection
- Outside endorsements and certifications

Examples of internal failure costs are

- Reaudit, retest, and rework
- Defects and their impact
- Unscheduled lost time
- Unscheduled overtime
- Excess inventory
- Obsolescence
- Scrap
- White-collar mistakes

Examples of external failures are

- Warranty
- Technical support
- Customer complaints
- Customer bad will costs
- Customer appeasement costs
- Lost business (margin only) due to poor quality
- Product liability
- Return/refunds
- White-collar mistakes

The COPQ can help to identify potential categories of waste embedded in the process.

Analyze Phase

The purpose of the Analyze phase is to analyze the data collected related to the VOC and the VOP to identify the root causes of the process problems, and to develop the capability of the process. The activities performed and tools applied during the Analyze phase are discussed in the following:

1. Develop cause and effect relationships.
2. Determine and validate root causes.
3. Develop process capability.

	Analyze Activities	Deliverables
1	Develop cause and effect relationships	Cause and effect diagrams Why–why diagram
2	Determine and validate root causes	Basic statistics, sampling, process analysis, failure mode and effects analysis, waste elimination, 5S, Kaizen
3	Develop process capability	DPPM/DPMO, process capability

FIGURE 2.23
Analyze phase activities and tools/deliverables.

Figure 2.23 shows the main activities mapped to the tools or deliverables most typically used during that step.

Develop Cause and Effect Relationships

There are several tools that can be used to generate the root causes of the problems identified in the Measure phase, including the cause and effect diagram, cause and effect matrix, and why–why analysis of the 5 Whys. The 5 Whys is more typically related to Lean, and it can be very effective in discerning the root causes of many problems.

Cause and Effect Diagram

The cause and effect diagram can be used to brainstorm and document root causes of an effect or problem. It is helpful to group the causes into categories or use categories to brainstorm the causes. Typical categories are people, machines, materials, methods, measurement, and environment. Transactional categories are places, procedures, policies, people, and information systems. The steps for creating a cause and effect diagram are as follows:

1. Define the problem.
2. Brainstorm all possible types of causes.
3. Brainstorm and organize causes by groups: people, machines, materials, methods, measurement, and environment. You can also add information systems.
4. Brainstorm/identify subcauses for each main cause.

An example cause and effect diagram is shown in Figure 2.24.

Why–Why Diagram

The why–why diagram is a powerful tool to generate root causes. It uses the concept of the 5 Whys, where you ask the question "Why?" several times

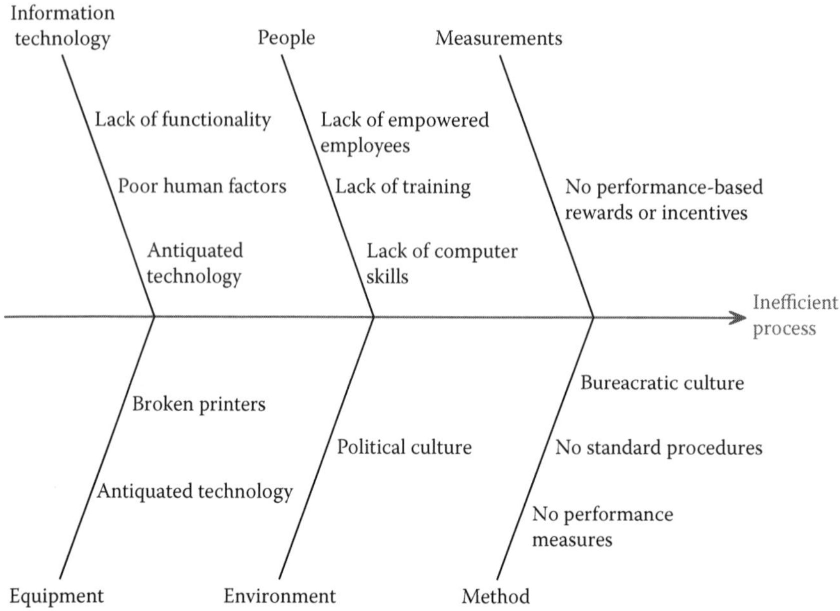

FIGURE 2.24
Cause and effect diagram created in Minitab.

until the root cause is revealed. Following are the steps to create a why–why diagram:

1. Start on the left with the problem statement.
2. State causes for the problem.
3. State causes of each cause.
4. Keep asking "Why?" five times.
5. Try to substantiate the causes with data.
6. Draw the diagram.

Figure 2.25 shows a sample why–why diagram for why potential customers leave a store without making a purchase.

It is critical that once you brainstorm the potential root causes of the problems, you collect additional data to substantiate the causes.

Determine and Validate Root Causes

Process Analysis

To determine and validate root causes, the Lean Six Sigma team can perform a process analysis coupled with waste elimination. A process analysis consists of the following steps:

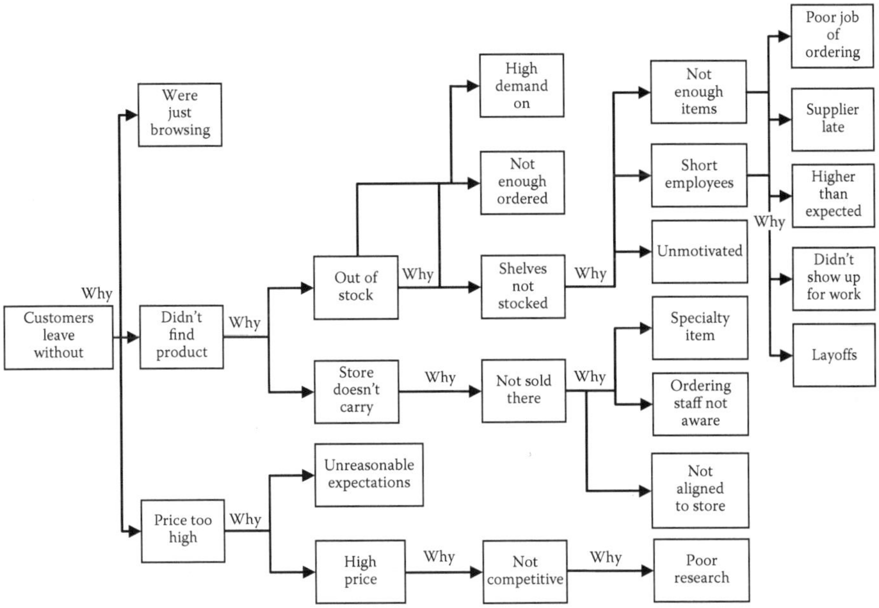

FIGURE 2.25
Why–why diagram.

1. Document the process (using process maps from the Measure phase).
2. Identify non-value-added activities and waste.
3. Consider eliminating non-value-added activities and waste.
4. Identify and validate (collect more data if necessary) root causes of non-value-added activities and waste.
5. Begin generating improvement opportunities.

Value-added activities are those activities that the customer would pay for and that add value for the customer. Non-value-added activities are those that the customer would not want to pay for or do not add value for the customer. Some are necessary, such as for legal, financial reporting, or documentation reasons, while others are unnecessary and should be reduced or eliminated. You can assess the percent of value-added activities as follows:

$$100 \times \left(\frac{\text{Number of value-added activities}}{\text{Number of total activities}}\right)\%$$

Value-added activities include operations that add value for the customer, and non-value-added activities include delays, storage of materials, movement of materials, and inspections. The number of total activities includes the value-added activities and the non-value-added activities.

You can also calculate the percent of value-added time as follows:

$$100 \times \left(\frac{\text{Total time spent in value-added activities}}{\text{Total time for process}} \right) \%$$

Typically, the percent of value-added time is about 1%–5%, with total non-value-added time equal to 95%–99%.

During the process analysis, the team can focus on areas to identify inefficiencies in the following areas (Process Flow Analysis Training Manual):

- Can labor-intensive processes be reduced, eliminated, or combined?
- Can delays be eliminated?
- Are all reviews and approvals necessary and value-added?
- Are decisions necessary?
- Why is rework required?
- Is all of the documentation, tracking, and reporting necessary?
- Are there duplicated processes across the organization?
- What is slipping through the cracks causing customer dissatisfaction?
- Activities that require accessing multiple information systems
- Travel—look at the layout requiring the travel
- Is it necessary to store and retrieve all of that information? Do we need that many copies?
- Are the inspections necessary?
- Is the sequence of activities or flow logical?
- Are standardization, training, and documentation needed?
- Are all of the inputs and outputs of a process necessary?
- How is the data and information stored and used?
- Are systems slow?
- Are systems usable?
- Are information systems user-friendly?
- Can you combine tasks?
- Is the responsible person at too high or low of a level?

Waste Analysis

Waste analysis is a Lean tool that identifies waste into eight different categories to help brainstorm and eliminate different types of wastes. The eight wastes are all considered non-value-added activities and should be reduced or eliminated when possible. Waste is defined as anything that adds cost to the product without adding value. The eight wastes are

- Transportation: Moving people, equipment, materials, and tools.
- Overproduction: Producing more product or material than is necessary to satisfy the customers' orders or faster than is needed.
- Motion: Unnecessary motion, usually at a micro or workplace level.
- Defects: Any errors in not making the product or delivering the service correctly the first time.
- Delay: Wait or delay for equipment or people.
- Inventory: Storing product or materials.
- Processing: Effort that adds no value to a product or service; incorporating requirements not requested by the customer.
- People: Not using people's skills, or mental, creative, and physical abilities.

5S Analysis

5S is a Lean tool that helps to organize a workplace. The 5S are as follows:

- Sort or Simplify: Clearly distinguish between what is necessary and what is unnecessary, disposing of the unnecessary. A red tag is used to identify items that should be reviewed for disposal.
- Straighten: Organize the necessary items so that they can be used and returned easily.
- Scrub: Fix the root cause of the dirt or disorganization.
- Stabilize: Maintain and improve the standards of the first 3S.
- Sustain: Achieve the discipline or habit of properly maintaining the correct 5S procedures.

Kaizen

Kaizen is a Lean tool where kai means "change" and zen "means" for the good." It represents continuous incremental improvement of an activity to constantly create more value for the customer by eliminating waste. A Kaizen consists of short-term activities that focus on redesigning a particular process. A Kaizen event can be incorporated into the Analyze or Improve phase of the Lean Six Sigma project to help design and/or implement a focused improvement recommendation.

The Kaizen event follows the Plan–Do–Check–Act (PDCA) cycle, including the following steps:

PLAN:

1. Identify need: Determine the purpose of the Kaizen.
2. Form Kaizen team: Typically 6–8 team members.

3. Develop Kaizen objectives: To document the scope of the project. The objectives should be SMART (Specific, Measurable, Attainable, Realistic, and Time-based).
4. Collect current state baseline data: From Measure phase or additional data as needed.
5. Develop schedule and Kaizen event agenda: Typically 1 week or less.

DO:

6. Hold Kaizen event using DMAIC

   Sample Kaizen event agenda:
   - Review Kaizen event agenda.
   - Review Kaizen objectives and approach.
   - Develop Kaizen event ground rules with team.
   - Present baseline measure and background info.
   - Hold event:
   - Define: Problem (derived from objectives), agree on scope for the event.
   - Measure: Review collected baseline measure.
   - Analyze: Identify root causes, wastes, and inefficiencies.
   - Improve: Create action item list and improvement recommendations.
   - Control: Create standard operating procedures to document and sustain improvements; prepare summary report and present to sponsor.
   - Identify and assign action items.
   - Document findings and results.
   - Discuss next steps and close meeting.
7. Implement: Implement recommendations, fine-tune, and train

CHECK/ACT:

8. Summarize results

   Kaizen summary report items:
   - Team members
   - Project scope
   - Project goals
   - Before Kaizen description
   - Pictures (with captions)
   - Key Kaizen breakthroughs

- After Kaizen description
- Results
- Summary
- Lessons learned
- Kaizen report card with follow-up date

9. Control: If targets are met, standardize the process. If targets are not met, or the process is not stabilized, restart Kaizen event PDCA cycle.

Failure Mode and Effect Analysis

A failure mode and effect analysis (FMEA) is a systemized group of activities intended to recognize and evaluate the potential failure of a product or process, identify actions that could eliminate or reduce the likelihood of the potential failure occurring, and document the entire process.

The FMEA process includes the following steps:

1. Document process, define functions.
2. Identify potential failure modes.
3. List effects of each failure mode and causes.
4. Quantify effects: severity, occurrence, and detection.
5. Define controls.
6. Calculate risk and loss.
7. Prioritize failure modes.
8. Take action.
9. Assess results.

A simple FMEA form is shown in Figure 2.26.

The risk priority number (RPN) is calculated by multiplying the Severity times the Occurrence times the Detection. The Severity is estimated for the failure, and given a numerical rating on a scale of 1 (low severity) to 10 (high severity). The Occurrence is given a numerical rating on a scale of 1 (low probability of occurrence) to 10 (high probability of occurrence). The detection scale is reversed, where a numerical rating is given on a scale of 1 (failure is easily detected) to 10 (failure is difficult to detect).

A Pareto diagram can be created based on the RPN values to identify the potential failures with the highest RPN values. Recommendations should be developed for the highest value RPN failures to ensure that they are incorporated into the improvement recommendations in the Improve phase.

Process Step	Potential Failure Mode	Potential Effects of Failure	S E V E R I T Y	Potential Causes of Failure	O C C U R R E N C E	Current Process Controls	D E T E C T I O N	R P N	Recommended Action

FIGURE 2.26
Failure mode and effect analysis form.

Customer Survey Analysis

Most surveys are attribute or qualitative data, where you are asking the respondent to answer questions using some type of Likert scale, asking about importance, the level of agreement, or perhaps, the level of excellence.

The following are some ways to analyze survey data:

1. Summarize percent or frequency of responses in each rating category using tables, histograms, or Pareto charts.
2. Perform attribute hypothesis testing use Chi-square analysis.

Unlike hypothesis testing with variable data, statistically with attribute data, we are testing for dependence, not a difference, but you can think "makes a difference."

We formulate our hypotheses as follows:

- H_o: "{Factor A} is independent of {factor B}."
- H_a: "{Factor A} is dependent on {factor B}."

In addition to the p value, we use contingency tables to help understand where the dependencies (differences) exist.

The customer survey analysis steps include the following:

1. State the practical problem.
2. Formulate the hypotheses.
3. Enter your data in Minitab or other statistical package.

4. Run the Chi-square.
5. Translate the statistical conclusion into practical terms.

If p, the significance level, is low, then reject H_0 (if p is low, H_0 must go). If you fail to reject the null hypothesis, H_0, that means that you fail to reject the hypothesis that the values are independent. If you reject H_0 that means that they are dependent, or dependencies or differences exist.

Develop Process Capability

To develop the process capability, you can calculate the defects per million opportunities (DPMO) and related sigma level, or you can calculate the capability indices. We first discuss DPMO.

DPMO

Six Sigma represents a stretch goal of six standard deviations from the process mean to the specification limits when the process is centered, but also allowing for a 1.5 sigma shift toward either specification limit. This represents a quality level of 3.4 defects per million. This is represented in Figure 2.27.

The greater the number of σs, the smaller the variation (the tighter the distribution) around the average. DPMO provides a single measure to compare performance of very different operations, giving an apples to apples comparison, not apples to oranges. Figure 2.28 shows a sigma to DPMO conversion.

DPMO is calculated as follows:

$$\text{DPMO} = \frac{\text{Defects} \times 1{,}000{,}000}{\text{Units} \times \text{opportunities}}$$

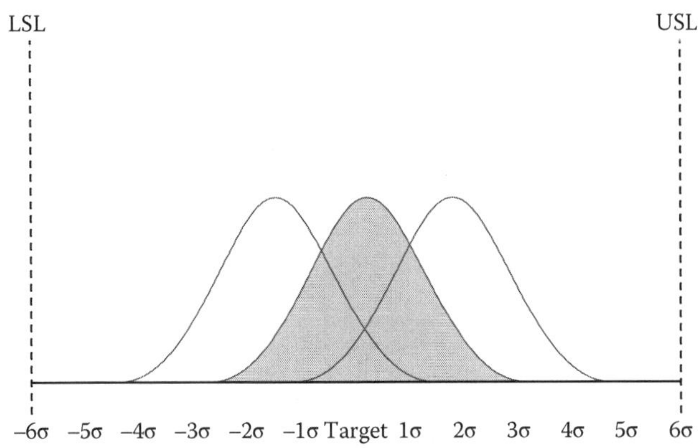

FIGURE 2.27
A 3.4 DPMO representing a Six Sigma quality level, allowing for a 1.5 sigma shift in the average. LSL, lower specification limit; USL, upper specification limit.

Sigma Level	DPMO
6σ	3.4 DPMO
5σ	233 DPMO
4σ	6,210 DPMO
3σ	66,810 DPMO
2σ	308,770 DPMO
1σ	691,462 DPMO

FIGURE 2.28
Sigma to DPMO conversion (assuming 1.5 sigma shift).

where defects is the number of defects in the sample, units is the number of units in the sample, and opportunities is the number of opportunities for error. For example, if we take a sample of 100 purchase orders with 30 fields each (opportunities for errors) and find 5 defects, we calculate a DPMO of 1667 or about 4.4 sigma.

Process Capability Study

Process capability is the ability of a process to produce products or provide services capable of meeting the specifications set by the customer or designer. You should only conduct a process capability study when the process is in a state of statistical control. Process capability is based on the performance of individual products or services against specifications. According to the central limit theorem, the spread or variation of the individual values will be greater than the spread of the averages of the values. Average values smooth out the highs and lows associated with individuals.

Following are the steps for performing a process capability study:

1. Define the metric or quality characteristic. Perform your process capability study for the metrics that measure your CTS characteristics defined in the Define and Measure phases.
2. Collect data on the process for the metric; take 25–50 samples.
3. Perform graphical analysis (histogram).
4. Perform a test for normality.
5. Determine if the process is in control and stable, using control charts. When the process is stable, continue to step 6.
6. Estimate the process mean and standard deviation.
7. Calculate the capability indices, usually Cp, Cpk.

$$\mathrm{Cp} = \frac{\text{Upper specification limit} - \text{Lower specification limit}}{6\sigma}$$

$$\text{Cpk} = \text{Minimum of}\{\text{CPU}, \text{CPL}\}$$

where

$$\text{CPU} = \frac{\text{Upper specification limit} - \text{Process mean}}{3\sigma}$$

$$\text{CPL} = \frac{\text{Process mean} - \text{Lower specification limit}}{3\sigma}$$

A process can be in control, but not meet the specifications established by the customer or engineering. You can be in control and not capable. You can be out of control or unstable, but still meet specifications. There is no relationship between control limits and specification limits. However, you must be in control before you use the estimates of standard deviation from your process to calculate process capability and your capability indices.

There are typically three scenarios regarding process capability:

1. Process spread is less than the specification spread. The process is quite capable. Figure 2.29 shows this scenario. Cp and Cpk are greater than 1.33.
2. Process spread is equal to specification spread, an acceptable situation, but there is no room for error. If the mean shifts, or variation increases, there will be nonconforming product. Figure 2.30 shows this scenario. Cp = Cpk = 1.
3. Process spread is greater than the specification spread. The process is not capable. Figure 2.31 shows this scenario. Cp and Cpk are less than 1.

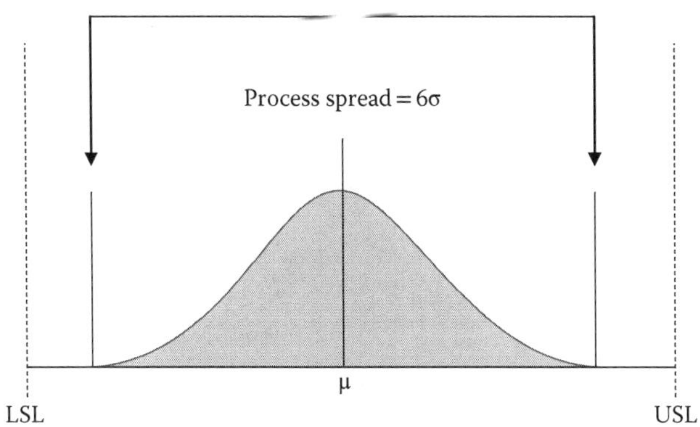

FIGURE 2.29
Process is quite capable.

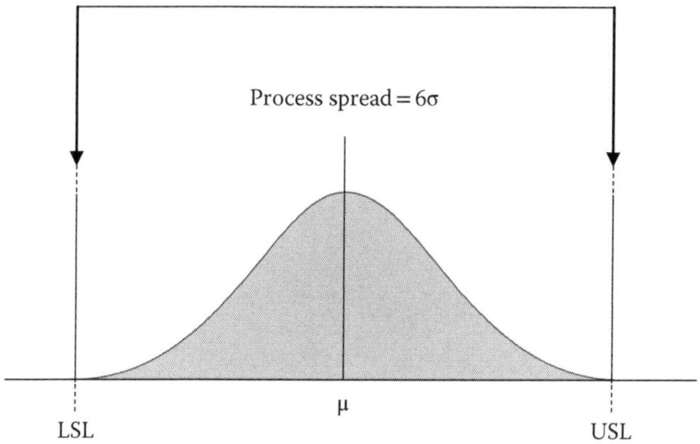

FIGURE 2.30
Process is just capable.

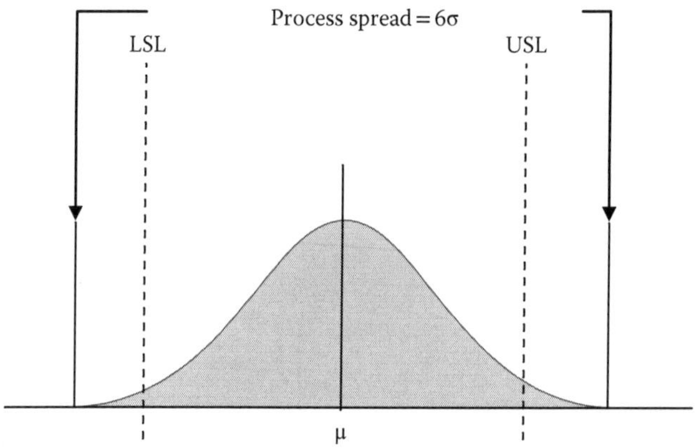

FIGURE 2.31
Process is not capable.

Improve Phase

The purpose of the Improve phase is to identify improvement recommendations, design the future state, implement pilot projects, and train and document the new processes. The activities performed and tools applied during the Improve phase are discussed in the following:

1. Identify improvement recommendations.
2. Perform cost/benefit analysis.
3. Design future state.

4. Establish performance targets and project scorecard.
5. Gain approval to implement and then implement.
6. Train and execute.

Figure 2.32 shows the main Improve activities mapped to the tools or deliverables most typically used during that step.

Identify Improvement Recommendations

The project team should use the data collected in the first three phases of the DMAIC to identify improvement recommendations. The recommendations should be designed to eliminate the root causes. The team should anticipate things that can go wrong or the type of resistance that you may have and design plans to work around these barriers.

Recommendations for Improvement

The recommendations for improvement can be a listing, with a description of the recommendation ideas. During the design of the recommendations, input should be obtained from the process owners to assess the reasonableness of the solutions. Information from the following previously used tools can be used to develop the improvement recommendations:

- Waste analysis
- Process analysis
- Cause and effect analysis (diagram and matrix)

	Improve Activities	Deliverables
1	Identify improvement recommendations	• Recommendations for improvement • Action plan
2	Perform cost/benefit analysis	• Cost/benefit analysis • Cost of poor quality
3	Design future state	• Future state process map • Future state value stream map • One-piece flow • Cellular design • SMED • Flow • Pull • Kanban
4	Establish performance targets and project scorecard	• Dashboards/scorecards • Revised VOP matrix
5	Gain approval to implement, and implement	• Project presentations
6	Train and execute	• Training plans, procedures

FIGURE 2.32
Improve phase activities and tools/deliverables.

- Why–why diagram
- 5S
- Kaizen
- Process map or value stream map

Action Plan

The action plan for the improvement recommendations can be divided into short-term and long-term recommendations. Short term is 3 months or less, and long-term recommendations are those that can be implemented in greater than 3 months.

The action plan should include project plans, with resources, timelines, and risk analysis.

Change management is important when planning the improvement recommendations pilot projects. The following steps can be used to enhance the probability of success for the pilot projects.

1. Create a vision of the future state and communicate with the stakeholders.
2. Understand what is in it for each stakeholder; include them in the solution.
3. Identify who will be resistant to change and who will be receptive. Gain some early adopters and engage them in the change process.
4. Implement the change, incorporating training and appropriate resources.
5. Monitor improvement and measure and assess results.
6. Implement new processes, systems, and organizational structures if needed.

Perform Cost/Benefit Analysis

Cost/Benefit Analysis

You can perform a cost/benefit analysis or a COPQ assessment to determine the return on investment or savings that are estimated by implementing the improvement recommendations.

In a cost/benefit analysis, we need to decide whether the benefits of the project outweigh the costs. We can use the benefit/cost ratio:

$$\text{B/C ratio} = \frac{\text{Present value of benefits}}{\text{Present value of costs}}$$

If the B/C ratio is greater than 1, then you can accept the project; the benefits outweigh the costs.

Cost of Poor Quality

The COPQ categories can be used to assess the costs that can be eliminated by implementing the improvement recommendations. The team can either use this in the cost/benefit analysis or compare the improvement in the COPQ before and after the recommendations are implemented.

Design Future State

It is important to design the new future state by developing a future state process map and/or a future state value stream map. The team should challenge the boundaries and incorporate quality and Lean principles.

Future State Process Map

The future state process map is simply a process map of the new process incorporating the improvement recommendations.

Establish Performance Targets and Project Scorecard

In this step, the team should identify the performance targets for the metrics identified in the Measure phase. They also should track pilot project status using project scorecards.

Dashboards/Scorecards

You should also create dashboards or scorecards to assess the performance of your process after trying out the improvement recommendations in the pilot projects.

The project scorecards or dashboards can be used to identify improvements in your process against the metrics that you have identified. The metrics should relate to your CTS characteristics.

Scorecards should include the following ways to present your metrics (Pyzdek, 2003):

- Assessment of improvement of central tendency and variation overtime. SPC average and range charts can be used to meet this objective.
- Graphical distribution using a histogram for the most recent time period.
- Assessment of quality or number of defects, using a SPC percent defective chart.
- Outliers showing a distribution of individual defectives using a dot plot.

Revised VOP Matrix

You can revise your VOP matrix from the Measure phase to relate your CTS, process factors, operational definition, metrics, and more realistic targets for your metrics.

Other tools that can be applied during the Improve phase can include cellular design, SMED, flow, pull, kanban, standard work, and visual management. Detailed discussion on these Lean tools is included Chapters 4, 5, 6, and 8.

Gain Approval to Implement, Then Implement

The project team should create a presentation and deliver it to the project sponsor, champion, and other management that must approve the improvement recommendations. The champion presentation should be a high-level executive summary.

The team can use the PDCA cycle to implement the pilot recommendation projects. Plan—you can plan your improvements; Do—implement your improvements, usually on a pilot scale; Check—verify that the improvement improved the process based on your metrics; and Act—if the improvements made a positive difference, implement them on a broader scale; if not, refine the improvements and try again. You may go through the PDCA cycle several times.

Train and Execute

The team should develop detailed procedures, as necessary, to ensure consistency of the new process. They should develop and roll out training. The train the trainer concept is sometimes used to reduce the resources needed to train. A core group of people are trained on the new process and then they train others in the organization and become subject matter experts. The process owners should be included in the change process, and changes should be communicated to appropriate stakeholders. The team can use the future state process map as a training guide. They should assess the effectiveness of the training as part of the control plan in the next phase.

Control Phase

The purpose of the Control phase is to measure the results of the pilot projects, manage the change on a broader scale, report scorecard data and the control plan, identify replication opportunities, and develop future plans for improvement. The activities performed and tools applied during the Control phase are discussed in the following.

1. Measure results and manage change.
2. Report scorecard data and create process control plan.

	Control Activities	Deliverables
1	Measure results and manage change	Hypothesis tests
2	Report scorecard data and create process control plan	Basic statistics, graphical analysis, mistake proofing/poka-yoke, FMEA, control plan, process capability, visual management
3	Apply PDCA process	Replication opportunities
4	Identify replication opportunities	Standard work, Kaizen
5	Develop future plans	Dashboards/scorecards, action plans

FIGURE 2.33
Control phase activities and tools/deliverables.

3. Apply PDCA process.
4. Identify replication opportunities.
5. Develop future plans.

Figure 2.33 shows the main Control activities mapped to the tools or deliverables most typically used during that step.

Measure Results and Manage Change

In the Control phase, the team should verify that the training and implementation was carried out correctly. They need to collect and analyze data to ensure process performance and improvements were made. The team needs to further manage the change for rollout of the pilot recommendations on a broader scale. The team also needs to keep all of the stakeholders in the loop by developing and implementing a communication plan. They will collect data after they improve the process, for the same CTS and metrics identified in the Measure phase. The team will then assess whether the changes implemented made a "statistically" significant difference, using hypothesis testing.

Hypothesis Testing

The purpose of hypothesis testing is to

- Determine whether claims on process parameters are valid.
- Understand the variables of interest, the CTS.
- Understand whether the improvements implemented have had a positive impact on the CTS and factors impacting the processes.

The hypothesis test begins with some theory, claim, or assertion about a particular characteristic (CTS) of one or more populations or levels of the X (independent variable).

The null hypothesis is designated as H_0 (pronounced "H-O") and defined as representing that there is no difference between a parameter and a specific value. The alternative hypothesis is designated as representing that there is a difference between a parameter and a specific value. The null hypothesis is assumed to be true, unless proven otherwise. If you fail to reject the null hypothesis, it is not proof that the null is true.

In hypothesis testing, there are two types of error. Type I error or alpha risk is the risk of rejecting the null hypothesis when you should not. The probability of a Type I error is referred to as alpha. Type II error or beta risk is the risk of not rejecting the null hypothesis when you should. The probability of a Type II error is referred to as beta.

When performing a hypothesis test, you select a level of significance. The level of significance is the probability of committing a Type I error or alpha and is typically a value of 0.05 or 0.01. Figure 2.34 shows the Type I and Type II errors.

The steps for performing a hypothesis test are

1. Formulate the null and alternative hypotheses.
2. Choose the level of significance, alpha, and the sample size *n*.
3. Determine the test statistic.
4. Collect the data and compute the sample value of the test statistic.
5. Run the hypothesis test in Minitab or a statistical package.
6. Make the decision. If the *p*-value is less than our significance level, alpha, reject the null hypothesis; if not, then there is no data to support rejecting the null hypothesis. Remember, if *p* is low, H_0 must go!

Some of the most common hypothesis tests are summarized in Figure 2.35.

Report Scorecard Data and Create Process Control Plan

In this step, the team should demonstrate the impact of the project's metrics and create or revise the process control plan. The process control plan helps to deploy the Lean Six Sigma approach across large areas and to coach groups through the major quality processes.

The purpose of the control plan is to maintain the gains. If a conscious plan and effort is not made to ensure that people continue to use the new process,

		Conclusion Drawn	
		Do not reject H_0	Reject H_0
Actual or true state	H_0 true	Correct conclusion	Type I error
	H_0 false	Type II error	Correct conclusion

FIGURE 2.34
Type I and Type II errors.

Test Statistics	Number of Variables	Test	Parameters
Mean	1	1 sample Z	Variance
Mean	1	1 sample t	Variance unknown
Mean	2	2 sample t	Variance unknown, assume equal variances
Mean	2	2 sample t	Variance unknown, do not assume equal variances
Mean	2	Paired t-test	Paired by subject (before and after)
Proportion	1	1 proportion	
Proportion	2	2 proportion	
Variance	1	1 variance (Chi-Square)	
Variance	2	Variance (F-test)	

FIGURE 2.35
Summary of hypothesis tests.

the gains can slip, and when push comes to shove, and people get pressured and busy, they can very easily slip back to their old ways and old processes. The control plan can include the following:

- Deploying new policies and removing outdated policies
- Implementing new standards
- Modifying procedures
- Modifying quality appraisal and audit criteria
- Updating prices and contract bid models
- Changing information systems
- Revising budgets
- Revising forecasts
- Modifying training

Useful tools that can be used to derive the information to create a control plan include

- Project planning for creating the control plan
- Brainstorming
- Failure mode and effects analysis (FMEA)
- SPC
- Process map

Process Step	Control Mechanism	Measure/ Metric	Criticality (High, Medium, Low)	Action to be Taken if Problems Occur	Responsibility

FIGURE 2.36
Control plan.

- Training
- Procedures
- Mistake proofing
- Statistics, graphical tools, sampling, process capability, and DPPM/DPMO

A control plan format is provided in Figure 2.36. For each major process step on the future process map, the control plan should identify how you will control the process step (control mechanism); how you will measure the process step; how critical it is to ensure control for that step; actions to be taken if problems occur; and who is responsible for monitoring control for each process step.

Mistake Proofing

Mistake proofing is a tool that helps to prevent errors in your process. Errors are inadvertent and unintentional, and accidental mistakes made by people because of the human sensitivity designed into our products and processes.

Mistake proofing, also called poka yoke, is the activity of awareness, detection, and prevention of errors, which adversely affect our customers and our people and result in waste.

Some of the underlying mistake proofing concepts are as follows:

- You should have to think to do it wrong, instead of right.
- Easy to perform inspection at the source.
- Reduces the need for rework and prevents further work (and cost) on a process step that is already defective.
- Simplifies prevention and repair of defects by placing responsibility on the responsible worker.

Apply PDCA Process

Applying the PDCA process to help people continually improve the process. There is the need to focus on the following:

- What are we trying to accomplish?
- How will we know that change is an improvement?
- What change can we make that will result in improvement?
- If the process is performing to plan, then standardize the activities; if not, then study why not and develop a new plan for improvement.
- Focus on the next most important root cause and implement additional improvements.

Identify Replication Opportunities

In this step, it is important to identify opportunities where you can replicate the same process in the organization. This will leverage the improvement effort across the organization and potentially save additional money for the company. Identifying replication opportunities can help to support organizational learning. An example of replication opportunities may be to implement 5S in other parts of the organization.

Develop Future Plans

The purpose of developing future plans is to recognize the time and effort that went into the Lean Six Sigma project by reflecting on the lessons learned and incorporating these into future projects. Some important questions and considerations are

- Have you identified lessons learned?
- Have you identified the next opportunity for improvement?
- Have you shared the learnings with others?
- Have you documented the new procedures?
- Has everyone been trained that needs to be?
- Take time to celebrate.

Dashboards and scorecards can be used to assess where you need to focus improvement efforts in the future. Also, the cause and effect analysis can be used to identify the next root cause to focus improvements on.

Summary

This chapter provided a comprehensive overview of the DMAIC problem-solving approach along with key Lean and Six Sigma tools that can be applied in each phase.

References

Alukal, G. Create a Lean, Mean Machine. *Quality Progress*. Milwaukee, WI: ASQ, 2003.

Brassard, M., and D. Ritter. *Sailing through Six Sigma: How the Power of People Can Perfect Processes and Drive Down Costs*. Marietta, GA: Brassard & Ritter, 2001.

Dubai Quality Group. *The Birth of Lean Sigma*. Dubai: The Manage Mentor, 2003.

Evans, J., and M. Lindsay. *The Management and Control of Quality*. Fifth Edition. Independence, KY: South-Western Thomson Learning, 2002.

Furterer, S. Total Quality Improvement Course Lecture Notes, ESI 5227, University of Central Florida, Department of Industrial Engineering and Management Systems, Orlando, FL, 2004.

Furterer, S. L. Invited Panelist at ASQ Conference on Quality in the Space and Defense Industries. *Critical Quality Skills of Our Future Engineers*, Cape Canaveral, FL, March 2006.

George, M., D. Rowlands, M. Price, and J. Maxey. *Lean Six Sigma Pocket Toolbook*. New York: McGraw-Hill, 2005. Print.

Gitlow, H. S., and D. M. Levine. *Six Sigma for Green Belts and Champions: Foundations, DMAIC, Tools, Cases, and Certification*. Upper Saddle River, NJ: Pearson/Prentice Hall, 2005.

Harrington, R. 11th International Conference on Industry, Engineering, & Management Systems, Cost of Poor Quality, March 14–16, 2005, Cocoa Beach, FL.

Kandebo, S. Lean, Six Sigma Yield Dividends for C-130J. *Aviation Week & Space Technology*, New York, July 12, 1999.

Malone, L. Class Notes Guest Lecture, ESI 5227, University of Central Florida, Department of Industrial Engineering and Management System, Orlando, FL, 2005.

McIlroy, J., and D. Silverstein. Six Sigma Deployment in One Aerospace Company. Six Sigma Forum, www.sixsigmaforum.com, 2002.

Process Flow Analysis Training Manual. Control Data, (internal publication) 1982.

Pyzdek, T. *The Six Sigma Handbook: A Complete Guide for Green Belts, Black Belts, and Managers at All Levels*. New York: McGraw-Hill, 2003. Print.

Sheridan, J. *Aircraft-Controls Firm Combines Strategies to Improve Speed, Flexibility and Quality*. Farmington Hills, MI Gale Group, Penton Media, Inc., 2000.

University of Wisconsin Focus Group Guidelines. Superior, WI: University of Wisconsin, 2005. http://www.uwsuper.edu/cipt/exsite/upload/Focus_Group_Guidelines.pdf.

Womack, J. P., and D. T. Jones. *Lean Thinking: Banish Waste and Create Wealth in Your Corporation*. New York: Simon & Schuster, 1996.

Bibliography

Certified Six Sigma Primer, Quality Council of Indiana, West Terre Haute, IN, 2001.

Section II

Lean Tools and Step-by-Step Implementation

3

Value Stream Mapping

Elizabeth A. Cudney

CONTENTS

Introduction

Once your organization has determined its strategic goals, you need to baseline the current state of the critical processes. Current state value stream mapping (VSM) is a valuable tool for understanding and documenting an end-to-end process. This chapter will provide you with a review of VSM so that you can see how value flows within your processes.

Purpose of Value Stream Mapping

VSM is a flowcharting method originally created in the Toyota Production System (TPS) to document the entire process (of a company or a department) on a single sheet of paper to encourage dialogue and understand the process better. First, you create a current state map of how the value presently flows in your organization; then, utilizing the principles of

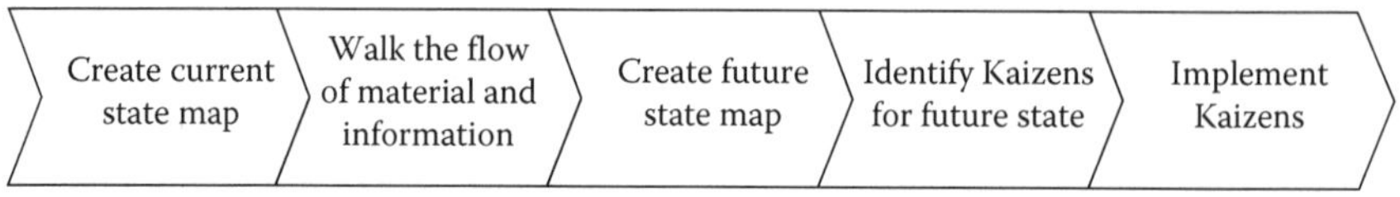

FIGURE 3.1
Value stream mapping (VSM) steps.

Lean, you envision a better state for how value "should" flow in an optimum manner in your organization: this is "future state VSM." Kaizen activities, which are events to overcome deficiencies in the current state that will allow the company to reach the future state, are identified for implementation on the current state map. The future state map illustrates the ideal state after the changes are implemented. It represents how your processes should flow.

When you use VSM, you can simultaneously analyze the flow of information and material to enable your organization to eliminate waste in both. By considering material flow and information flow concurrently, it is easy to identify and correct whether material flow is hampering information flow or vice versa. The overall steps of VSM are shown in Figure 3.1.

VSM captures the flow of a product from the point that raw material enters the process to the point where a final product is delivered to the customer. This includes all value-added and non-value-added activities that have been accepted as necessary to produce the product. You can apply VSM to any product or service; therefore, it is finding widespread application in nonmanufacturing processes, such as designing a product, and in business processes, such as purchasing, billing, and selling.

Also, because VSM enables visualization of an entire process as a holistic set of operations rather than as a disjointed movement of material through discrete operation steps, it enables interaction between the operators to help see how they can optimize an entire process, rather than seeking out only local optimization for specific operations. In doing so, the following eight wastes—which have the acronym of DOWNTIME—become evident:

1. *Defects*: Inspection or repair of a product or service to fulfill customer requirements. Root causes of defects include poor process control, insufficient preventive maintenance, lack of training, and inadequate tools or equipment.
2. *Overproduction*: Producing more than needed or faster than needed. Root causes of overproduction include incapable processes, long changeovers, just-in-case logic, insufficient preventive maintenance, and lack of level schedules.

3. *Waiting*: Idle time produced when people wait on machines or machines wait on people or people wait on people. Root causes include unbalanced workloads, unplanned downtime, long change-overs, insufficient preventive maintenance, and quality problems.
4. *Nonutilized resources/talent*: Poor utilization of the existing abilities, talents, skills, and ideas of employees. Root causes include lack of training, lack of teamwork, poor management involvement and understanding, and poor communication.
5. *Transportation*: Any material movement not part of a value-added operation. Root causes include multiple storage locations, large lot processing, poor plant layout, unleveled scheduling, and poor work-place organization.
6. *Inventory*: Any supply in excess of the customer's just-in-time demand. Root causes include incapable processes, incapable suppliers, unleveled schedules, poor communication, long changeovers, insufficient preventive maintenance, and low uptime.
7. *Motion*: Any movement of people or machines that does not contribute added value to the product or service. Root causes include poor workplace organization, poor plant layout, and inconsistent work methods.
8. *Excess processing*: Human or machine effort that adds no value (only cost) to the product or service. Root causes include product changes without process changes, just-in-case logic, undefined or unclear customer requirements, and lack of training.

The waste of nonutilized talent/resources is not part of the seven classical wastes in the original Japanese wastes; however, recognizing that people are the most important part of Lean has led to the addition of this waste.

VSM comprises all the value-added and non-value-added activities required to manufacture a product. The essential process flow can include bringing a product from raw materials through delivery to the customer or designing a product from concept to job production. A "value-adding activity" is any activity that transforms raw material to meet your customers' requirements. "Non-value-adding activities" are those activities that take time, resources, or space but do not add to the value of your product or service. The customer is not willing to pay for non-value-adding activities. Unfortunately, typically, 95% of process time is spent in non-value-adding activities, and only 5% of the time is spent in value-adding activities, adding value to the product or service.

VSM also provides a common platform to apply various Lean principles and tools and allows an organization to create an integrated plan to follow for implementation. Further, because it captures the delay caused by material and information flow on the same page, it is possible to tackle the two together, to create an optimized process such that the lack of one is not hampering the flow of the other.

Benefits of Value Stream Mapping

VSM, used as a tool, provides several benefits to a process. The first main benefit is enabling people to see the entire process rather than just a single step. The flow of the entire process becomes apparent after using VSM. This also makes the sources of waste in the value stream evident. With a clear process flow and identified sources of waste, the decisions needed to improve the flow are also visualized.

Another major benefit to the VSM technique is that it utilizes a format that provides a common language for the manufacturing process, which ties together Lean concepts and techniques. VSM is also the only tool that currently provides a link between information flow and material flow. VSM is conducive in any environment that contains a process to meet a desired objective.

As mentioned at the beginning of this chapter, VSM was taken from the TPS method called "Material and Information Flow Mapping." The TPS used this method more as a means of communication by individuals learning through hands-on experience. It is used to illustrate the current and future states of a process to implement Lean systems. The focus at Toyota is to establish a flow, eliminate muda (the Japanese word for waste), and add value. Toyota teaches the following three flows in manufacturing:

1. Flow of material
2. Flow of information
3. Flow of people and processes

VSM is based on the first two of these three, that is, the flow of material and information.

Product Families

A product family is a group of products that use similar processing steps and/or common equipment. In VSM, the goal is not to create a value stream map for every individual product. The goal is to improve operational issues. Therefore, grouping similar products enables management to act on the most important operational issues that affect a group of products to gain the

	Processes								
Product									

FIGURE 3.2
Product family matrix.

biggest improvement impact. This is achieved by first defining the product families through a product family matrix as shown in Figure 3.2.

To create a product family matrix, you start by listing your products down the left column. The process steps are then listed at the top of the matrix by working backward from the customer. The product in the matrix is then reviewed for each process step listed. If the product goes through that step, an "X" is marked in the portion of the matrix. To determine the product families, you review the overall matrix for several products that are processed using common process steps. These products are then grouped into product families.

Creating Current State and Future State Maps

The initial step is to map the current process in a brainstorming process. This step is often more effective when involving several cross-functional stakeholders who offer varied degrees of opinions and perspectives. Taking an initial gemba walk is generally helpful to document the process and kick-start creative ideas. As a team, in evaluating the current state of a process, you can typically identify several improvements, including cellular manufacturing, one-piece flow, single-minute exchange of dies (covered in more detail in Chapter 5), and Kaizen wherever possible.

Next, the team should map the improved process to represent the desired future state. The purpose of the current state map is to make a clear representation of the production situation by drawing the material and information flows. As such, if the team is having difficulty mapping the process flow, a helpful tip is to start at the customer, in the upper right portion of the value stream map, and detail the process moving upstream to clearly define the process.

The goal of Lean is to get one process to make only what the next process needs when it needs it by linking the processes from raw material to the final customer. Value can be defined by the customer as a product that is delivered at the right time, with the defined specifications and at the right price.

The purpose of VSM is to identify sources of waste and eliminate the waste by implementing a future state value stream. The first pass of implementing a future state value stream should ignore the inherent waste from product

Symbol	Description
	Manufacturing process
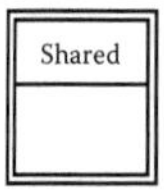	Shared manufacturing process
	Outside supplier
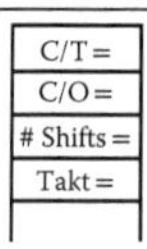	Data box
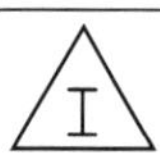	Inventory box
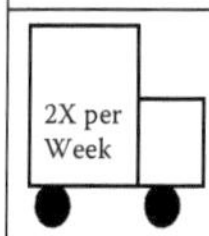	Truck shipment
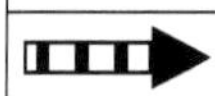	Push arrow
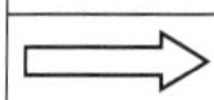	Finished goods to customer
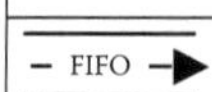	First-in-first-out sequencing

FIGURE 3.3
VSM icons.

	Supermarket
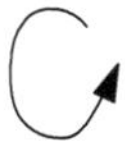	Physical pull
	Move by forklift
	Manual information flow
	Electronic information flow
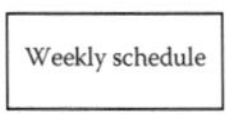	Schedule
	Load leveling
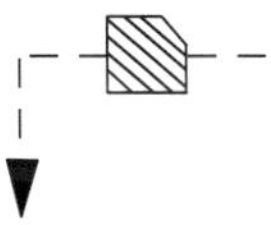	Withdrawal kanban
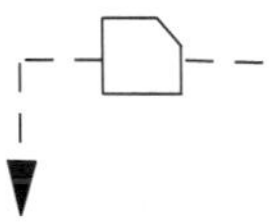	Production kanban
	Signal kanban
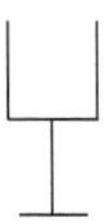	Kanban post

FIGURE 3.3 ***(Continued)***
VSM icons.

	Sequenced pull ball
	Go see production scheduling
	Kaizen lightening burst
	Buffer or safety stock
	Operator
	Quality problem

FIGURE 3.3 (*Continued*)
VSM icons.

design, current processing machinery, and the location of some activities because these changes may require a great deal of work and will not be changed immediately. You should address these features in later iterations of your maps. Remember that your first future state mapping activity will not be your last. You should continue to expand on your future state mapping through iterative activities held periodically.

The common VSM symbols are given in Figure 3.3.

Steps of Value Stream Mapping

In VSM, it is important to walk the process to start with a process map of how the process actually works today. Walk a product's production path from "beginning to end" and draw a "visual representation" of every transportation, storage, and overall product process in the "material and information flows." When starting out, it is preferable to hand draw/sketch a VSM first when walking a product's production path. Once everyone is in agreement of the appropriate process, then formally draw the VSM icons. The future state map will be drawn later and shows how value should flow.

As you walk the process, you should collect data on the process to add to the map. Make sure to differentiate between a human's effort and a machine's effort. During the assessment, make sure to answer a few critical questions:

- How long does each step take?
- What is the queue time between each step?
- What are the inventory levels?
- What are the scrap and rework rates?

There are 10 main steps to VSM:

1. Select the product family.
2. Define the boundaries.
3. Define the value.
4. Walk the process—go to the gemba.
    a. Identify tasks and flows of material and information between them.
5. Gather data.
    a. Identify resources for each task and flow.
6. Create the current state map.
    a. Understand how the process currently operates.
7. Analyze current conditions.
    a. Identify value added and waste.
    b. Reconfigure process to eliminate waste and maximize value.
8. Visualize the ideal state.
9. Create the future state map.
    a. Design a Lean flow.
10. Develop and track action plans.

Observe and record the flow of orders, materials, and information for a product family. Then identify non-value-added activities. It is important to visualize areas of waste and areas that generate value.

In VSM, it is important to understand your takt time. Based on your takt time, the focus of the future state map is on developing continuous flow, sending the schedule to only one process, leveling the production mix evenly, and producing every part every day.

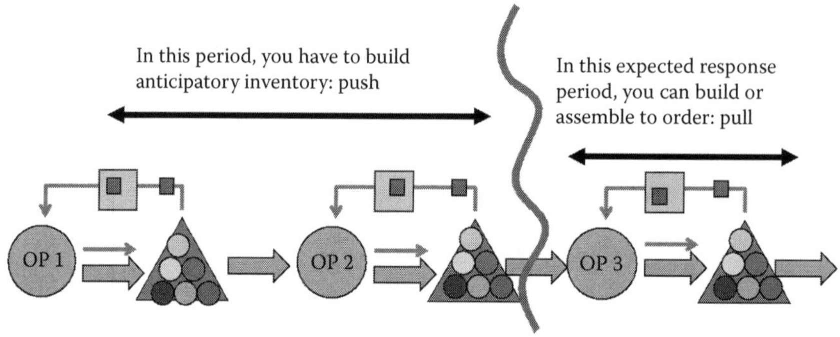

FIGURE 3.4
Pacemaker process.

Future State Maps

In developing future state maps, it is critical to revisit the five Lean fundamentals developed by Womack and Jones (1996):

1. *Specify value*: Value is defined by customer in terms of specific products and services.
2. *Identify the value stream*: Map out all end-to-end linked actions, processes, and functions necessary for transforming inputs to outputs to identify and eliminate waste.
3. *Make value flow continuously*: Having eliminated waste, make remaining value-creating steps "flow."
4. *Let customers pull value*: Customer's "pull" cascades all the way back to the lowest level supplier, enabling just-in-time production.
5. *Pursue perfection*: Pursue continuous process of improvement striving for perfection.

The goal is to make value flow. To create flow, focus on the product or service that is flowing through the process. It is important to not be limited by organizational or functional boundaries. In addition, you should rethink work practices and the tools used to get the job done by reducing or eliminating bottlenecks and stoppages. Finally, you should focus on controlling the process.

Part of determining and achieving the future state is locating the pacemaker. The pacemaker is the point in the process where the customer's order enters the process, as shown in Figure 3.4. The location of the decision point is governed by customer-expected response time and build time. If response time is very long, you do not need any work-in-process (WIP) inventory and can have full build to order. By cutting waste, you should be able to move the decision point upstream.

Macro Value Stream Mapping

Macro value stream maps extend beyond plant level maps. There are three key reasons to create macro value stream maps. First, purchased materials are the single largest cost driver of the process. Second, costs from downstream activities can potentially negate any internal cost savings realized from process improvements. Finally, macro value stream maps identify who does what where. This information can be used for major asset reconfigurations that can impact the entire supply chain.

The current state macro VSM should focus on a product or partner that is important to your business and include practical limits. You should begin drawing the current state by mapping the facility closest to the customer and then work backward to create the map. Just as in the process or plant level value stream map, you should collect data for each step.

For the current state map, you should calculate the takt time and focus on leveling the demand upstream from the customer. To create flow and pull, it is useful to create a supermarket downstream from each facility and implement cells in each facility.

It is important to note that VSMs should extend beyond a single facility at some point in a Lean enterprise. Once VSM initiatives have gained success within companies, it is important to transplant VSM efforts across a single company and eventually beyond the company walls. Known as an "extended value stream map" (Jones and Womack, 2009), VSM should eventually engage multiple sites and even suppliers upstream and downstream from the targeted facility. This transparency and partnership among suppliers allow multiple mapping activities to be conducted among several companies all working together to produce a good or service. Jones and Womack (2009) describe the conditions for an extended VSM effort:

> Every facility along the stream needs to be aware of the end rate of consumption to calculate facility-specific takt time. Production at every upstream stage should run on average at the same rate, as adjusted for the available amount of production time at each step and the need to make multiple copies of some products to incorporate in products downstream. Any time we see a chronic pattern of imbalanced production rates in different facilities, we know we do not have a Lean value stream.

Conclusion

Now we have reviewed the basics of VSM, and Chapter 4 will discuss 5S and visual management. Chapters 11 through 21 provide case studies of value stream maps for manufacturing, service, and healthcare applications.

Homework Problems

1. Define VSM.
2. Describe how a value stream map is tied to eliminating the eight wastes.
3. Fill in the blank: "VSM is based on the flow of ____________ and the flow of ____________."
4. What is a non-value-added activity? Provide a specific example.
5. A value stream map does not normally provide data on
   a. Changeover time
   b. Cycle time
   c. WIP inventory
   d. Supplier's finished good inventory
6. What is the purpose of VSM? Explain the main benefits.
7. Describe the key steps of VSM.
8. What are the three flows in manufacturing that Toyota teaches?
9. What is the Japanese word for waste?
10. Which waste is defined as making more than is required by the next process?
11. Explain the following statement: "Waste is really a symptom, rather than a root cause of the problem."
12. What are the key aspects of properly planning a Kaizen event?
13. Lean is most concerned with which of the following? Explain your answer.
    a. Reducing waste
    b. Reducing people
    c. Reducing management layers
    d. Eliminating bottlenecks in a process
14. Which waste is defined as any movement of people or machines that does not contribute added value to the product or service?
15. Where should the pacemaker be located?
16. What is a macro value stream map? Where should it start?
17. Explain how product families are selected.
18. Define the term "gemba."
19. Should a value stream map be drawn by a team in a conference room? Why?

20. What type of information listed below should be used in the data boxes for a value stream map? Explain your reasoning.
    a. Engineering standards.
    b. Average measurement for a fiscal year.
    c. Measurement for an ideal day.
    d. What you observe as you draw your map.
21. Typically, how much of the total lead time is non-value-added?
22. Should the current state value stream map include value-added and non-value-added activities? Explain your answer.
23. Observe a repetitive process in action that generates a product or service. This may be someone stocking fruit in a grocery store or action behind the counter of your local fast food restaurant, for example. Roughly diagram the process and document three "value-added" activities and three "non-value-added" activities associated within the process.
24. Develop the current state value stream map for the following organization:
    a. Production Processes
        i. XYZ's process for this product family (Products 1 and 2) involves Plants 1, 2, and 3 using the same processes and transportation methods.
        ii. Switching between Product 1 and Product 2 requires no changeover time at Plant 1, 2 hours at Plant 2, and an 8 hour changeover in Plant 3. Transportation time is 5 days by barge between Plants 1 and 2, 3 days by rail from Plant 1 to Plant 2, 5 days by rail and truck from Plant 1 to the customer, 6 days by rail from Plants 2 and 3, and 1 day by truck between Plant 3 and the customer.
        iii. Raw materials at each plan are supplied by various vendors. Raw materials are delivered on a daily basis.
    b. XYZ Production Control Department
        i. Receives ABC Carpet's 90/60/30 day forecasts and enters them into manufacturing resource planning (MRP).
        ii. Issues XYZ's 6 week forecast to ChemCo Supplier via MRP.
        iii. Secures raw materials by weekly faxed order release to ChemCo Supplier.
        iv. Generates MRP-based, weekly departmental requirements based on customer orders.
        v. Issues weekly build schedules to Plants 1, 2, and 3.
        vi. Issues daily shipping schedule to shipping department at Plants 1, 2, and 3.

c. Process Information
    i. All processes occur in the following order with Products 1 and 2 going through all processes.
d. Plant 1 Receiving
    i. Raw material is shipped in by pipe, rail, and truck.
    ii. Raw material is stored on-site in tanks.
    iii. Production of some raw materials is on-site.
e. Plant 1
    i. The plant makes the same material for Products 1 and 2.
    ii. Cycle time is 12 days.
    iii. No changeover time is required.
    iv. Process reliability is 90%.
    v. Yield is 95%.
    vi. First pass yield is 87%.
    vii. Shut down for 3 weeks every 18 months to refurbish the process.
    viii. Observed inventory:
        A. 5 days of hydrogen cyanide (HCN) (in tank).
        B. 0 days of propane (in pipe).
        C. 30 days of platinum (catalyst).
f. Plant 1 Shipping
    i. Products 1 and 2 are shipped out by barge and rail to Plant 2 (83%) and to customers (17%) by rail and truck.
    ii. Products 1 and 2 are stored on-site in tanks before shipment.
g. Plant 2 Receiving
    i. Products 1 and 2 are shipped in by barge and rail.
    ii. Products 1 and 2 are stored on-site in tanks before use.
h. Plant 2
    i. The plant makes the same material for both Products 1 and 2.
    ii. Cycle time is 30 days.
    iii. Two hour changeover time is required.
    iv. Process reliability is 82%.
    v. Yield is 85%.
    vi. First pass yield is 68%.
    vii. Shut down for 2 weeks every 24 months to refurbish the process.

        viii. Observed inventory:
            A. 25 days of Product 1 (in tank).
            B. 12 days of Product 2 (in tank).
            C. 30 days of mixture 3Z4 (in tank).

    i. Plant 2 Shipping
        i. Products 1 and 2 are shipped out by rail and truck.
        ii. Products 1 and 2 are stored on-site in tanks before shipment.

    j. Plant 3 Receiving
        i. Products 1 and 2 are shipped in by rail and truck.
        ii. Products 1 and 2 are stored on-site in tanks before use.
        iii. Observed inventory:
            A. 35 days of Product 1 (in tank).
            B. 32 days of Product 2 (in tank).

    k. Plant 3
        i. The plant makes a different item for both Products 1 and 2.
        ii. Cycle time is 30 days.
        iii. Eight hour changeover time is required.
        iv. Process reliability is 72%.
        v. Yield is 75%.
        vi. First pass yield is 62%.
        vii. Shut down for 3 weeks every 18 months to refurbish the process.
        viii. Observed inventory:
            A. 35 days of Product 1 (in warehouse).
            B. 32 days of Product 2 (in warehouse).

    l. Plant 3 Shipping
        i. Products 1 and 2 are shipped out by rail and truck.
        ii. Products 1 and 2 are stored on-site in warehouses before shipment.

References

Jones, D.T., and Womack, J. Seeing the Whole. Cambridge, MA: Lean Enterprise Institute, 2009.

Womack, J., and Jones, D.T. *Lean Thinking*. New York: Simon & Schuster, 1996.

4

Using 5S and Visual Management to Create a Clean and Manageable Work Environment

Elizabeth A. Cudney

CONTENTS

Introduction

While value stream mapping (described in Chapter 3) is a Lean tool that improves product quality by reducing process variation, 5S is another Lean tool that improves quality and productivity in a different way. 5S builds a foundation for a company to deliver high-quality products and services in the right quantity at the right time to satisfy and exceed customer requirements. It is key to creating a work environment that focuses on quality. 5S also creates a safer and more pleasant place to work for employees. It makes it possible to increase productivity and improve quality. The term 5S is derived from five Japanese terms for practices that lead to a clean and manageable work environment.

Overview of 5S

There are five steps in 5S, defined briefly here and discussed in more detail in the rest of this chapter:

- *Step 1 is* Seiri *or Simplify.* In this step, the team must determine what is necessary and unnecessary. The unnecessary must then be disposed of properly.
- *Step 2 is* Seiton *or Straighten.* The team organizes the necessary items so they can be found easily, used, and then easily returned to the proper location.
- *Step 3 is* Seiso *or Scrub.* This includes not only cleaning the equipment but also cleaning the floor and furniture in all areas of the workplace.
- *Step 4 is* Seiketsu *or Stabilize.* In this step, the first three steps are maintained and improved.
- *Step 5 is* Shitsuke *or Sustain.* The discipline or habit of properly maintaining the correct 5S procedures is maintained.

In order for a 5S program to be successful, it must be fully supported by management. They must give guidance, coordination, support, and proper communication. Management must provide a suitable environment for employees to utilize their skills properly. An important element is management's support to encourage the team to focus on value-added activities to expose problems and respond accordingly.

Benefits of 5S

There are numerous benefits from implementing 5S principles:

- Quality is improved by clearly identifying all necessary objects in an area.
- 5S reduces waste by eliminating unnecessary steps to search for necessary materials, tools, or equipment.
- Safety is also improved by providing a place for everything and having everything in its place.
- Maintenance is easier by having all the needed tools in a specific location.
- 5S improves profitability by reducing waste, improving quality, and eliminating all unnecessary equipment.

5S also provides benefits that directly relate to the employee:

- Employees take more pride in their workplace when the area is clean and organized.
- Respect is gained from associates and customers after they visit the area.
- Employees also keep a more positive mental attitude working in this environment.
- 5S can build trust, remove frustration, and improve morale.

The 5S Steps

Step 1: *Seiri* or Simplify the Workplace

The first step of 5S is to Simplify (*Seiri*). In the first step, all items in the work area should be sorted. The team clearly distinguishes between items that are necessary and items that are unnecessary. The unnecessary items are then disposed of properly. *Seiri* means separating the necessary equipment such as tools, parts, and instructions from the unneeded materials and removing the latter.

An example of how simplifying improves quality can be seen with two different but similar parts that are used to make two different products. Occasionally, the wrong part was used in assembly. The parts were color coded along with their container and work order to reduce confusion. Simple color coding can be an easy way to eliminate unnecessary confusion in daily activities.

This step is essential because it removes waste, which improves quality. It also makes the workplace safer by eliminating clutter. Floor space is gained by disposing of all the unnecessary items. This makes the necessary items that remain much easier to visualize. Simplifying the area reduces crowding and clutter, which improves the workflow. The wasted time searching for tools is eliminated and worker safety is improved due to the removal of debris and potential trip hazards. Decreasing the unneeded inventory and equipment also reduces costs. Excess stock can hide other problems such as quality defects. Unneeded items make improving process flow difficult.

Separating Necessary from Unnecessary Items

The team must thoroughly examine the area. All unnecessary items should be discarded. Items thought to be necessary should be considered for their use, why it was used, and how often it is used. In a production area, your team should evaluate items such as components, documentation, supplies, tooling, gauges, parts, and machines. When using 5S in an office area, your team should evaluate items such as records, forms, books, supplies, equipment, computers, shelving, reports, and parts. When using 5S in a health-care environment, your team should evaluate items such as medical equipment, supplies, furniture, beds, storage cabinets, forms, computers, printers, office equipment, pharmaceuticals, and instruments.

Using Red Tags for Identification

An important part of the "Simplify" step is red-tagging any items that are unnecessary or in question. The red tag should be a large piece of red paper, typically 8.5″ × 11″. The lower portion of the tag should be perforated to keep as a record of the location of the tag. An example is provided in Figure 4.1.

To organize the red tag materials, create a red tag board for tracking purposes. Then move all red-tagged items into a temporary red tag area and set up a time to dispose of the items. This will allow items to be removed by other teams for their use instead of them buying new material.

After the predetermined time period is up, promptly dispose of all remaining items. Continue this process on a regular basis to ensure that more unnecessary items do not begin to clutter the area again.

Area:	**Red Tag**	**Tag No.**
Category (Circle): Supplies Office Materials Furniture Books/Magazines/Files	WIP Raw Material Finished Goods	Tools Equipment Other: ______
Date Tagged:		Tagged By:
Item:		Quantity:
Reason:		
Disposition (Circle): Discard Store in Dept. Long Term Storage		Sell Transfer Other: ______
Action Taken:		Date:
- - - - - - - - - - - - - - - - - - - -	- - - - - Perforated Line - - - - -	- - - - - - - - — - - - - - - - - -
Area:	**Red Tag Locator**	Tag No.

FIGURE 4.1
Red tag.

The steps in a red tag effort include

1. Identify red tag target/area.
2. Establish criteria for evaluation.
3. Make the red tags.
4. Attach tags and separate lower portion for traceability.
5. Evaluate the tagged items.

Step 2: *Seiton* or Straighten Up the Workplace

In the second step, the entire workplace is organized. *Seiton* means arranging and identifying parts, materials, and tools for ease of use. Items should be placed in the best location for point of use and visually organized. This improves quality by visually identifying all products and creating a specific location for that item, which makes it easier for employees to find, use, and easily return all items to their proper location.

First, the team should decide where items should be placed for easier organization and the locations and systems are readily understood. When selecting a storage method, the team should pay careful attention to these details: their goal should be to minimize inventory and space and improve visual management of items. Again, the locations of items are visually indicated so that items can be easily returned. This labeling also promotes readily identifying missing items.

Here is a list of items that should be identified using visual controls:

- Shadow boxes for tools
- Tooling/fixtures
- Movable objects
- Documentation control
- Storage area for common tools
- Raw material
- Finished material
- Discrepant material
- Packaging material

Step 3: *Seiso* or Scrub the Workplace

In the third step, all the areas of the workplace are thoroughly cleaned, including floors, equipment, and furniture. *Seiso* means performing a cleanup campaign, which is important because a clean work environment promotes quality work.

While cleaning, the team should consider how the area became dirty to help maintain a clean environment. The team should inspect for safety hazards or leaks, and before continuing with the cleaning, the team should repair any items that need to be fixed. Cleaning equipment should also have an identified location to ensure that it can be found readily. These activities should be integrated into daily maintenance.

The areas in the workplace that should be cleaned include

- Floors
- Ceilings
- Walls
- Computer equipment
- Furniture
- Cabinets
- Desks
- Production equipment
- Unnecessary computer files

Step 4: *Seiketsu* or Stabilize the Workplace Standards

Stabilize consists of maintaining and improving the first three 5S standards by implementing needed changes. *Seiketsu* means performing *Seiri, Seiton,* and *Seiso* at specified frequency intervals to maintain and improve the well-organized and clean work environment. This is an essential step in any type of quality improvement effort.

Seiketsu can be performed by creating a daily checklist of cleaning and organizing activities. The team can also perform 5S patrols and spot checks to monitor the progress of the 5S efforts. During this step, you should make changes to existing equipment to make cleaning quicker and easier, and you can use the "five-why" method to determine the root source. The five-why is a simple method of asking "why" five times to get to the root cause of a problem. A 5S question could be, "Why does the workplace become so cluttered?" You can then modify equipment and systems to improve the cleanliness of the area. You should also eliminate flat storage areas and excess storage areas.

During Step 4, you should also schedule time and resources for 5S activities. You can create model areas to demonstrate the 5S philosophy, and you can use check sheets to provide feedback. The team should arrange for management walk-throughs to be performed to maintain the 5S culture. Finally, you should require training on 5S to occur on a specified schedule.

Step 5: *Shitsuke* or Sustain

In the final step of 5S, you must instill the discipline or habit of properly maintaining the 5S procedures: 5S must become a habit in your team's daily life. *Shitsuke* means forming a habit of following the first four S's. The team must develop a habit of simplifying, straightening, scrubbing, and stabilizing the work environment. The entire team must be focused on the goal of 5S.

This is the most difficult "S" to obtain. It means maintaining self-discipline and practicing 5S until it is a way of life. Focusing on continuous improvement is essential. An important part of any 5S program is also sharing the lessons learned. By sustaining the effort, employee spirit is improved. Management involvement is critical to sustain the success of the 5S effort.

Using Visual Boards to Document 5S Status

There are several forms that you can use to visually document the 5S status of your workplace. These forms may exist in paper form or be displayed using digital monitors posted in workplace areas:

1. The first form is the 5S wheel, which gives a visual overview of the level of 5S an area has obtained. As each level is obtained, that "S" is filled in and levels not yet achieved are left unfilled. Figure 4.2 shows an example of a 5S wheel.
2. The second form is a series of questions that the team should address to obtain each level. Figure 4.3 is a checklist of the types of questions that can be asked for each S. Post these forms in each area on the shop floor or work area.

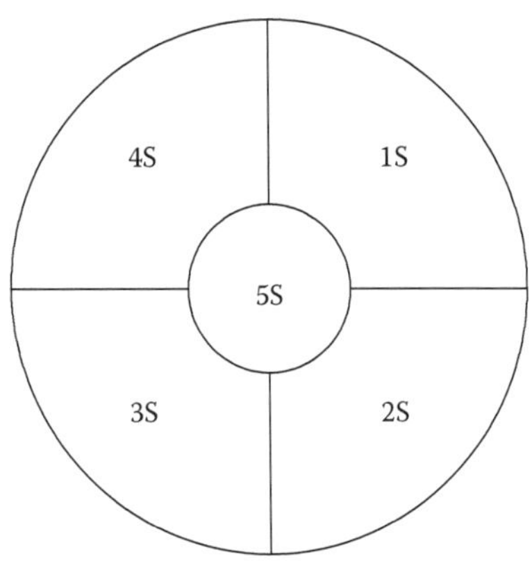

FIGURE 4.2
5S wheel.

5S Evaluation Sheet		Year:				Area:																			
Manufacturing/Shop Floor		JAN		FEB		MAR		APR		MAY		JUN		JUL		AUG		SEP		OCT		NOV		DEC	
		Y	N	Y	N	Y	N	Y	N	Y	N	Y	N	Y	N	Y	N	Y	N	Y	N	Y	N	Y	N
1S Simplify (*Seiri*)—clearly distinguishing between what is needed and what is unneeded.																									
1	Are only necessary items present?																								
2	Are defective materials identified?																								
3	Are unneeded items marked with a red tag?																								
4	Are red-tagged items in the red tag area?																								
5	Is standard work present and current?																								
2S Straighten (*Seiton*)—organizing the necessary items, making it easy for anyone to find and use them.																									
6	Are areas and walkways marked for location?																								
7	Are materials visually controlled and at the point of use?																								
8	Are maximum and minimum allowable quantities on materials indicated?																								
9	Is nonconforming material segregated, identified, and documented?																								
10	Is product clearly identified?																								

3S Scrub (*Seiso*)—cleaning floors and equipment; finding ways to prevent dirt and debris from accumulating in the work area.																									
11	Are floors and areas free of trash, liquids, or dirt?																								
12	Is the work area lighting sufficient, ventilated, and free of dust and odors?																								
13	Are cleaning operations visually managed?																								
14	Are cleaning supplies and equipment easily available and stored in a marked area?																								
4S Stabilize (*Seiketsu*)—maintaining the first three S's.																									
15	Are machines equipped with removal systems for waste or chips?																								
16	Are there records defective material entries with corrective action?																								
17	Are equipment TPM procedures in place and current?																								
18	Are waste and recyclable material receptacles emptied regularly?																								
19	Are scrap and product nonconformance areas cleared on a regular basis?																								
5S Sustain (*Shitsuke*)—adopting the discipline of properly maintaining correct 5S procedures.																									
*	The fifth S (Sustain) is awarded when a minimum of three consecutive months of 1-S, 2-S, 3-S, and 4-S have been sustained.																								

FIGURE 4.3
5S checklist.

Conclusion

5S is a powerful tool for improving workplace organization. You can use the forms provided to implement 5S in your workplace. Chapter 5 will take you through single-minute exchange of dies and changeover. Chapters 11 through 21 will take you through case studies of improvement related to application of the 5S tool.

Homework Problems

1. What should management do to ensure a 5S program is successful?
2. List and describe the five benefits from implementing 5S principles.
3. Which step in 5S uses the red tag?
4. What are the steps in the red tag effort?
5. Describe each of the steps in 5S. Provide a specific example of how each step can be applied.
6. The 5S methodology provides five specific steps. Why is the order of the steps important?
7. Review the following 5S elements and identify the step that is being referenced.
   a. Create a name and location for everything.
   b. Use aisle and material placement markings.
   c. Use labels, tool boards, and color codes.
8. Which step focuses on creating a daily checklist of cleaning and organizing activities? What should be included in this checklist? Give an example of a checklist for your daily activities.
9. What is the five-why method? Give a specific example of how you could use it.
10. Which 5S step is primarily the responsibility of management? Why?
11. Which 5S step is associated with the statement, "A place for everything and everything in its place?"
12. Which 5S step involves operators cleaning their own equipment and working area and performing basic preventive maintenance?
13. Find an area of your house/apartment that needs a significant 5S project. You will be looking for an area that is physically disorganized, but used often. Typical examples include a garage, basement, or storage shed. Next, move through the sequence of steps for 5S at

each phase documenting the progress with photos. Be sure to use as many of the 5S tools as possible including red tagging, separating, sorting, shadow boxes, visual management, and so on. After you complete the first four phases, document how you intend on sustaining this new system with a sustainment plan.

14. Analyze the work/office area of your home/apartment where that you typically work in for homework and studying. This area should include your computer. Follow the 5S system below for the office area:

    a. *Sort:* Take a before photo of your work area and then segregate your items in groups: physical items should include files, letters, sticky notes, pens, notebooks, paper clips, stapler, and so on. Next, conduct the same procedure on your computer's file system. Move/rename file folders, combine file folders, and so on.
    b. *Straighten:* Take a before photo of your area and then organize your area so that everything has a place. On your computer, organize the file structure format accordingly.
    c. *Shine:* Take a before photo and then dust, clean, and polish your work area. On your computer, clean your hard drive, defrag, virus scan, and optimize your registry.
    d. *Standardize:* Develop a time-constrained checklist to review to keep your area clean. Design the workstation area to make it difficult to disorganize it in the future by standardizing the setup. Take an image snap of your computer's file format structure and print out the image to place in your work area to set a standard reference.
    e. *Sustain:* Adhere to the time-constrained checklist by incentivizing yourself to maintain the workspace area. Set up a maintenance schedule on your computer to run your activities automatically and set up reminder appointments in your calendar to remind you to reorganize your computer system.
    f. Take an after photo and compare and contrast the different photos.

5

Using Single-Minute Exchange of Dies and Total Productive Maintenance to Reduce Setup Time and Downtime

Elizabeth A. Cudney

CONTENTS

Introduction

In addition to the 5S process of creating a clean and manageable work environment, other Lean tools that improve the workplace are single-minute exchange of dies (SMED) and total productive maintenance (TPM), which are methodologies that focus on reducing setup time and downtime, respectively. Though rarely achievable, the theoretical goal of SMED is zero setup, in which changeovers are instantaneous and do not affect continuous flow. In practice, SMED setup should take fewer than 10 minutes, which is where the term "single minute" originated. Here are the main benefits of SMED (which are described in more detail in this chapter):

- Reduced inventory
- Improved flexibility
- Increased capacity
- Better customer service

Achieved by detailed analysis of a continuous process, the purpose of SMED is to make setup a standard, routine operation.

Origin of SMED

Initially developed to improve die press and machine tool setups, Shigeo Shingo introduced SMED after 19 years of observing setup operations in factories. Setup time is calculated as the time span between the completion of last good piece of product A and the time the first good piece of product B is completed. This includes the time to change materials, tools, fixtures, and dies, and to stabilize the process again to make good parts.

In developing his methodology, Shingo discovered two main points. First, setup operations can be divided into two categories:

1. Internal setups, which are performed while the machine is not operating. An example is changing the tooling in the machine.
2. External setups, which are performed while the machine is running. An example of this is setting tools for a different product.

The second point Shingo discovered is that converting internal setups to external setups can substantially reduce changeover times. This also increases the time the machines can operate. Shigeo Shingo believed any setup time could be reduced by 59/60ths.

Benefits of SMED

As mentioned, implementing SMED provides many benefits:

1. There are fewer physical adjustments in process changes, which means less of a chance for error.
2. Flexibility in scheduling is increased because setup is less than takt time.
3. The expense of excess inventory is avoided. This, in turn, increases capacity by reducing the amount of lost time for changeover.
4. There is also less material that is wasted due to scrap during setup.
5. Product quality is also improved by reducing variation between each setup.
6. Also in terms of quality, by producing smaller runs, there is a decreased chance of a large-scale defect in inventory being discovered after the part has already been fabricated.
7. SMED also reduces defects from setup errors and eliminating trial runs.
8. Customer service is improved because of the ability to changes over quickly and meet changing customer needs. SMED enables companies to produce more cost-effective products with less waste in smaller quantities.
9. Faster changeovers increase productivity by reducing the amount of downtime.
10. SMED also results in a safer work environment. By simplifying setups, the risk of physical strain or injury is reduced.
11. Dovetailing with the 5S theme, the physical space, or footprint, of an area is reduced, thereby providing a more efficient use of floor space and reducing the amount of clutter.

SMED Methodology

All setup operations consist of five main steps:

1. Before beginning SMED, the setup operation must be determined.
2. The next step includes preparation, after-process adjustments, and checking of materials and tools.
3. The third step is mounting and removing blades, tools, and parts.

4. Next, measurements, settings, and calibrations are performed.
5. The final step is trial runs and adjustments.

Next, let us look at each step in more detail.

Step 1: Determining the Setup Operation

In observing the current setup, the first step is to select the area for improvement. In selecting an area, do not think of the machines as independent from each other. Instead, you should consider setup time reduction as part of a complete flow of production. The manufacturing process should consist of the continuous flow from raw materials to finished product. This includes the following four basic phases of manufacturing processes: processing, inspection, transport, and storage. When evaluating process SMED candidates, often process improvement specialists locate the largest monument machine within the production facility and use that as a starting point to assess upstream and downstream impacts of the machine operations. The largest machine may, or may not, be the SMED to be analyzed; however, it provides a starting point for process evaluation.

Step 2: Preparation, After-Process Adjustments, and Checking of Materials and Tools

Once the SMED evaluation candidate is selected, the purpose of this next step is to ensure that all necessary tools and materials are in place and functional. You should perform this step external to the machine operating. This is the primary source of streamlining the setup process.

Step 3: Mounting and Removing Blades, Tools, and Parts

Once Step 2 is complete, the next step consists of removing parts and tools after processing a part and then attaching the parts and tools for the next part. To perform this step, typically, you must stop the machine. Therefore, it should be identified as an internal setup.

Step 4: Measurements, Settings, and Calibrations

In this step, you should perform the necessary measurements and calibrations during production operations. These operations are typically internal, but they can be converted to external.

Step 5: Trial Runs and Adjustments

The final step in a changeover is to make adjustments after machining the first piece. If you have taken accurate measurements and calibrations during the previous step, that will make these adjustments easier. During traditional setup, this is considered to be an internal setup element because changeover is not complete until good parts are produced.

Analyzing Setup

There are three stages of SMED (outlined in Figure 5.1) that simplify changeover:

1. Separating internal and external setup
2. Converting internal setup to external setup
3. Streamlining the setup operation

The preliminary step to implementing SMED improvements is to analyze the current operation. First, observe the setup by using video capture techniques for easier reference. Pay attention to each machine operator's hand, eye, and body movements. Then show the video to the setup person and others involved with the operation. Finally, using the video, record the time and motion on a setup analysis chart, as shown in Figure 5.2.

Then, chart each time category in Pareto format. An example of the format is provided in Figure 5.3. There are several time categories, including searching, fixture change, walk time, first piece inspection, gaging, tool change, and programming.

The key to successfully implementing SMED is to properly distinguish between internal and external setups. You should perform elements such as preparation and transport while the machine is running; this can typically reduce internal setup time by 30% to 50%. Another method of reducing internal setup is to reexamine operations that are assumed to be internal. You may assume that some steps are internal, but they can be performed externally. You can also convert internal setup to external setup by identifying the function of those steps during the setup process. Another means of further reducing setup time is to analyze the internal setup steps to shorten the time needed.

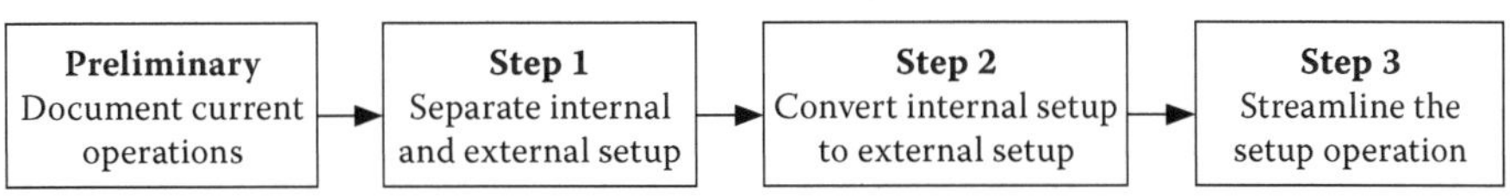

FIGURE 5.1
Stages of SMED implementation.

Setup Analysis Chart			Machine 1000 From: A		To: B				Date
		Time	Changeover Categories				Goal of Improvement Plan		
Step No.	Changeover Element	Element	Internal	External	Waste	Improvement Plan	Eliminate	Internal to External	Reduce
1	Remove fixture (fixture change)	12	12			Eliminate bolts—install quick release			X
2	Search for tools (search)	8	8		X	Mount tools at machine	X		
3	Clean fixture (fixture change)	10	10			Standardize procedure to externalize step		X	
4	Put on fixture (fixture change)	17	22			Eliminate bolts—install quick release			X
5	Load tools (tool change)	15	15			Standardize tooling on all models	X		
6	Machine first part (first piece)	3		3					
	Totals	65	62	3					

FIGURE 5.2
Setup analysis chart.

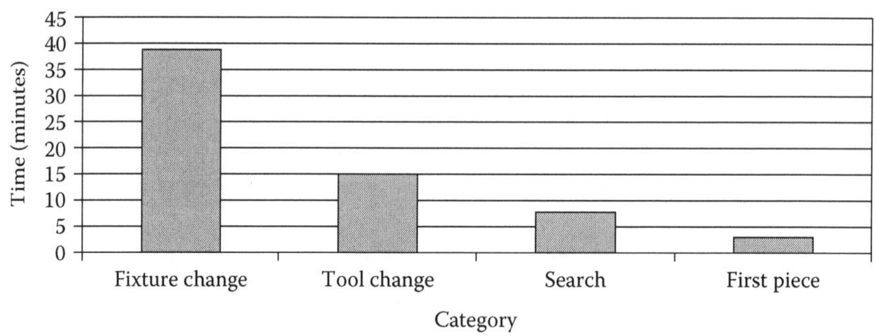

FIGURE 5.3
Setup Pareto analysis.

The next sections of this chapter describe the three steps of analyzing setup in more detail.

Step 1: Separate the Elements

You must separate internal and external elements to identify areas for improvement. The goal of SMED is to have all the necessary tools at the station so the operator never leaves (and never needs to leave) to perform any of the external setups. You should also identify the separation of internal and external elements on your setup analysis chart under changeover categories, as shown in Figure 5.2. Internal activities are performed while the machine is not operating. External elements are performed while the machine is running.

Machine operators can easily perform several tasks while the machines are running: for example, they can call the proper personnel, set tools, and get parts. However, many operators often do not begin these tasks until the machine stops. As mentioned earlier in this chapter, by performing these tasks external to the setup, you can typically reduce setup time in your organization by 30%–50%.

Practical techniques to separate internal and external tasks include checklists, function checks, and improving transport.

Using Checklists

Your checklists should include everything required to setup and run the next operation, including tools, specifications, people, operating conditions, and measurements. By checking off items on the list before the machine is stopped, operators can correct missed steps and errors prior to the internal setup. The checklist should also be specific for each machine or operation, because general checklists can be confusing and therefore, are often ignored.

Using Function Checks

Function checks are also important before beginning the internal setup because they verify that the parts are in working order and they allow time for repair before changeover. If your operators do not find broken equipment until setup, there may be a large delay during internal setup.

Improving Transport

Another means of reducing the amount of time the machine is down is to move the parts and tools to the machine during external setup. Machine operators should transport all necessary tools and equipment to the machine or operation before they shut down the machine for changeover. On the other hand, employees should not move the parts and tools from the previous operation to storage until they have installed the new parts and the machine is running.

Step 2: Convert Internal Setup to External Setup

In the previous step, you separated the internal tasks from the external tasks. To reduce the setup time, you need to convert internal setup to external setup. The first stage is to evaluate the true functions and purposes of each task in the current internal setup. The second stage is to convert the internal steps to the external setup.

There are three methods to convert internal setup to external setup, described in the following paragraphs.

Technique 1: Prepare the Operation Conditions in Advance

This means having the parts, tools, and conditions ready before starting the internal setup. An example is preheating the die molds in advance, instead of heating the molds after the setup.

Technique 2: Standardize the Essential Functions

This means keeping characteristics the same from one operation to another. If tools or parts vary between operations, the operators must typically make adjustments while the machine is down, which tends to be very time consuming. In contrast, by standardizing the parts and tools, you can reduce internal setup time.

Functional standardization focuses on standardizing only the elements essential to the setup, such as securing the mold or fixture, centering, and dimensioning. This involves two steps:

1. First, evaluate each function in the setup process and determine which can be standardized.
2. Next, evaluate each function again to determine if a more efficient means is possible by using fewer parts. An example is standardizing the tooling for each model to eliminate changing the tools during setup.

Technique 3: Use Intermediary Jigs

These are standard dimension plates or frames that can be removed from the machinery. The purpose is to externalize as much of the setup as possible. The current fixture is attached to an intermediary jig on the machine. The next fixture is also installed onto an intermediary jig as an external setup procedure. During the setup, the next fixture is already attached and ready to be installed onto the machine. An example is standardizing the fixture subplate so that all fixtures are installed the same way. All fixtures are designed and built with an intermediary jig.

Step 3: Streamline the Setup Operation

In the final step of SMED, you can improve the remaining internal and external setup operations by evaluating each task's functions and purpose again. Methods for implementing improvements are separated into external setup improvements and internal setup improvements. In developing the improvement plan, consider the following seven forms of waste/*muda*:

1. Overproduction
2. Waiting

3. Transportation
4. Excess processing
5. Inventory
6. Motion
7. Defects

(Refer back to the wastes discussed in Chapter 3 for a description of each type of waste.) Your improvement plan should convert internal to external setup, eliminate or reduce both internal and external setups, and eliminate adjustments.

Streamlining External Setups

The external setup improvements focus on streamlining the storage and transportation aspect of setup. To refine these areas, tool and part management are key. 5S is essential to a successful changeover. 5S will ensure that you and your employees do not lose time searching for tools and materials, because the required materials will be in the proper place and they will be clean and working. You can often cut setup times in half just by organizing your materials.

Streamlining Internal Setups

There are a few techniques for streamlining the internal setup:

- *Implementing parallel operations*: Certain changeovers require tasks be performed at the front and back of the machine. Parallel operations reduce the time lost by walking back and forth from the front to the back of the machine. Instead, divide the setup operations between two people, one for each side of the machine; this eliminates the walk time and reduces the internal setup time. However, you must carefully develop a detailed procedural chart to maintain safe and reliable operations during changeover. Your procedural chart should list the task sequence for each person, the time it will take, and when safety signals are required. The signal should be a buzzer, a whistle, or a light to clearly notify the other person.
- *Using functional clamps*: Another technique to streamline internal setups is to use functional clamps. Bolts are considered an enemy in SMED because they slow down internal setups. Bolts often get lost, get mismatched, and take too long to tighten. In contrast, functional clamps attach items in place with a minimal amount of effort and can be loosened or tightened quickly. Also, because they are typically attached to the machine, operators cannot lose or mismatch

them. Types of functional clamping systems include one-turn, one-motion, and interlocking.

- *Eliminating adjustments*: Eliminating adjustments also reduces the time spent during internal setup. Trial runs and adjustments typically account for 50% of the total time in a changeover. By eliminating these adjustments, you avoid any time lost due to machine downtime. The key is to have the proper settings before starting the machine for the new operation. Trial runs and adjustments depend on the accuracy of centering, dimensioning, and condition setting.

Eliminating adjustments can be achieved in several ways:

- Using numerical scales and standardized settings: for example, you can make a graduated scale with marks that indicate the proper settings.
- Identifying imaginary reference planes and centerlines: for example, by placing V-blocks and rods on the machine table parallel to the centerline and then aligning the center of the cutter.
- Using the least common multiple system, which takes into account operations that have elements in common but are different with respect to dimensions, patterns, or functions.
- *Mechanization*: This is the final attempt to streamline setups. This is because it will not significantly reduce the setup time as much as the other techniques. Another reason is that mechanization will reduce an inefficient operation, but it will not make the process better. Techniques in mechanization include the following:
    - Using forklifts for inserting large dies or molds into machines
    - Moving heavy dies
    - Tightening and loosening dies by remote control
    - Using the energy from presses to move heavy dies

Document the New Setup

The final step of SMED is to document the new setup on a new setup analysis chart. Videotape the setup procedure again and record the time for each element. Develop a new improvement plan. This process will continue until the setup is eliminated or the setup is within takt time. The setup analysis chart is now the basis for the setup procedure because it contains the necessary steps for successful changeover. Any other detailed work instructions not included must also be posted and personnel trained. Remember that continuous improvement is a cycle. Upon completion, repeat the cycle by attempting to streamline the setup process further.

Total Productive Maintenance

Traditional manufacturing operates under the "we fix it" mentality in which the maintenance department performs all maintenance activities. These activities are "firefighting" maintenance that occurs when a machine breaks down. The maintenance department performs some preventive maintenance; however, there is limited time for this because it is performed around normal machine operating times. In the traditional manufacturing environment, the manufacturing department functions under the "we operate" and "run it until it breaks" mentality. The operators generally do not perform any maintenance activities. Instead, the operators contact maintenance when a machine breakdown occurs. In addition, the operators are inactive during the maintenance activities.

TPM is an innovative approach to equipment maintenance involving maintenance personnel and operators working in teams focusing on eliminating equipment breakdowns and equipment-related defects. There are three main goals of TPM:

1. Reduce unplanned equipment downtime.
2. Eliminate barriers between departments.
3. Reduce equipment-related defects.

In addition, there are three main objectives including

1. Total employee involvement
2. Hands-on approach
3. Improve the organization's competitiveness

TPM methodology consists of four key phases as outlined in Figure 5.4. The methodology starts by returning equipment to almost new condition. Next, the focus is on zero breakdowns through proper maintenance. The third phase focuses on consolidating information for future use. The final phase of TPM is zero defects.

Complete TPM implementation centers on autonomous maintenance, equipment improvement, maintenance prevention systems, and quality maintenance as shown in Figure 5.5. Autonomous maintenance involves developing preventative maintenance practices.

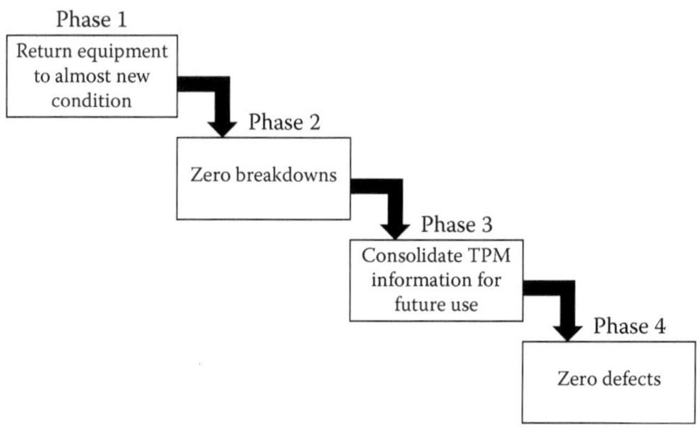

FIGURE 5.4
Total productive maintenance (TPM) phases.

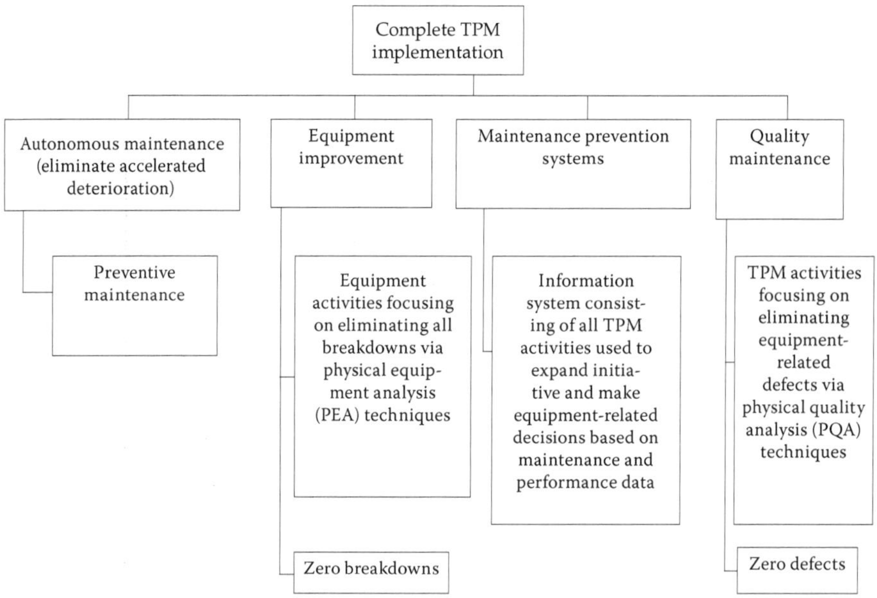

FIGURE 5.5
TPM implementation.

For equipment improvement, the equipment activities should focus on eliminating all breakdowns through physical equipment analysis techniques to target zero breakdowns. For maintenance prevention systems, the organization should develop an information system that consists of all TPM activities. Equipment-related decisions are then based on the maintenance and performance data. Finally, quality maintenance focuses on eliminating equipment-related defects through physical quality analysis with the goal being zero defects.

Preventive maintenance (PM) is a time- or usage-based method of maintaining equipment. Much like maintenance of oil changes in your automobile, maintenance activities are performed on equipment based on defined time and/or usage intervals to prevent equipment breakdowns from occurring. Examples of PM are PM schedules and team activities.

Predictive maintenance is a situation-based method of maintaining equipment. Maintenance activities are performed on equipment based on visible signals or diagnostic techniques to prevent equipment breakdowns from occurring. Examples of predictive maintenance include vibration analysis, ultrasound, thermography, laser measuring, generator analysis, and oil analysis.

Overall Equipment Effectiveness

Overall equipment effectiveness (OEE) is the measure of the percent of time a piece of equipment is producing quality product. OEE measures the effect of the six big losses, which are breakdowns, setup, idling, minor stoppages, quality factors, and rework, as shown in Figure 5.6.

These six big losses fall under the three categories of availability level, operating level, and quality level. The availability level consists of breakdowns and setup, when the equipment is not available to run production. The operating level consists of idling and minor stoppage, which are the times when the equipment is not running; however, these are normal stoppages during the operation for activities such as unloading a part from a machine or malfunctions that are less than 10 minutes. The quality level consists of quality

Availability Level	Operating Level	Quality Level
Breakdowns	**Idling**	**Quality Factors**
Malfunctions causing equipment to stop processing product more than a set time (i.e., 10 minutes)	Load/unload time for product on a piece of equipment and/or operators waiting for activities out of their control	Activities related to ensuring the quality of the product produced on the equipment
Setup	**Minor Stoppages**	**Rework**
Any activity related to equipment changeover, setup, and adjustments	Malfunctions causing equipment to stop processing product less than a set time (i.e., 10 minutes)	The time taken to process product for rework

FIGURE 5.6
Major equipment losses.

factors and rework, which relate to activities to ensure the product meets the customer's quality requirements or rework.

Overall equipment utilization and utilization are related concepts. Utilization shows only the black-and-white utilization of an asset from a financial standpoint. However, OEE further defines the effectiveness of equipment by monitoring the effect of load/unload time and minor stoppages (operating level) and quality factors and rework (quality level). It indicates the actual time the equipment is producing a quality product. OEE takes utilization to a more detailed level.

OEE is important because it helps to prioritize improvement projects and reflect results. It also combines the utilization, operation, and quality aspects of the equipment into one measure. OEE provides a measure for changes to capacity, productivity, and quality. Figure 5.7 provides the OEE calculation and Figure 5.8 provides a sample calculation.

Loading Time (LT)	Total hours available for equipment operation		
Availability Level (AL)	$\frac{\text{LT - Administrative + Setup + Breakdown}}{\text{LT}}$	Breakdowns and Admin.	Setup and Adjust
Operating Level (OL)	$\frac{\text{AL hours} - (\text{Idling} + \text{Minor Stoppages})}{\text{AL hours}}$	Idling	Minor Stoppages
Quality Level (QL)	$\frac{\text{OL hours} - (\text{Quality} + \text{Rework})}{\text{OL hours}}$	Quality Factors	Rework
OEE	AL × OL × QL		

FIGURE 5.7
Overall equipment effectiveness (OEE) calculation.

Loading Time (LT)	Total hours available for equipment operation = 168 hours		
Availability Level (AL)	$\frac{168 - (12.3 + 12.4 + 12.5)}{168} = 78\%$	Breakdowns and Admin.	Setup and Adjust
Operating Level (OL)	$\frac{131 - (15.4 + 1.1)}{131} = 87\%$	Idling	Minor Stoppages
Quality Level (QL)	$\frac{114 - (3.1 + 0.1)}{114} = 97\%$	Quality Factors	Rework
OEE	78% × 87% × 97% = 66%		

FIGURE 5.8
Sample OEE calculation.

Preventive Maintenance

The initial OEE calculations for equipment identify the current baseline. OEE will reflect improvements as the TPM initiative evolves. The initial focus of TPM is to return the equipment to like-new condition and prevent any further deterioration. Implementing PM schedules is the first step toward eliminating deterioration. A four-phased approach to PM schedules is presented in Figure 5.9.

In four-phase PM schedule development, the first phase consists of inspecting target equipment, utilizing customized checklists, and tagging and documenting problems. The second phase involves prioritizing identified problems and identifying causes of the highest priority problems. Commonly used tools in this phase include brainstorming, data collection, and maintenance/operator experience. In the third phase, the focus is to develop inspection standards and build PM schedules. The inspection standards should be written standards. These are essential for developing accurate PM schedules. They should communicate the procedure necessary to carry out effective PM. Finally, the fourth phase is to deliver training, implement PM schedules, and monitor and adjust. Effective training should be developed and delivered to targeted operators and maintenance personnel. This is required for effective PM implementation. In addition, implementing monitoring measurements provides the vehicle to adjust and change the performance of the equipment. OEE is an effective monitoring tool for these reasons.

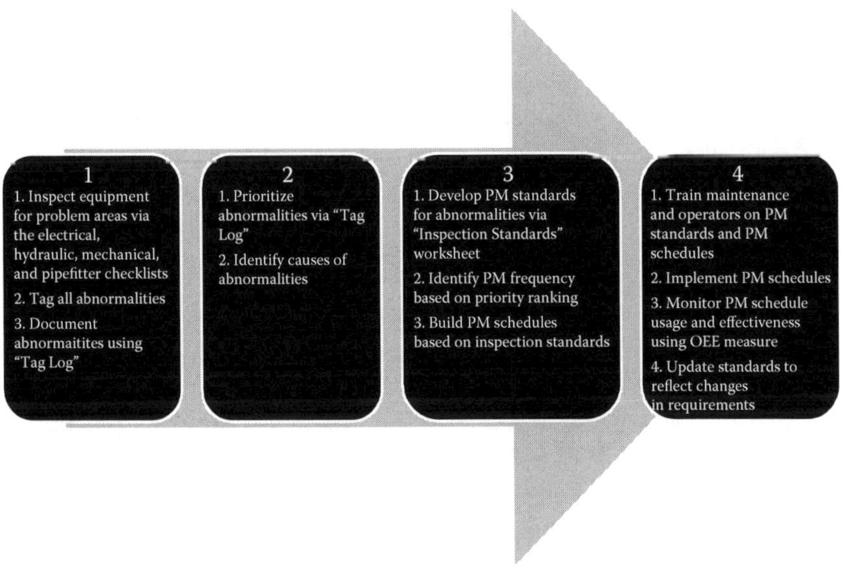

FIGURE 5.9
Four-phase preventive maintenance approach.

Conclusion

SMED enables significant reductions in changeover time, which also enables organizations to be more flexible for changes in customer demands and requirements. TPM is an innovative approach that focuses on eliminating equipment breakdowns and equipment-related defects. Chapter 6 discusses flow, pull, and kanban in detail so that you can implement pull in your organization. Chapters 11 through 21 provide in-depth case studies of implementing SMED in various industries.

Homework Problems

1. In a SMED system, what is the real meaning of the single minute concept?
2. What are the main benefits of SMED? Explain each benefit.
3. Describe the origins of SMED. What were the two main points that led to the development of SMED?
4. How is setup time calculated? Give a specific example.
5. Define an external setup. Give a specific example.
6. What are internal changeover operations? Give a specific example.
7. List and describe the five main steps in SMED.
8. What are three stages of SMED that simplify changeover?
9. Video capture techniques are very useful in a SMED event. Explain why.
10. The largest gains from SMED are realized by separating __________ operations. Explain your answer.
    a. Preparation from replacement
    b. Location from adjustment
    c. Internal from external
    d. Pre-setup from post-setup
11. Performing elements such as preparation and transport while the machine is running typically reduces internal setup time by how much?
12. How are function checks used and why are they important?
13. Describe the three methods to convert internal setup to external setup.

14. Explain the two steps in functional standardization.
15. While streamlining the setup operation what should be considered when developing the improvement plan?
16. Why is 5S important when streamlining the external setups?
17. List and describe the four techniques for streamlining internal setups.
18. Describe how adjustments can be eliminated.
19. Why should mechanization be the final attempt to streamline setups? Describe at least two techniques for mechanization in detail.
20. Define TPM. What are the three goals of TPM?
21. What are the three categories of major equipment loss?
22. Define OEE. How is OEE different from utilization?
23. What is the initial focus of TPM?
24. What is autonomous maintenance? Give an example.
25. Explain the difference between PM and predictive maintenance.
26. You work for a major dishwasher company. You have received customer complaints centered around a long process changeover during dishwasher batch processing. Although you do not know of any specifics, you have heard rumors that your competitor has come up with a number of clever features that reduce dishwasher loading cycle time. As a team, detail the steps involved in processing a load of dishes in a dishwasher using SMED (you may use any team member's dishwasher as an example). Estimate the time involved in each activity and use the setup analysis chart in Figure 5.2 to detail the processing of the dishes by classifying internal and external setups. After you have detailed the steps involved your team will propose changes to your design team. Detail the setup analysis chart and proposed design changes.

6

Flow, Pull, and Kanban

Elizabeth A. Cudney

CONTENTS

Introduction

This chapter will take you through the history and progression of Lean from traditional manufacturing to one-piece flow. These major transitions often mirror the process improvement milestones you will undertake in your own efforts.

Lean emphasizes the elimination of waste and creation of flow within an enterprise. The primary focus of Lean is on the customer, to identify, from a customer perspective, value-added and non-value-added tasks and minimize and eliminate non-value-added tasks. Value-added tasks are the only operations for which the customer is ready to pay. On the contrary, non-value-added tasks are processing steps that produce a form of waste in which the customer would not want to pay to have these activities performed. These types of non-value-added tasks fall into two main categories: (1) non-value-added activities that are considered necessary to produce a product or service and (2) non-value-added activities that are not necessary to produce a product or service.

The idea in creating flow in Lean is to deliver products and services just-in-time (JIT), in the right amounts, and at the right quality levels at the right place. This necessitates that products and services are produced and delivered only when a pull is exerted by the customer through a signal in the form of an order or purchase. A well-designed Lean system allows for an immediate and effective response to fluctuating customer demands and requirements.

Lean History and Progression

Manufacturing has continued to evolve over time from traditional manufacturing to heijunka with several steps in between. Each period had various methods of controlling production: instruction, production control, material, humans, machine, and lead time.

Prior to the 1900s, products were handcrafted by artisans. Typically, they were made one at a time from start to completion. Because of this sequential process, delivery to the customer was typically slow. Quality was high because the focus was on that one product. However, there was a wide variation in product and process because a process step from one product to the next product might be several weeks apart. Artisan training was not consistent as it largely remained an apprenticeship style of learning structure. As a result of this type of customized training, product variation in the form of dimensional control, structural quality, and speed of delivery among different artisans remained erratic. In many cases, the same artisan could not produce the same part repeatability relative to dimensional control and structural characteristics across the products that they made.

Batch production was used in Henry Ford's assembly line. There was one operator at each machine or station and, therefore, single-task orientation. The machines were high volume with a focus on high machine utilization. Subsequently, there were large production lots of sequential steps. There was little variation in the process or product. Lead times in batch production were long because each product in a batch must wait for the other products in that batch to be completed before it can continue to the next process. The following sections describe each manufacturing environment, up to current practices.

Traditional Manufacturing Environment

Traditional manufacturing is characterized by verbal production instructions and is still commonly used today. This type of manufacturing system is completely reactionary to forecasting and, as a result, changed often by manufacturing planners. Product is made at many locations and is scheduled. Inventory and material are decentralized and difficult to control. Material is bought and stored in large quantities. Processes are individualized for specialized operators. Similar to traditional accounting practices, production is aimed toward maximum capacity and machine utilization with equipment being dedicated to a specific function. To justify the large batch style machine utilization, long setup times are often associated with the equipment. The lead time to manufacture a product becomes highly variable and difficult to predict.

Continuous Flow Manufacturing Environment

Continuous flow manufacturing is similar to traditional manufacturing in that the production orders are verbal and forecasted, and they change often. Product in continuous flow manufacturing is manufactured at limited locations. Production is scheduled and decentralized. The focus of improvements is now on controlling processes, and material has changed from large undefined quantities to limited work-in-process. The operator responsibilities have also changed from specialized tasks to related tasks within processes. Equipment is now arranged according to the process flow rather than grouped by function with product lead times becoming more predictable with reductions in lead time occurring.

Standardized Work Environment

In a standardized work environment, production instruction is similar to continuous flow with verbal instructions, forecasting, and limited production locations. In addition, product is manufactured to customer order with a defined work-in-process. The process is well defined to the work sequence for the operator with work instructions clearly defined for a flexible workforce. The idealized state of the standard work environment should be seamless substitution of workers, regardless of skill level, to perform a specific process function based on standardized work instructions. In addition to manual labor, machines are synchronized to approximately the same process speed, and the lead time for a product is predictable.

Pull System of Manufacturing

The pull system is based on a concrete order for customer requirements. Production control is visible and disciplined, typically with the use of kanbans (signals for production). Material is also controlled with the use of kanbans to replenish the system and determine proper inventory levels. With systems to control inventory, the operator's indirect work becomes manageable. Lead time now is based on the customer rather than the process.

The customer lead time is predictable as a product is only constructed on a specific customer demand. This leads to the customer "pulling" the product through the value chain, or the sequence of activities that create a product or provide a service.

Manufacturing Environments That Use Small Lots

The method of using small lots is similar to the pull system; however, customer requirements change frequently. Material requirements are reduced due to the reduced size of the lots. Kaizen is used to improve the processes and reduce the lead time. The equipment is frequently changed over for different products. Single-minute exchange of dies (SMED) is used whenever possible to reduce changeovers. With the use of Kaizen and SMED, the lead time for an order is reduced.

Heijunka Method of Leveling Production Work

Heijunka is a method focused on leveling production. Production instruction and control is similar to that of the small-lot method. Within this environment, material usage is uniform with a leveled production. The operator also is leveled with respect to his or her workload, with operating equipment efficiency and machine utilization held in a uniform condition. Again, the lead time for production is reduced and predictable.

Just-in-Time Manufacturing

The JIT production system is based on the philosophy of providing the right product or the right service in the right quantity or amount at exactly the right time based on customer requirements; as such, JIT drives to eliminate waste. The seven forms of waste are overproduction, inventory, transportation, processing, motion, waiting, and defective parts. An eighth waste of unused creativity and intellect of people can also be considered. Derivative wastes have also been identified and evolved over time as more industries have become involved in Lean efforts. Implementing JIT concepts to eliminate waste drives improvements in safety, quality, delivery, and cost.

The key to the JIT system is the customer. The customer in the past could only select high quality, good service, or low price. At best, the customer was able to get two out of the three. Now, to satisfy customer demands organizations must provide for all three. Therefore, the profit equation has changed drastically over time, changing the dynamic perspective of business profitability. The old calculation was to add the cost to the profit desired to determine your selling price:

$$\text{Cost} + \text{Profit} = \text{Selling price}$$

In contrast, the new method is to calculate profit by subtracting cost from selling price:

$$\text{Profit} = \text{Selling price} - \text{Cost}$$

One-Piece Flow and Cellular Manufacturing

Taiichi Ohno at Toyota developed Lean production in which operators are multiskilled and run several machines or steps. Lean production may incorporate cellular design; however, flow production is based on customer orders. Other characteristics include one-piece flow and flexible setups.

One-piece flow in a cellular layout has many benefits:

- It significantly reduces transportation, inventory, and waiting time.
- It improves quality, delivery, and cost.
- It lowers lead times.
- It improves product distribution.
- It reduces scrap and rework.
- It makes scheduling easier.
- It reduces floor space used.
- It reduces material handling.
- It uses labor better and increases productivity.
- It exposes problems.

Cellular manufacturing involves a group of machines or processes connected by a process sequence in a pattern that supports the efficient flow of production. There are several elements that are characteristic of a good cell design:

- The process will determine the layout.
- Quality is built into the process at each step rather than at final inspection.
- One operator can run the cell.
- Machines are in close proximity to each other.
- The cell is a U-shaped design that flows counterclockwise.
- The operators are multiskilled.

The cell design should focus on operators and their multiskilled training. This is based on the philosophy that people *appreciate* over time, whereas machines *depreciate*. In other words, people are more valuable than machines.

The cell should also be designed around the process. Similarities between products can be determined by using part/quantity analysis to show model/volume relationships and process route analysis to show process relationships.

Several different flows are shown in Figure 6.1. Figure 6.1a is a small U-shaped layout, which provides flexibility and visual management. Figure 6.1b is the traditional straight-line manufacturing, which does not provide as much flexibility as a U-shaped cell. In addition, the straight-line layout hinders communication and visual management since the operators are not able to see all of the machines easily. Figure 6.1c is also an example of a U-shaped cell. In Figure 6.1c, the material enters and leaves the cell at the same point. This layout promotes the most flexibility since as few as 1 and as many as 12 operators can work in the line depending on demand. In addition, the operators can see all machines and easily communicate. Finally, Figure 6.1d shows a dedicated machine in which one operator is stationed at one machine.

The key to determining the proper layout is to be product focused. The equipment should be laid out in the process order to promote flow. The machines should be close together to facilitate one-piece flow and reduce non-value-added walking. The U-shaped configuration allows volume flexibility with staffing, reduces non-value-added walking, and improves communication and visual management. It is also critical to be customer focused. Modules of capacity should be matched to major customer sites; this allows production to match customer needs.

For example, a traditional manufacturing process has one operator per machine as shown in Figure 6.2. There is inventory before and after each machine. The focus is on the individual machine and labor productivity.

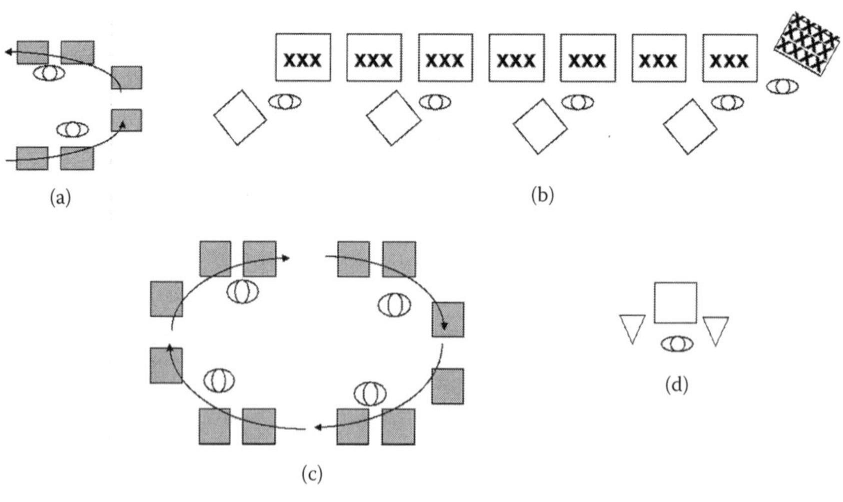

FIGURE 6.1
Types of layouts.

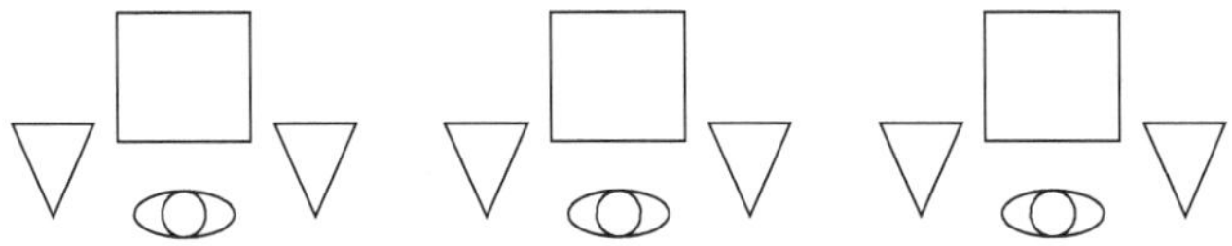

FIGURE 6.2
Traditional manufacturing layout.

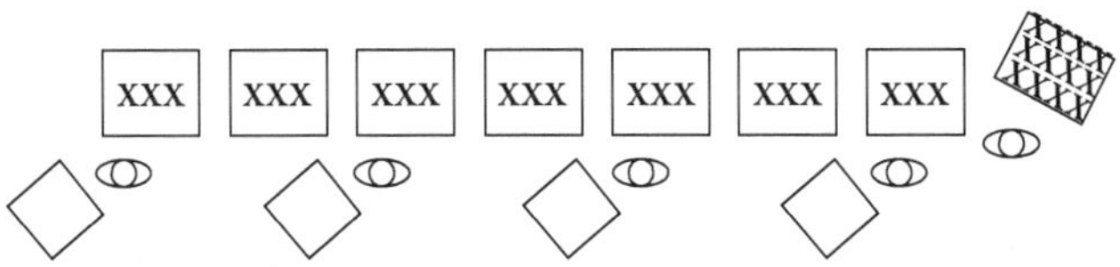

FIGURE 6.3
Transfer line layout.

This is similar to a transfer line. The flexibility is minimal in terms of labor and changeover, as shown in Figure 6.3. In the figure, the process is currently run with five operators. If there are only four operators, the process is very difficult to run due to self-imposed stations. In addition, there is no difference in the output if the process is run with five operators or six operators due to the lack of flexibility. The changeover from one part to another also presents a serious issue. If one station is down for changeover, then the entire line is not producing the product. Therefore, the total changeover time is significantly higher in this case than in a U-shaped cell.

Moving from batch and queue processing also significantly reduces the lead time and opportunities for defects. Figure 6.4 presents the difference between a batch and queue process and one-piece flow. With the batch and queue process, parts are batched in quantities of 10. Therefore, the parts do not move to the next step in the process until all 10 parts are complete with each process step. The total lead time for the order is 30 minutes. The first piece is not complete for 21 minutes. In addition, if a defect occurs it may not be caught until the end of the batch is complete and the entire production of 30 pieces is at risk. In continuous flow processing, each part moves to the next step as soon as it is complete. Therefore, the total time for the first piece is reduced from 21 to 3 minutes and the total time to complete the order is reduced from over 30 minutes to 12 minutes. In addition, defects or quality issues are caught quicker and a lesser number of parts are at risk.

Figure 6.5 shows a U-shaped cell with multiskilled operators working on a multiprocess line. The operators are assigned stations in a way that balances their workload and reduces walking. For example, instead of assigning operator 3 machines 4, 5, and 6, the operator is assigned machines 4, 7, and 8 to level load the operators and reduce walking. All of the operators are located inside the U-shaped cell to promote communication and visibility. The layout is flexible to increase or decrease the number of operators depending

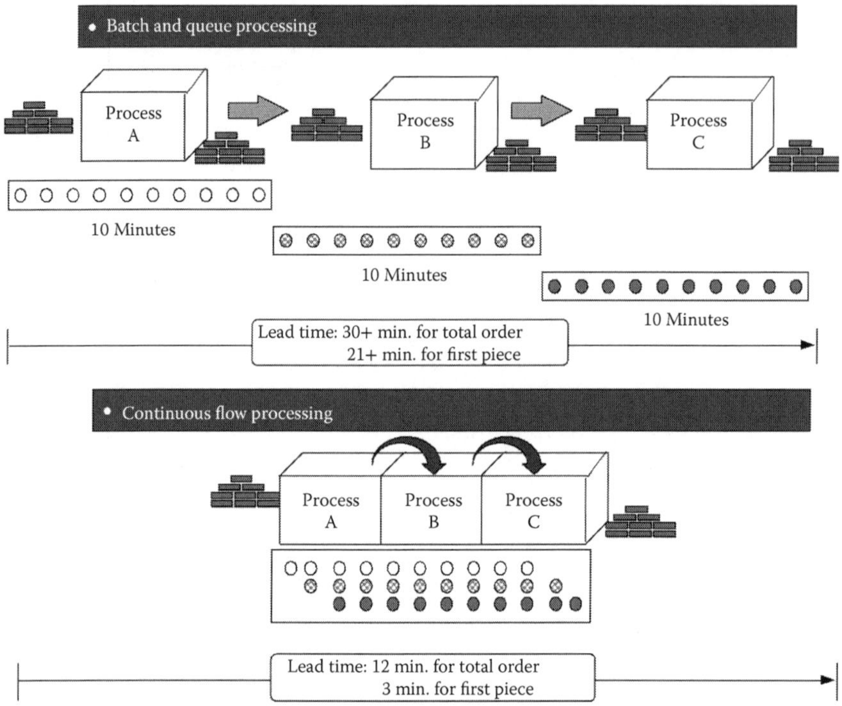

FIGURE 6.4
Batch and queue versus continuous flow processing.

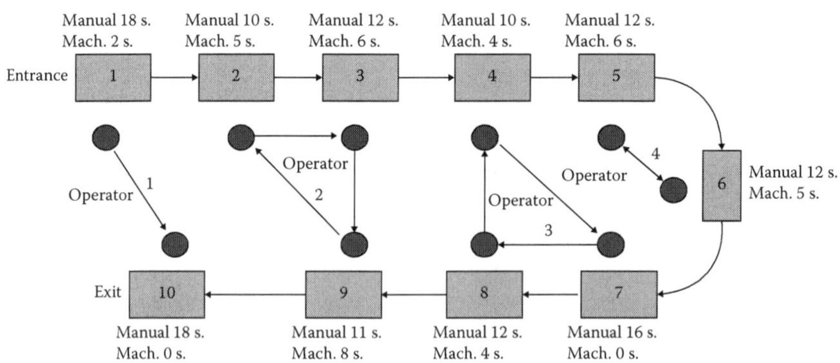

FIGURE 6.5
U-shaped cell with multiskilled operators.

on the demand. This layout can support from 1 to 10 operators. Internal machine time can be used such that the machine waits for the operator and the operator does not watch the machines run. Ideally, this is balanced with standard work (covered in Chapter 8) to reduce the machine and operator waiting times.

Spaghetti Diagrams

One method to understand the flow of a product or service is the use of spaghetti diagrams. An example spaghetti diagram is shown in Figure 6.6. The spaghetti diagram shows the flow of the product from each station or machine to the next, usually showing an erratic and wasteful flow.

In addition, the path of the operator can also be traced to show their movement. One method is to document the flow of the product and the operator using different colors to illustrate or highlight potential areas of waste.

By grouping common processes together, the flow shown in Figure 6.7 is achieved. Using production lines, the process is streamlined, as shown in Figure 6.8.

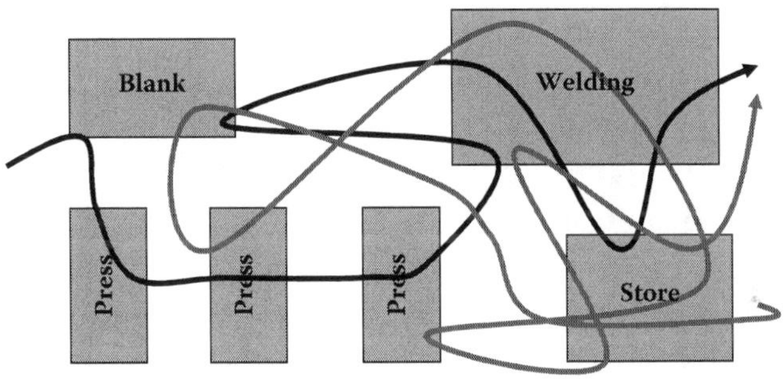

FIGURE 6.6
Spaghetti diagram.

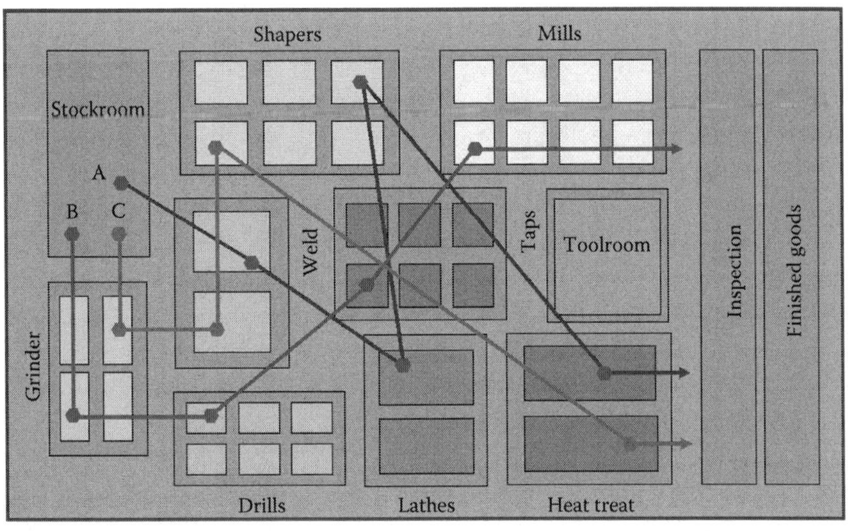

FIGURE 6.7
Process layout.

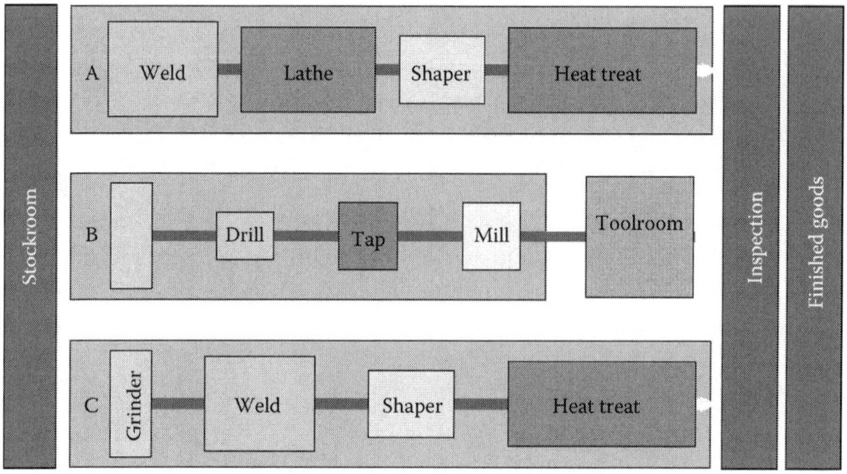

FIGURE 6.8
Process layout based on the product.

Pull Systems and Kanban

In a push system, each operation is considered independent. The work schedules for each operation are based on a forecast. This typically requires large, cumbersome computer-based scheduling systems that generate detailed operation schedules. Expediting is required to identify real priorities. This method also requires looking for and counting actual material versus computer record balance. Schedules at the upper level seldom agree with schedules at the lower level. Push systems operate under "make all we can just in case we need it." Push systems have the following characteristics:

- Large lots, high inventory
- Poor product availability
- Hidden problems
- High levels of waste
- Poor communication
- Large enterprise resource planning software configurations

A pull system is a customer-controlled system that utilizes signals (kanbans) to authorize the replenishment of material. It is a method of controlling the flow of material by replacing what has been consumed. Pull systems operate under "make what we need when we need it." In a pull system, each operation is considered part of the process. The final assembly schedule is the schedule for each operation in the process. Work is based on consumption, and product is only produced to replace what has been consumed. The computer systems are much simpler because the focus is on planning

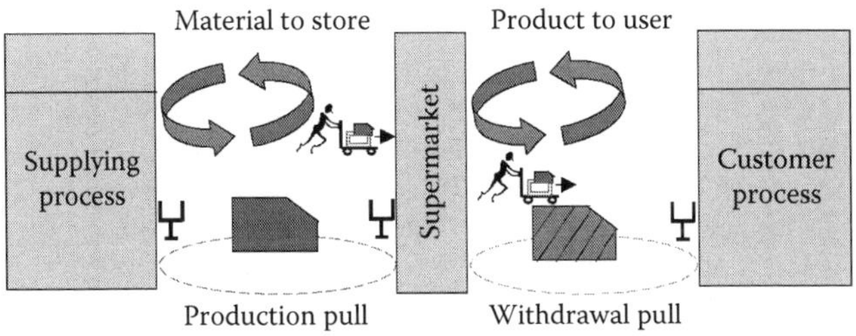

FIGURE 6.9
Production and withdrawal pull.

systems. In addition, expediting is generally not necessary because what is required is obvious. Pull systems have the following characteristics:

- Small lots, lower inventory
- Better product availability
- Better visibility of problems
- Much less waste
- Good communication

There are two main kinds of pull systems as shown in Figure 6.9. A withdrawal pull authorizes movement of material. A production pull authorizes production of material.

The term kanban originates from a Japanese word meaning card or visible record; it refers to cards used to control the flow of production through a factory. Kanbans act as a signaling method to produce a specific amount of product to be consumed downstream in the process. In a simplified system, kanbans take the physical shape of simple cards or the container itself. In a more complex signaling system, kanbans may be electronic signals programmed into production planning software. When adjusting a production line, or setting up a new production line, you may need to determine the amount of kanban cards needed to manage the signaling of production. To determine the appropriate amount of kanban containers needed, n_k, the following equation is used:

$$n_k = \frac{\left[\left(D_{\text{Units}} \times t_l\right) + s\right]}{c}$$

$$s = p\left(D_{\text{Units}} \times t_l\right)$$

where D_{Units} is the average unit demand over a given period, t_l is the lead time it takes to restock an order, s is the percentage of safety stock during demand lead time, p is the specified management buffer percentage, and c is the container size.

Example

An industrial engineer is determining the amount of kanban containers that may be necessary for a new production line being set up as a U-shaped production cell. The cell is required to produce 100 parts per hour. Each container has a single kanban card attached to it, and the maximum number of parts a container can hold is 15. Management has determined that a safety stock of 10% should be kept at all times. It takes about 15 minutes to forklift new parts from the previous workstation. Determine the number of kanbans needed.

SOLUTION:

$$D_{\text{Units}} = 100 \text{ parts per hour}$$

$$t_1 = 15 \text{ minutes} = \left(\frac{15}{60}\right) = 0.25 \text{ hour}$$

$p = 10\%$
$s = 0.10\ (100 \times 0.25) = 2.5$ parts
$c = 15$ parts

$$n_k = \frac{[(100 \times 0.25) + 2.5]}{15} = 1.833 \text{ or 2 kanban cards are needed}$$

Kaizen

Kaizen is the philosophy in which one seeks to continuously improve. It is a state of mind to never accept the status quo. *Kai* means change, and *zen* means for the good. Kaizen is a constant process. What is done today to improve should be done tomorrow as second nature. It drives us to improve. Kaizen should also be done with the imagination before turning to costly improvements. Chapter 22 will take you through planning and executing a Kaizen event.

The improvement cycle begins with exposing and quantifying problems. The next step is determining the root cause. After uncovering the root cause, solutions should be implemented. After implementation, the focus is on standardization and adherence. Because this is a cycle, the process continues for further improvements.

Conclusion

Flow, pull, and kanban are powerful tools for improving responsiveness and flexibility in your organization. Chapter 7 will take you through mistake proofing to prevent defects from occurring in your processes.

Homework Problems

1. Give an example of a necessary non-value-added activity.
2. Where was batch production first used?
3. Describe continuous flow manufacturing. How is it different from traditional manufacturing?
4. Define a pull system. Give an example of a pull system that you have experienced.
5. Define a withdrawal pull. Give a specific example.
6. The management philosophy of make all we can just in case we need it is an example of what kind of system?
7. Describe heijunka.
8. What is the philosophy of the JIT production system?
9. What is the key to the JIT production system?
10. Name at least five benefits of one-piece flow in a cellular layout.
11. Name five elements that are characteristic of a good cell design.
12. Why is it important to be product focused when determining the proper layout?
13. What is a spaghetti diagram? Why is it useful?
14. Compare the different characteristics between a pull system and a push system.
15. The kanban method is most closely associated with which of the following? Explain your answer.
    a. The elimination of non-value-added activities in the process
    b. The development of a future state process stream map
    c. Making problems visible in a process, thus clarifying targets for improvement
    d. The control of material flow
16. What is the ideal batch size in a continuous flow operation? Why?
17. What does Kaizen mean?
18. Work as a team and map out an area for a small spaghetti diagram project. It is often helpful to target a work process that occurs frequently and can be easily tracked. Commonly used examples include restaurants, bookstores, coffee baristas, libraries, and convenience stores. Again, remember that you are identifying a process that repeats and is measureable. Follow the following sequence of steps to assist in your spaghetti diagramming:

a. First, map the layout of the service process. Next, watch a worker walk through the process. If possible, try to receive measureable distances using a pedometer on the worker or approximate the distances of travel. Record this information on a spreadsheet format that maps distances to work function. Also, note any barriers, safety hazards, or major impediments that the worker faces when conducting routine tasks. When observing the worker's motion, draw a line on your drawing following the worker as he or she moves about to perform the routine tasks.
b. Hold a meeting to communicate the current state layout of the spaghetti diagram. Discuss how wasted steps can be eliminated or reduced. Discuss how physical barriers can be relocated, or how process sequencing may be altered to suggest an improved state.
c. Map out a few scenario options of future state layouts that were discussed, and estimate the savings for each scenario based on the metrics gathered. If possible, implement the most attractive scenario option and validate the option by following a worker through the future state process. If a standard work exists for the current state process, edit the standard work to reflect the new layout.

19. A machining facility is developing a new factory layout for its standard part business. As a newly hired engineer, you are determining the amount of kanban cards needed between work cells. You start at cell B for the analysis. Cell B is required to produce 115 parts per hour. Each container that moves between work cells has a single kanban card attached to it, and the maximum number of parts a container can have is 12. Management has determined that a safety stock of 8% should be kept at all times. It takes roughly 6 minutes to move inventory from cell A. Determine the number of kanbans needed.

7

Mistake Proofing (aka Poka-Yoke*): Preventing Defects by Monitoring Process Conditions and Correcting Errors at the Source*

Elizabeth A. Cudney

CONTENTS

Introduction

The next tool is poka-yoke, a methodology that focuses on preventing defects and improving quality. The goal of zero defects is achievable. The Zero Quality Control system approach of mistake proofing (*poka-yoke*, in Japanese) prevents defects by monitoring process conditions and correcting errors at the source. It is human nature to make mistakes. In this approach, poka-yoke devices are used to perform 100% inspections and give feedback about each part or operation. Installing a poka-yoke device has a considerable effect on quality, but it can also give you a false impression that these devices alone will eliminate defects. However, if you combine poka-yoke systems with self-checks or successive checks, you can effectively obtain 100% inspections, prompt feedback, and action. To do that, it is essential for you to understand how to use poka-yoke systems, their functions, the types of systems, and various detection methods—all of which are covered in this chapter.

Overview of Poka-Yoke/Mistake Proofing

Poka-yoke is a technique to prevent simple human error. Preventing the defect before it is produced is, of course, the most effective means of reducing defects. However, implementing poka-yoke devices to detect errors and immediately stop the action is also a valuable part of the continuous improvement effort.

The concept of poka-yoke has existed for a long time in various forms. Shigeo Shingo, a Japanese manufacturing engineer, developed the idea into a tool to achieve zero defects. The idea behind poka-yoke is focused on taking overrepetitive tasks and actions to free the worker's time and mind for more creative and value-adding activities.

Zero Quality Control is an approach for achieving zero defects. It is based on the idea that controlling process performance can prevent defects even when mistakes are made by a machine or a person. Zero Quality Control is a blameless approach, which recognizes that people sometimes make mistakes.

Focusing on producing zero defect products is essential to maintaining customer satisfaction and loyalty. Cost is another reason to focus on eliminating defects. Defects result in costs from scrapping a product, reworking, or repairing damaged equipment. Zero defects are also key for a company to achieve Lean production and smaller inventories. Reducing defects allows a company to decrease buffer inventories that are built with anticipation of problems. Companies are able to produce the exact quantity of products ordered by the customer.

Inspection Techniques

Three major inspection techniques exist in the field of quality control:

1. *Judgment inspection:* With this technique, the operator separates defective parts from good parts after processing. This method of inspection prevents defects from reaching the customer; however, it does not lower the company's internal defect rate.
2. *Informative inspection*: With this technique, you investigate the cause of the defect and relay that information back to the appropriate process so that the operator can take action to reduce the defect rate.

3. *Source inspection:* Because defects are typically caused by simple mistakes, this approach to inspection focuses completely on the source so that the mistake can be corrected before it even becomes a defect. Zero defects can be achieved using source inspection.

Zero Quality Control consists of three main methods leading to eliminating defects. First, source inspection checks for the factors that *cause* an error, rather than inspecting for the *resulting defect.* It ensures that certain conditions exist to perform a process properly. An example of source inspection is adding a locator pin to prevent a part from being misaligned in a fixture. Source inspection differs from judgment inspection and informative inspection in that it catches errors and provides feedback so that the errors can be corrected before processing the product.

Second, inexpensive poka-yoke devices, called 100% inspection devices, can provide 100% inspection automatically for errors or defective operating conditions. Zero Quality Control varies from statistical quality control inspections in that it inspects every single product produced. Statistical quality control only gives an idea of whether a process is in control and does not prevent defects. A limit switch or inexpensive sensing device is an example of a 100% inspection device.

The third component is taking immediate action. The operations are stopped immediately when a defect or mistake is made and will not resume until the mistake is corrected. An interlocked circuit that automatically shuts down a machine when a mistake is made is an example of taking immediate action.

Types of Errors

There are several different types of errors:

1. The first type of human error is *forgetfulness,* which can occur when an operator is not concentrating. A safeguard to prevent forgetfulness is setting inspections for the operators to perform at regular intervals.
2. *Errors due to misunderstanding* can happen when people make conclusions before they are familiar with the situation. Training and standardizing work procedures can help avoid these situations.
3. *Identification errors* are another type of human error. Situations can be misjudged when viewed too quickly or from too far away to be clearly visible. Training and attentiveness can avoid these.
4. Errors are also made by newer operators due to *lack of experience.* Skill building through formal training programs can help avoid these types of mistakes.

5. The fifth type of human error is *willful errors*, when operators decide they can ignore certain rules. Experience, basic education, and corrective actions from management are safeguards against these errors.
6. *Inadvertent errors* can also be made when people are absentminded or lost in thought. Through proper discipline and work standardization, these defects can be avoided.
7. *Delays in judgment* can also cause errors. Again, skill building and work standardization are aspects to avoiding these defects.
8. *Lack of suitable work standards or instructions* can cause mistakes. Work instructions and work standardization can help avoid these errors.
9. Occasionally, equipment and machines will run differently than expected, resulting in *surprise errors.* Using total productive maintenance can avoid these errors.

Types of Defects

There are various types of defects, including

- Omitted processing, for example, a missed process
- Processing errors, for example, a broken tool
- Errors in setup, for example, the wrong machine setting
- Missing parts, for example, a missing component in an assembly
- Wrong parts, for example, assembling the wrong component
- Processing the wrong part, for example, picking up the wrong part
- Misoperation, for example, variation in the process
- Adjustment errors, for example, changing the setting incorrectly

Defects typically occur during one of five situations:

1. Defects can occur because of inappropriate procedures or standards during process planning. Proper planning to ensure correct standards can avoid this situation.
2. Defects can also occur because of excessive variability in a process. Maintenance can prevent these types of defects.
3. Defects can occur when material is damaged or varies excessively. A means of eliminating this situation is implementing inspection on receipt of the materials for defects and variation.

4. Defects can also be caused by worn equipment and tools. Again, regular maintenance can prevent these defects.
5. The final situation can occur even when the above situations do not exist: simple human mistakes occur that result in the production of defective products.

Defects are either about to occur or already exist. Poka-yokes have three main functions: shutdown, control, and warning. *Prediction* is recognizing that a defect is going to happen. *Detection* is recognizing that a defect has occurred.

Quality Improvement Activities

The traditional quality improvement cycle consists of *Plan, Do, Check,* as shown in Figure 7.1. The processing conditions are determined in the Plan stage. These planned actions then occur in the Do stage. Finally, the Check stage performs the quality monitoring, where information regarding defects is relayed back, so that you can take corrective action in the next Plan stage to improve conditions during the next Do stage.

However, you cannot completely *prevent* defects even when continuously repeating the functions in this cycle, because feedback about defects is given only after a defect occurs. The Plan, Do, Check cycle suggests there is no means of preventing the error before it happens.

Zero Quality Control addresses this problem by integrating the Do and Check stages, as shown in Figure 7.2. Errors can occur between the Plan and Do functions. In the Zero Quality Control approach, inspection is carried out at the point where the error happens. When an error is detected, the operator is able to correct the problem before the work is done, using source inspection.

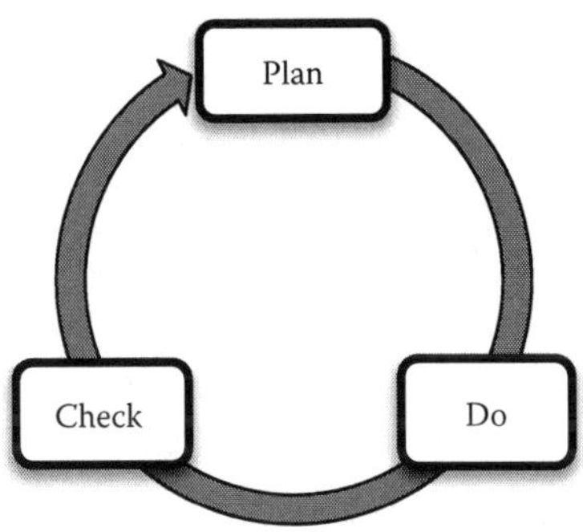

FIGURE 7.1
Plan, Do, Check cycle of traditional quality improvement cycle.

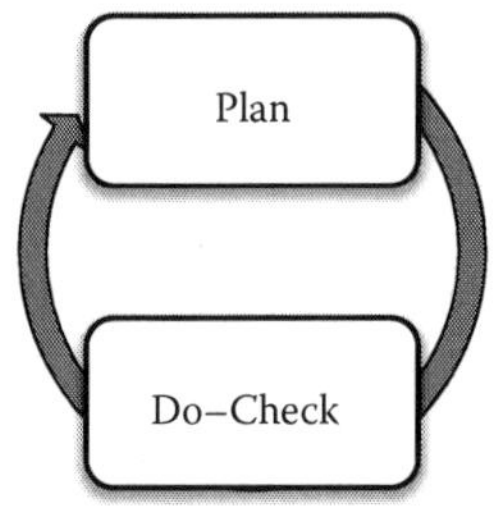

FIGURE 7.2
Integrated Do and Check in the Zero Quality Control approach.

Poka-Yoke Detection Systems

There are several methods to identify the proper poka-yoke for a situation. Based on the situation, there are appropriate systems to prevent defects.

- The first method is to identify an item by its characteristic. This can be the weight, dimension, or shape of the product.
- Another method is to determine how the defect can deviate from procedures or omitted processes by following the process sequence. Using basic work standardization methods of travelling documentation often provides a living history of the process in question.
- A third method is detecting deviations from fixed values. This can be achieved by using a counter, odd-part-out method, or critical condition detection.

In Zero Quality Control, true poka-yokes are used as source inspections to detect errors before the production process creates a defective product. The poka-yoke system detects an error and automatically shuts down the equipment or gives a warning. Control systems are used to stop the equipment when an error is detected. This is a more effective means of achieving zero defects because it does not depend on the operator. This type of system may not always be possible or convenient; therefore, a warning system may also be used. A warning system is used to get the attention of the operator and could be in the form of a flashing light or a sound. Control systems, however, also typically use lights and noise to direct the attention of the operator to the equipment with the problem. Nonautomated warning systems can also be effective, such as color-coding parts and part holders.

Poka-yoke detection devices can be categorized into three main methods: contact methods, fixed-value methods, and motion-step methods. These methods can be used with either control systems or warning systems, but each uses a different approach to detect defects or errors.

Using Contact Methods to Detect Defects or Errors

Contact methods detect if a product makes a physical or energy contact with a sensing device. Microswitches and limit switches are the most commonly used contact devices. However, contact methods do not have to be very technical. Inexpensive contact devices such as guide pins or blocks will not allow a part to be loaded into a fixture in the wrong position. These types of contact methods are *passive devices* and take advantage of a part's design or uneven shape.

Using Fixed-Value Methods to Detect Defects or Errors

You should use a fixed-value poka-yoke method to detect defects or errors when you have a process with a fixed number of parts that must be attached or assembled in a product. You can also use this method when you have a process that requires a fixed number of repeated operations to be performed at a station.

With fixed-value methods, you use a device to count the number of times a task is done; the device then signals or releases the part only when the required number is reached. Limit switches can be used with each movement sending a signal to the counter that, in turn, detects when the required number of movements is complete. The device will only release the part when the preset number of signals is reached. For example, consider an assembly with three bolts that must be set at a specific torque. The assembly fixture would hold the part in place until the count (in this case three) was reached, and the device would be preprogrammed such that it would only count up when the torque level is reached.

Using Motion-Step Methods to Detect Defects or Errors

Another approach is to use the motion-step method to sense whether a motion has been completed within a certain expected time. You can also use this method to detect whether tasks are performed according to a specified sequence, which is often a helpful tool for assembling the proper parts in a particular product model.

Poka-Yoke Product Design/Redesign

In some cases, it may be necessary to design or redesign a product to be mistake proof during assembly or installation. This concept also is known as *design for manufacturing, design for assembly,* or simply, product optimization. The idea behind this concept includes designing parts so that if mistakes are made during fabrication and assembly, the part will still be functional and will not be considered scrap. This practice uses the simple design rules of symmetric design. For example, consider a flat piece of sheet metal that may be formed into a 90° bend. By making the sheet metal symmetric, the bracket may be bent in either direction to avoid simple fabrication mistakes.

Another design rule includes reducing the number of left-hand/right-hand-specific designs to eliminate part count numbers and fabrication errors, thus resulting in scrap reduction. In addition, designing parts to be multifunctional and multiuse eliminates unnecessary assembly steps that may result in defects and scrap. Also, designing an assembly fixture for a single assembly direction within the assembly process prevents an operator from wasting time and motion and prevents operator fatigue and the propensity for mistakes.

Conclusion

Mistake proofing improves the overall quality of a product or service. Chapter 8 will discuss standard work, which is a critical Lean tool for standardizing processes. Chapters 11 through 21 provide in-depth case studies illustrating the implementation of mistake proofing devices in manufacturing, service, and healthcare.

Homework Problems

1. Who is the founder of poka-yoke? What was the main idea behind the development of poka-yoke?
2. Define poka-yoke. Give a specific example.
3. Describe the three major inspection techniques that exist in the field of quality control.
4. Which of the following is the best way to error proof activities? Explain your answer.
    a. Corrective action
    b. Preventive action
    c. Containment action
    d. Temporary action
5. How is source inspection different from judgment inspection and informative inspection?
6. Give a specific example of a 100% inspection device.

7. Describe at least five different types of errors. Give a specific example for each.
8. How can you prevent willful errors?
9. Describe at least five different types of errors. Give a specific example for each.
10. What are the five situations where defects typically occur?
11. What are the three methods to identify the proper poka-yoke for a situation?
12. What kind of systems are used to stop equipment when an error is detected? Why is this more effective?
13. Describe the three main methods of poka-yoke detection devices.
14. Which type of poka-yoke detection device can be used to detect whether tasks are performed according to a specified sequence? Explain how this type of device works.
15. How can poka-yoke concepts be used to design or redesign a product to be mistake proof during assembly or installation? Give a specific example.
16. You are the operations manager for a small company that assembles fans for computer systems. Your takt time is one fan assembly every 2 minutes with 480 minutes available for working per day. The assembly consists of a single injection-molded fan housing, a stainless steel shaft that is slid into a hole in the fan housing, and a plastic injection-molded fan blade as shown in Figure 7.3, which is press fitted onto the shaft and capped with a small plastic injection-molded cap as shown in Figure 7.4. The shaft freely rotates in the housing and is fixed to the blade. In addition, the end of the shaft is geared to match the gearing of a motor that is installed downstream of the production line.

FIGURE 7.3
Fan and housing assembly.

FIGURE 7.4
Fan in assembled condition.

After the housing and blade assembly is assembled into a subassembly, it travels to the motor installation department, which installs and wires the motor onto the assembly. The quality control department has contacted you with a significant reoccurring scrap issue after the parts have been made. You have discovered that during the assembly process the fans are being assembled backward and as a result the airflow is reversed. On further investigation, you learn that the motor assembly department is wiring the direction of the motor in reverse to compensate and get the product out of the door. However, you learn that this practice reduces the life span of the motor and causes significant customer complaints. After learning of the success of poka-yoke, you decide to remedy the scrap issue at the point of assembly.

As a team, develop a solution to eliminate the scrap by implementing poka-yoke techniques. This solution may be in the form of a change in design or process or a combination of each. Three-dimensional computer-aided design drawings are not required, but your solution must be sketched if it is a design change. If your solution is a process change, diagram the change. In addition to the diagramming and sketches include a description of the change.

8

Standard Work: Documenting the Interaction between People and Their Environment

Elizabeth A. Cudney

CONTENTS

Introduction

So far, we have covered five Lean tools: value stream mapping (described in Chapter 3); 5S and visual management (described in Chapter 4), which improves productivity by creating a clean and manageable workplace; single-minute exchange of dies (described in Chapter 5), which improves productivity by reducing setup time; flow, pull, and kanban (described in Chapter 6), which helps create flow of products and services based on customer demand; and mistake proofing (described in Chapter 7), which helps eliminate defects. Standard work is another Lean tool: it defines and documents the interaction between people and their environment.

Standard work provides a routine for consistency of an operation and a basis for improvement by detailing the motion of the operator and the sequence of action. First, you document your current process to provide a basis or standard for continuous improvement. After you have seen improvements, then you should revise your organization's standard work to incorporate these improvements.

Overview of Standard Work

Standard work consists of three elements: takt time, standard work sequence, and standard work-in-process (which are discussed in detail in this chapter). Used as a tool, standard work accomplishes the following:

- It establishes a routine for repetitive tasks.
- It makes managing resource allocation and scheduling easier.
- It establishes a relationship between a person and the environment (both the machine and materials).
- It provides a basis for improvement by defining the normal process and highlighting areas for improvement (by making problems visual and obvious).
- It prohibits backsliding, or relapse into previous bad habits.

The goal of the implementation project is to first record your existing process. Then, using time observations of your initial process, you should make changes to the line to improve working conditions, flow, and the level load of your operators. Finally, you should document your improved process to provide a baseline for your team to use in the future.

Standard work is a tool to determine maximum performance with minimum waste through the best combination of operator and machine with the goal to improve. In addition, standard work helps eliminate variability from the process and functions as a diagnostic device. Furthermore, standard work exposes problems and facilitates problem solving for both operator and process control monitors. Finally, it helps us to identify waste and drives us to Kaizen the process.

Prerequisites for Standard Work

To implement standard work, the operation must be observable, repetitive, and based on human motion. In addition, the process must be standardized with all variable processes reduced. Moreover, the floor supervisor must be responsible for the implementation of standard work.

Three Elements of Standard Work

The elements of standard work are takt time, work sequence, and standard work-in-process.

Takt Time

Takt time is how frequently a product must be completed to meet customers' expectations. It is calculated using customer demand and available time. Takt time sets the rhythm for standard work.

Operator cycle time (CT_O) is the total time required for an operator to complete one cycle of operation, including time needed for walking, loading and unloading, and inspecting products. The *machine cycle time* (CT_m) is the time between the instant an operator presses the "on" or "start" button and the point at which the machine returns to its original position after completing the target operation.

Takt time is equal to the total daily operating time divided by the total daily requirements. The variables include customer demand and available work time. Therefore, you should recalculate takt time when your customer demand available work time changes.

$$\text{Takt time} = \frac{\text{Available time}}{\text{Customer demand}}$$

The *total cycle time* (TCT) is the rate of completion of a process or operation. This is a summation of operator cycle time and machine cycle time for all processes in the operation.

$$\text{TCT} = \sum^{n} i = (CT_o + CT_m)$$

Where i is the machine number and n is the total number of machines. If takt time is known, computing TCT will provide an understanding of how many operators may be needed for line balancing needs.

$$\text{Number of operators needed} = \frac{\text{TCT}}{\text{Takt time}}$$

Work Sequence

The work sequence is the specific order in which an operator performs the manual steps of a process. The *work sequence* may be different from the *process sequence*. Focusing on the work sequence identifies waste and stabilizes the process.

The work sequence requires multiskilled operators. A complete operator cycle is from the time the operator begins the sequence to the time the operator returns to that same point.

Standard Work-in-Process

Standard work-in-process is the minimum amount of parts on the line that will allow an operator to flow product efficiently. Keeping the number of parts standard is key to allowing work to continue without the operator waiting.

Documenting Standard Work

Several standard forms are used for documenting standard work. These forms help standardize the process and are described in detail in this section.

Takt Time Sheet

The takt time sheet documents all of the following information:

- Minutes available
- Pieces per shift
- Allotted break time
- Allotted wash time
- Allotted cleanup time

Figure 8.1 shows the takt time sheet. Figure 8.2 provides an example of takt time calculation. Figure 8.3 shows the total cycle time calculation. Equation 8.1 shows the number of operators needed for calculation.

$$\text{Number of operators needed} = \frac{\text{Total cycle time}}{\text{Takt time}} = \frac{270}{67} = 4.03 = 4 \text{ operators needed} \qquad (8.1)$$

Time Observation Sheet

The time observation sheet focuses on manual and walk time elements. You should fill out a separate sheet for each operator. The time observation sheet has three key steps:

- Step 1: Identify work elements.
- Step 2: Determine the observation points.
- Step 3: Time each element with a running clock.

Calculating Takt Time				
____	Hours =	______	Minutes (based on standard work shift)	
		– ______	Minutes (break time)	
		– ______	Minutes (wash time)	
		– ______	Minutes (cleanup)	
		– ______	Minutes (team meetings)	
	Total	– ______	Available minutes per shift	
		=		
______	Minutes available × 60 =	______	seconds per shift	
______	Seconds per shift × 2 shifts = ______ seconds per day			
______	Seconds divided by	______	pieces/day = ______	seconds
Takt time =		______	seconds per piece	

FIGURE 8.1
Takt time calculation.

Calculating Takt Time				
8	Hours =	480	Minutes (based on standard work shift)	
		−30	Minutes (break time)	
		−10	Minutes (wash time)	
		−0	Minutes (cleanup)	
		−5	Minutes (team meetings)	
	Total	435	Available minutes per shift	
435	Minutes available × 60 =	26,100	seconds per shift	
26,100	Seconds per shift × 2 shifts = 52,200 seconds per day			
52,200	Seconds divided by	775	pieces/day = 67	seconds
Takt time =		67	seconds per piece	

FIGURE 8.2
Example of takt time calculation.

	Operation 1 Machining		Operation 2 Drilling		Operation 3 Deburring		
	Operator cycle time (CT_o)	Machine cycle time (CT_m)	Operator cycle time (CT_o)	Machine cycle time (CT_m)	Operator cycle time (CT_o)	Machine cycle time (CT_m)	**TCT**
Seconds	100	30	40	10	60	30	270

FIGURE 8.3
Example of total cycle time calculation.

The first step in appraising the current process is to time the process. Begin by outlining the process steps. Then, time the existing process for walking time, loading time, unloading time, standard inspection time, and machine cycle times. You should take 10 observations of the process and record them on your time observation form (refer to Figure 8.4).

When making your observations, you can use a running clock. In this case, you should subtract the times between observations to obtain the time for each step. Then, determine the lowest elemental time for each step and then add the lowest elemental times to obtain the time for one cycle. You should also add together the times for individual steps to calculate the time for one cycle. You can then make adjustments to the steps to make the total of the lowest elemental times equal to the lowest cycle time of your actual observations.

Process for Observation		**TIME OBSERVATION FORM**									**Part No.**		**Part Type**	
											Part Name		**Daily Demand**	
No.	Component Task	1	2	3	4	5	6	7	8	9	10	Low Elem. Time	Adj.	Adj. Elem. Time
	Time for One Cycle													

FIGURE 8.4
Time observation form.

If the total time for one cycle of your actual observations is greater than the takt time for the operator performing the machining responsibilities, you will know that your process must be improved to meet takt time. Figure 8.4 provides the time observation sheet, and Figure 8.5 is an example of a completed time observation sheet.

Process Capacity Table

The process capacity table documents the machine capacity per shift. Use one sheet for each cell. The table focuses on the total machine time, which includes any load and unload times. Document only the load, unload, and cycle start when calculating the manual time. You should also take into account tool changes in your calculations. Do not include any abnormalities that are not standard.

You should perform process capacity calculations for each process step. List each process step with its associated manual time, machine time, and walking time. In addition, evaluate tool changes. Then add together the manual time, machine time, and walking time to obtain the total time to complete the process step. Finally, divide this time into the available operating time per shift: this calculation results in the processing capacity per shift. Figure 8.6 provides the process capacity table, and Figure 8.7 is an example of a completed process capacity table.

Standard Work Combination Sheet

The next step of outlining the existing process is to fill out the standard work combination sheet. The standard work combination sheet combines manual, automatic machine, and walk elements. Plot these against the takt time. Use one sheet for each operator, and post the sheet at the starting point of each operator sequence. Figure 8.8 shows the standard work combination sheet, and Figure 8.9 provides an example.

Fill out a standard work combination sheet for each operator. Use the same steps to complete this form as you used when completing your process capacity sheet. Then list the manual, machine, and walking times for their respective process step:

- Drawing a straight line first, draw the manual time on the combination sheet chart.
- Then add the machine cycle time from the manual time by drawing a dotted line.
- Finally, draw a curved line from the end of the manual time for the observed walking time to the next process step.

After you have determined whether the current process is within takt time, evaluate the process to identify areas for Kaizen.

Process for Observation		TIME OBSERVATION FORM										Part No.		Part Type	
												Part Name		Daily Demand	
No.	Component Task	1	2	3	4	5	6	7	8	9	10	Low Elem. Time	Adj.	Adj. Elem. Time	
1	Unload, Blow off chips, Load, Clamp, Start	:33	:37	:38	:43	:40	:43	:56	:72	:57	:52	:33	:10	:43	
		:33	:37	:38	:43	:40	:43	:56	:72	:57	:52				
	Walk	:10	:11	:10	:10	:11	:09	:19	:12	:11	:15	:09	:00	:09	
		:43	:48	:48	:53	:51	:52	1:15	1:24	1:08	1:07				
2	Unclamp, Unload, Blow off chips, Load, Clamp, Start	:32	:38	:37	:41	:30	:36	:36	:33	:41	:41	:30	:10	:40	
		1:15	1:26	1:25	1:34	1:21	1:28	1:51	1:57	1:49	1:48				
	Walk	:18	:16	:14	:14	:17	:17	:24	:19	:17	:17	:14	:00	:14	
		1:33	1:42	1:39	1:48	1:38	1:45	2:15	2:16	2:06	2:05				
3	Unload, Blow off chips, Load, Start	:27	:35	:32	:29	:23	:29	:29	:34	:40	:43	:23	:05	:28	
		2:00	2:17	2:11	2:17	2:01	2:14	2:44	2:50	2:46	2:48				
	Walk	:10	:08	:07	:08	:09	:09	:09	:08	:11	:07	:07	:00	:07	
		2:10	2:25	2:18	2:25	2:10	2:23	2:53	2:58	2:57	2:55				
4	Unload, Blow, Load, Start	:24	:33	:28	:28	:30	:29	:35	:38	:44	:43	:24	:05	:29	
		2:34	2:58	2:46	2:53	2:40	2:52	3:27	3:36	3:41	3:38				
	Walk	:08	:12	:08	:11	:09	:10	:12	:09	:09	:09	:08	:00	:08	
		2:42	3:10	2:54	3:04	2:49	3:02	3:39	3:45	3:50	3:37				

FIGURE 8.5
Example of time observation form.

Process for Observation		TIME OBSERVATION FORM										Part No.		Part Type	
												Part Name		Daily Demand	
No.	Component Task	1	2	3	4	5	6	7	8	9	10	Low Elem. Time	Adj.	Adj. Elem. Time	
5	Unload, Blow off chips, Load, Start	:31	:30	:41	:33	:32	:33	:29	:33	:36	:40	:29	:05	:34	
		3:13	3:40	3:35	3:37	3:21	3:35	4:08	4:18	4:26	4:17				
	Inspect	:13	:13	:14	:13	:13	:14	:13	:14	:14	:13	:13	:00	:13	
		3:26	3:53	3:49	3:50	3:34	3:49	4:21	4:32	4:40	4:30				
	Walk	:08	:08	:07	:08	:07	:07	:07	:08	:07	:07	:08	:00	:08	
		3:34	4:01	3:56	3:58	3:41	3:56	4:28	4:40	4:47	4:37				
6	Unload, Blow off chips, Load, Start	:31	:30	:30	:32	:33	:34	:34	:36	:29	:49	:29	:05	:34	
		4:05	4:31	4:26	4:30	4:14	4:30	5:02	5:16	5:16	5:26				
	Inspect	:16	:13	:19	:14	:13	:15	:13	:16	:14	:14	:13	:00	:13	
		4:21	4:44	4:45	4:44	4:27	4:45	5:15	5:32	5:30	5:40				
	Walk	:03	:06	:04	:04	:03	:05	:03	:04	:04	:03	:03	:00	:03	
		4:24	4:50	4:49	4:48	4:30	4:50	5:18	5:36	5:34	5:43				
7	Unload, Blow off chips, Load, Start	:33	:29	:26	:29	:30	:31	:33	:35	:28	:30	:28	:05	:33	
		4:57	5:19	5:15	5:17	5:00	5:21	5:51	6:11	6:02	6:13				
	Inspect	:08	:12	:10	:09	:08	:10	:09	:09	:08	:09	:13	:00	:13	
		5:05	5:31	5:25	5:26	5:08	5:31	6:00	6:20	6:10	6:22				
	Walk	:12	:08	:10	:08	:08	:09	:08	:10	:09	:09	:08	:00	:08	
		5:17	5:39	5:35	5:34	5:16	5:40	6:08	6:30	6:19	6:31				

FIGURE 8.5 ***(Continued)***
Example of time observation form.

Process for Observation		TIME OBSERVATION FORM									Part No.		Part Type	
											Part Name		Daily Demand	
No.	Component Task	1	2	3	4	5	6	7	8	9	10	Low Elem. Time	Adj.	Adj. Elem. Time
8	Unload, Load, Blow off chips, Start	:33	:20	:19	:23	:20	:25	:25	:27	:27	:32	:19	:05	:24
		5:49	5:59	5:54	5:57	5:36	6:05	6:33	6:57	6:46	7:03			
	Walk	:05	:05	:05	:07	:07	:06	:05	:06	:06	:15	:05	:00	:05
		5:54	6:04	5:59	6:04	5:43	6:11	6:38	7:03	6:52	7:18			
9	Blow out, Put gloves on	:57	:63	:67	:65	:63	:65	:66	:64	:65	:67	:57	:00	:57
		6:51	7:07	7:06	7:09	6:46	7:16	7:44	8:07	7:57	8:25			
	Plug, Blow off, Remove gloves	:78	:85	:89	:80	:79	:85	:84	:86	:82	:85	:78	:00	:78
		8:09	8:32	8:35	8:29	8:05	8:41	9:08	9:33	9:19	8:50			
	Walk	:20	:13	:09	:17	:22	:29	:32	:28	:30	:29	:09	:00	:09
		8:29	8:45	8:44	8:46	8:27	9:10	9:40	10:01	9:49	9:19			
	Time for One Cycle	**8:29**	**8:45**	**8:44**	**8:46**	**8:27**	**9:10**	**9:40**	**10:01**	**9:49**	**9:19**	**7:40**	**:50**	**8:30**

FIGURE 8.5 ***(Continued)***
Example of time observation form.

Department Manager		PROCESS CAPACITY FORM				Part No.		Part Type		Operating Time per Shift in Seconds
Supervisor						Part Name		Daily Demand		
Step No.	Process Description	Machine No.	Base Time (Seconds)		Tool Change		Time (Seconds)		Processing Capacity	Remarks
			Manual	Machine	No. of Pieces per Change	Replacement Time	Tool Change Time	Total Time to Complete		
1										
2										
3										
4										
5										
6										
7										
8										
9										
10										
		Total								

FIGURE 8.6
Process capacity table.

Department Manager		PROCESS CAPACITY FORM				Part No.		Part Type		Operating Time Per Shift in Seconds
Supervisor						Part Name		Daily Demand	55	26,100
Step No.	Process Description	Machine No.	Base Time (Seconds)		Tool Change		Time (Seconds)		Processing Capacity	Remarks
			Manual	Machine	No. of Pieces per Change	Replacement Time	Tool Change Time	Total Time to Complete		
1	Machine Z-plane	CNC 1	53	294	800	30	360	347	75	
2	Drill bolt holes	CNC 2	33	414	1,500	30	90	447	58	Bottleneck operation
3	Rough drill tube bores	CNC 3	28	349	750	30	90	377	69	
4	Finish drill tube bores	CNC 4	27	196	750	30	90	223	117	
5	Rough and finish drill center bore	CNC 5	32	423	375	30	90	455	57	Bottleneck operation
6	Mill clearance cut	CNC 6	34	378	750	30	360	413	63	
7	Drill oil hole	CNC 7	30	343	2,500	30	60	373	70	
8	Hand tap	CNC 7	22	229	75	90	0	252	103	
9	Assemble gear housing	Assy.	59	0	0	0	0	59	442	
		Total	318	2,626						

FIGURE 8.7
Example of process capacity table.

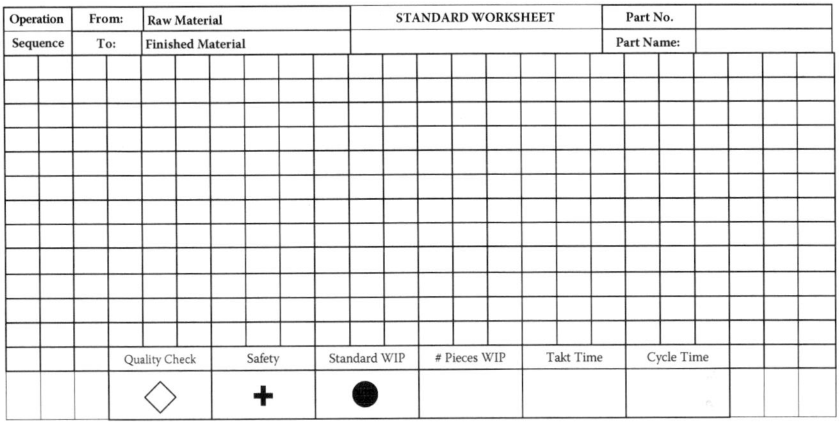

FIGURE 8.8
Standard worksheet.

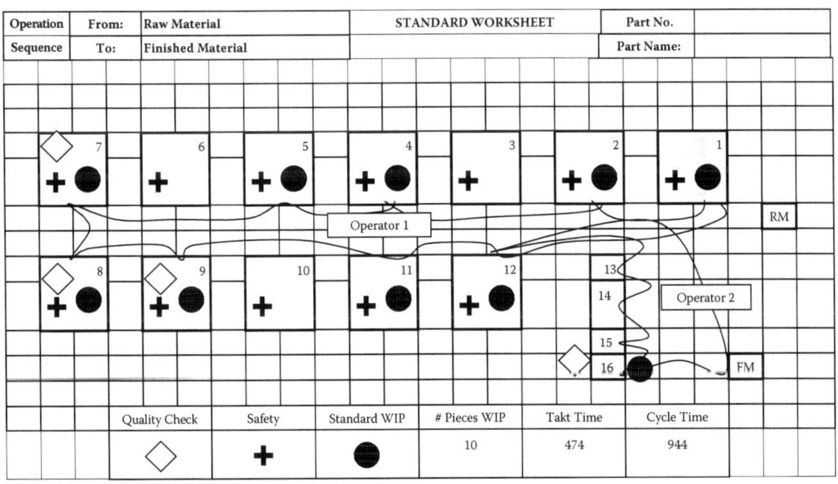

FIGURE 8.9
Example of standard worksheet.

Standard Worksheet

After documenting your process capacity and your time observations, the next step is to draw a standard worksheet to depict the process flow and machine layout of the current process. The standard worksheet is an overhead view of the cell or operations that illustrates the process and work sequence. It documents the standard work-in-process, safety precautions, and quality checks. Again, you should complete a sheet for each operator, and you should post the standard worksheet at the starting point of each operator sequence. Figure 8.10 shows the standard worksheet, and Figure 8.11 shows an example.

Draw the operation sequence from raw material to finished material. First, draw the machine and assembly stations to illustrate your floor layout. Then add the operator flow to show the steps the machining and assembly operator takes to complete the process. You should number these steps in sequence for each operator. Then mark each station for quality checks, safety precautions, and work-in-process.

Operator Loading Chart

The final charting of the standard work process is to graph the operator loading chart. The operator loading chart documents the time allocated for all operators in the cell. It also documents how many operators are in the cell. The chart is a bar chart. Figure 8.12 shows an example.

This graph depicts the operators versus their respective takt time. It shows if the operators are able to do their assigned tasks within takt time. If the operator's time is above takt time, then they will not meet customer demand. However, if the operator time is significantly below takt time, then this is an indication that there is waste in the process (operator or machine is waiting or idle). For example, in Figure 8.12, operator A has a cycle time of 460 seconds and operator B has a cycle time of 458 seconds. The takt time is 474 seconds, which indicates both operators can complete their respective tasks within takt time.

Least Operator Concept

Since the purpose of Lean is to drive continuous improvement, there is a technique within standard work that helps drive continuous improvement. The least operator concept states that the cell should be front-loaded, and you should allocate all waiting time to the least operator. The least operator should be the last operator in the sequence. The other operators should be fully loaded to takt time. This makes the waste, the waiting time, visible. It also exposes the opportunity for improvement at the last operation.

In contrast, natural work has the operator at various loads. The traditional work setup has the operators at equal loads; however, this is usually not at takt time. The front-loaded concept has all the operators except the last operator loaded close to takt time.

Resource requirements can be calculated. The total number of operators required is equal to the sum of individual operator cycle times divided by the takt time.

Model		STANDARD WORK COMBINATION SHEET	Date Prepared		Quota Per Shift		Manual Work Machine Work Walking
Work Seq.			Dept.		Takt Time		

Step No.	Description of Operation	Time			Operation Working Time (Seconds)																			
		Manual	Auto	Walk	25	50	75	100	125	150	175	200	225	250	275	300	325	350	375	400	425	450	475	500
1																								
2																								
3																								
4																								
5																								
6																								
7																								
8																								
9																								
10																								
	TOTAL																							

FIGURE 8.10
Standard work combination sheet.

STANDARD WORK COMBINATION SHEET

Model		Date Prepared		Quota Per Shift	55	Manual Work ——
Work Seq.		Group		Takt Time	474	Machine Work – – – / Walking ∿

Step No.	Description of Operation	Time: Manual	Time: Auto	Time: Walk
1	Unload, Blow off chips, Load, Clamp, Start	53	294	9
2	Unclamp, Unload, Blow off chips, Load, Start	33	414	15
3	Unload, Blow off chips, Load, Start	28	349	7
4	Unload, Blow off chips, Load, Start	27	196	13
5	Unload, Blow off chips, Load, Start	32	423	21
6	Unload, Blow off chips, Load, Start	34	378	19
7	Unload, Blow off chips, Load, Start	30	343	20
8	Unload, Blow off chips, Load, Start	22	229	7
9	Blowout, Plug, Blow off, Unload	59	0	19
	TOTAL	318	×	130

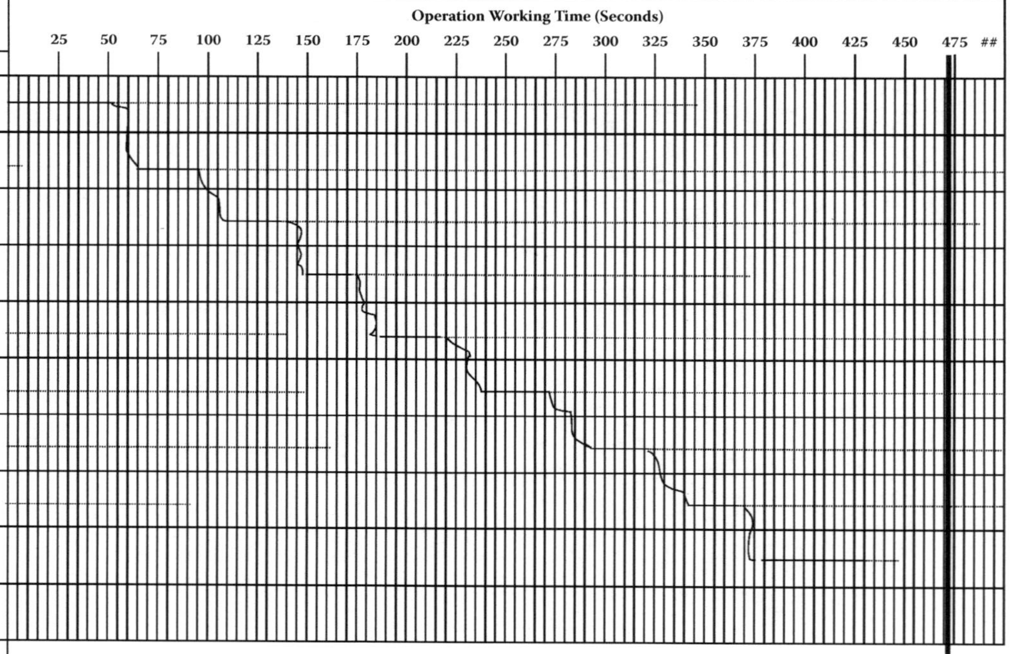

FIGURE 8.11
Example of standard work combination sheet.

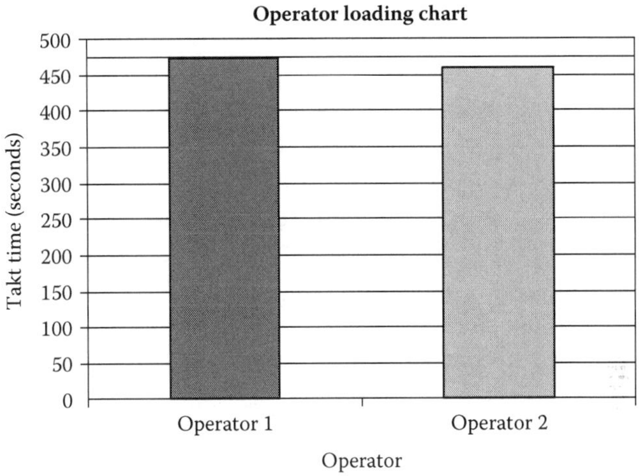

FIGURE 8.12
Operator loading chart.

Conclusion

Standard work documents a process so that all operators follow the same process to reduce process variability. Chapter 9 will take you through systems thinking and the theory of constraints. Chapter 10 will describe Hoshin Kanri, and Chapters 11 through 21 provide in-depth case studies of the Lean tools in healthcare, manufacturing, and service.

Homework Problems

1. Define standard work.
2. What are the three elements of standard work? Describe each element.
3. Describe at least three benefits of standard work.
4. To implement standard work, what are the three prerequisites?
5. Define takt time. Give a specific example showing your calculation. Provide detailed information on the example such that anyone else could calculate the takt time.
6. When calculating takt time:
   a. Include lunches and breaks in the available working time and reduce them afterward.

b. Include lunches and breaks in the available working time but do not include planned machine downtime.
c. Do not include lunches and breaks but include machine downtime as available working time.
d. Do not include lunches and breaks, machine downtime, and any other unavailable production time.

7. Given the following information, calculate the takt time. The operation has an 8 hour shift, 20 minutes for break, 5 minutes for cleanup, 5 minutes for a daily stand-up team meeting, and a customer demand of 175 pieces per shift.
8. What is standard work-in-process? Why is it important?
9. What is the purpose of the time observation sheet? Describe the three key steps.
10. What information does the process capacity table document?
11. Explain how standard work creates process repeatability.
12. What information does the standard work combination sheet combine?
13. What does the curved line on the standard work combination sheet represent?
14. What information does the standard worksheet provide? How many standard worksheets should you develop?
15. What is the least operator concept? Describe how it is used.
16. You are the managing director of a large amount of convenience stores/fueling stations. You believe you can offer a competitive advantage over other convenience store competitors by offering faster service. Customer feedback indicates a demand to reduce the process to fuel an empty tank, acquire a soda, and pay at the cash register before leaving.

    You gather data and determine most people watch their gas being pumped. It takes the customer 1 minute to prepare the pump. The pump averages 4.5 minutes from start to finish. Once finished with pumping fuel, customers enter the store and move to the soda fountain area where they spend about 1 minute searching for cups, lids, and choosing the flavor of soft drink. The soda fountain discharges for an average of 30 seconds. Next, the customer moves to the store clerk to pay. This takes approximately 1 minute of labor for the store clerk to determine what to ring up and 30 seconds to enter the charge in the register, swipe a credit card, and complete the sale.

    a. What is the TCT of this operation? Diagram this process.

b. Each store clerk works 4 hour shifts with a 30 minute break time and a 5 minute team meeting. Management is targeting servicing 100 customers per shift. Assuming a steady rate of customers, what is the current takt time?

c. Given the information from (a), is this a reasonable takt time?

d. How many store clerks (operators) are needed?

9

Systems Thinking and Theory of Constraints

Elizabeth A. Cudney, David M. Dietrich, and Sandra L. Furterer

CONTENTS

Introduction

During Lean implementation, it is critical for you to consider the organization as a system of integrated processes. What do we mean by a system? A system is defined as (Evans and Lindsay, 2008)

> A set of functions or activities within an organization that work together for the aim of the organization.

The system provides the enterprise-wide view of the organization. It aligns with Deming's (1982) view of a production system, by connecting customers and suppliers to the inner workings (processes) of the organization.

We want to understand the system's components and the relationship of the components to each other and to the environment. We should incorporate the value streams of the enterprise. The system and value streams connect the organization to its external customers and suppliers as well as the processes and internal customers (employees). It incorporates processes identified as workflows and events. Lean helps us through value stream mapping to understand the activities that are performed to achieve value for the customer. By mapping the value streams, we can begin to understand where the constraining activities and bottlenecks exist in our processes.

Once the value streams are identified through value stream mapping, the organization must understand how to identify and manage bottlenecks in the processes, how to relate the capacity and performance measures of one process to another, and how to use that information to determine the firm's best service or product mix. Some improvement efforts will only impact small portions of the business while other efforts can impact the entire business. The goal of this chapter is to get you thinking about the entire business as interconnected processes so that you can select improvement activities with the greatest impact on your organization.

The theory of constraints (TOC) is a methodology that focuses on profit. The basis of TOC is that every organization has at least one constraint that limits it from getting more of whatever is the goal—typically, profit. TOC defines a set of tools that can be used to manage constraints.

Most organizations can be defined as a linked set of processes that take inputs and transform them into saleable outputs. TOC models this chain of linked processes. This system is based on the theory that a chain is only as strong as its weakest link.

A constraint is defined as any restricting factor that limits the performance of a system in terms of output (Krajewski et al., 2010). Capacity is defined as the maximum rate of output of a system. Constraints may be categorized into different types. The first type of categorization is known as *physical*. This type of constraint may include physical space, product quality, machine performance, work area capacity, or material shortages.

A second constraint is categorized as *market quantity demand*. In this type of scenario, market demand for a product is less than process capacity, leading to process underutilization. Finally, the third constraint includes *managerial* constraints that tend to inhibit the performance of a system through inappropriately issued metrics, inflexible worker policies, and inflexible cultural mind-sets that stymie optimized system utilization.

A *bottleneck* is a specific type of constraint that inhibits the performance of a complete system by slowing down an entire system at a single resource point, thereby governing the performance of a complete system.

Five Steps to Strengthening Your Weakest Link

There are a number of key principles of TOC. A day lost at a bottleneck is a day lost for the entire system. As a contrast, a day saved at a non-bottleneck is deceiving due to the fact that it does not improve the productivity performance of an entire system. Eliyahu Goldratt has defined a five-step process that a change agent can use to strengthen the weakest link:

- *Step 1: Identify the system constraint.* A constraint is anything limiting a system from achieving higher performance. This link can be either a physical or a policy constraint. The focus should be on balancing the flow, not on balancing capacity.
- *Step 2: Determine how to exploit this constraint.* The change agent should obtain as much capability from the constraining link as possible.

As with any other type of continuous improvement, these changes should be inexpensive measures.

- *Step 3: Subordinate the nonconstraint components.* Maximizing the output and efficiency of every resource may not maximize the throughput of an entire system. As such, subordinating the nonconstraint components will allow the constraint to operate at the level of maximum effectiveness. The overall system should then be reviewed to determine if the constraint has moved to another component. If the constraint is eliminated, the change agent skips Step 4 and continues to Step 5.
- *Step 4: Elevate the constraint.* In this step, action must be taken to eliminate the constraint. These actions may include major changes to the existing system. This step is only necessary if Steps 2 and 3 were not successful.
- *Step 5: Return to Step 1 with the next constraint.* TOC is a continuous improvement process.

Three Ways to Measure Change

TOC also defines three essential measurements to drive changes in the process:

1. *Throughput* is defined as the rate money is generated through sales of a product or service. This represents all the money coming into an organization.
2. *Inventory* is the money invested in a product or service that an organization intends to sell. According to Goldratt, inventory includes all of the following:
    - Facilities
    - Equipment
    - Obsolete items
    - Raw material
    - Work-in-process
    - Finished goods
3. *Operating expense* is the third measurement; it is defined as the money used to turn inventory into throughput. Operating expenses include items such as the following:
    - Direct labor
    - Utilities
    - Consumable supplies
    - Depreciation of assets

The three measurements are interdependent on each other. Because one measurement will change one or both of the other measurements, TOC has defined the following heuristic for the change agent:

Maximize Throughput while Minimizing Inventory and Operating Expense

All improvement efforts should be prioritized by how each affects the three measures. Therefore, throughput is especially critical because the only limit on how much it can be increased is the market size.

Conclusion

Systems thinking is critical to making process improvements that impact the overall organization rather than a subset. You need to think about the big picture and focus on improvements that are tied to the strategic objectives of the organization. Understanding the TOC methodology and concepts provides a different outlook on how systems interact. Chapter 10 presents Hoshin Kanri, which focuses on creating a core strategy for the organization and driving this information throughout all levels of the organization.

Homework Problems

1. Using the definition by Evans and Lindsay as a basis, create your own definition of a system.
2. Define TOC.
3. What is a constraint? Define the three categories of constraints.
4. What is a bottleneck? Give a specific example.
5. Describe the five-step process that a change agent can use to strengthen the weakest link.
6. Why will maximizing the output and efficiency of every resource not necessarily maximize the throughput of an entire system?
7. What are the three measurements used to measure change in the system? Describe each measurement.
8. What is the heuristic measurement? Why was it developed?
9. Explain why throughput is the most critical measurement. Give an example to support your explanation.

Illustrated in Figure 9.1 is a process for customers arriving at a local barber shop. After signing in, (a) customers are routed to a waiting area (b) to queue

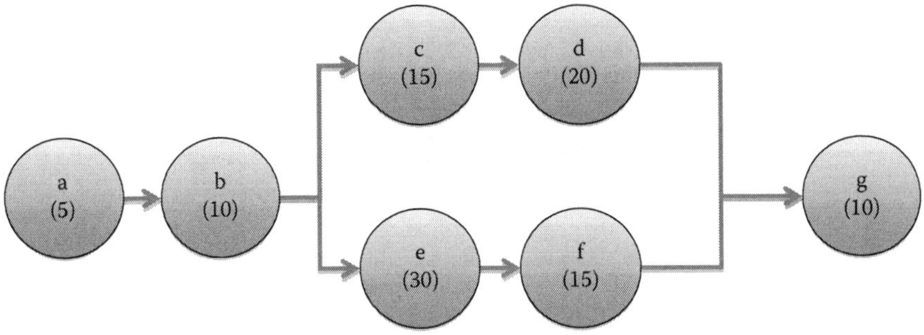

FIGURE 9.1
Hair salon process.

for two barbers cutting hair. Stylist (e) generally takes twice as long to cut hair compared to stylist (c), but washes hair (d,f) 5 minutes faster. (g) includes the cleanup/hair-dry and payment. The numbers in parentheses are the time in minutes for each step in the process.

10. Using the information in Figure 9.1 and what you have learned regarding TOC, what is the throughput time for the (a, b, c, d, g) routing?
    i. 35 minutes
    ii. 60 minutes
    iii. 50 minutes
    iv. 105 minutes
11. Using the information in Figure 9.1, how many customers can be processed through the (a, b, c, d, g) routing during a 4-hour evening hair appointment?
    i. 24
    ii. 48
    iii. 12
    iv. 32
12. Where is the constraint? a, b, c, d, e, f, g?

References

Deming, W. Edwards, *Out of the Crisis*, MIT Press, Cambridge, MA, 1982.

Evans, James R., and William M. Lindsay, *Managing for Quality and Performance Excellence*, 8th Edition, Independence, KY: South-Western Cengage Learning, 2008, 2011.

Krajewski, Lee J., Larry P. Ritzman, and Manoj K. Malhotra, *Operations Management: Processes & Supply Chains*, 9th Edition, Upper Saddle River, NJ: Prentice Hall, 2010.

10

Hoshin Kanri

Elizabeth A. Cudney

CONTENTS

Introduction

This chapter provides a detailed description of *Hoshin Kanri*, also known as policy deployment. Hoshin Kanri is a strategic decision-making tool that focuses resources on critical initiatives to accomplish organizational objectives. This process tool links major objectives with specific support plans throughout an organization. This chapter also presents an integrated approach to using Hoshin Kanri and value stream mapping to strategically select Lean initiatives.

Value stream mapping is the first building block in Lean. The purpose of value stream mapping is to understand the big picture. The current value stream consists of all actions necessary to deliver a product or accomplish a task including value-added and non-value-added activities. Value stream mapping must be conducted first to provide an effective blueprint for implementing an improvement strategy. A key step in creating the current state map is to calculate takt time.

The Lean philosophy can be used in almost any area of the business process, be it marketing, manufacturing, design, or even human resources. The range of implementation depends only on the motivation and creativity of the instigating administration. To make the best of Lean principles, the organization must look for new opportunities. Lean as a philosophy is not about just doing better than competitors; it is about going beyond and being the best in every process and product.

It is necessary to have a strategy that ties together the long-term vision of an organization with the Lean philosophy. This method provides a means for the strategic vision to flow down through all levels of the organization into its daily management activities. This enables the proper Lean technique to be used for optimal results that impact the organization as a whole rather than just a business unit.

This strategy consists of a five-phase methodology for implementing the enhanced future state value stream map, which is expedited using Hoshin Kanri (Cudney, 2009). A graphical representation of the five phases is provided in Figure 10.1. In phase 1, you start by deploying formalized Lean and variation reduction (or Six Sigma) training. Formal Lean training should include training on the technique followed by an implementation project.

At the same time as deploying the formalized training, you begin phase 2 to capture the strategic goals of the organization. Your goals should then be driven down through the organization and integrated into the daily activities of everyone in the organization.

In phase 3, value stream mapping is performed to identify all value-added and non-value-added steps required to bring a product from raw materials to the customer. Next, you map the current state to identify how the process

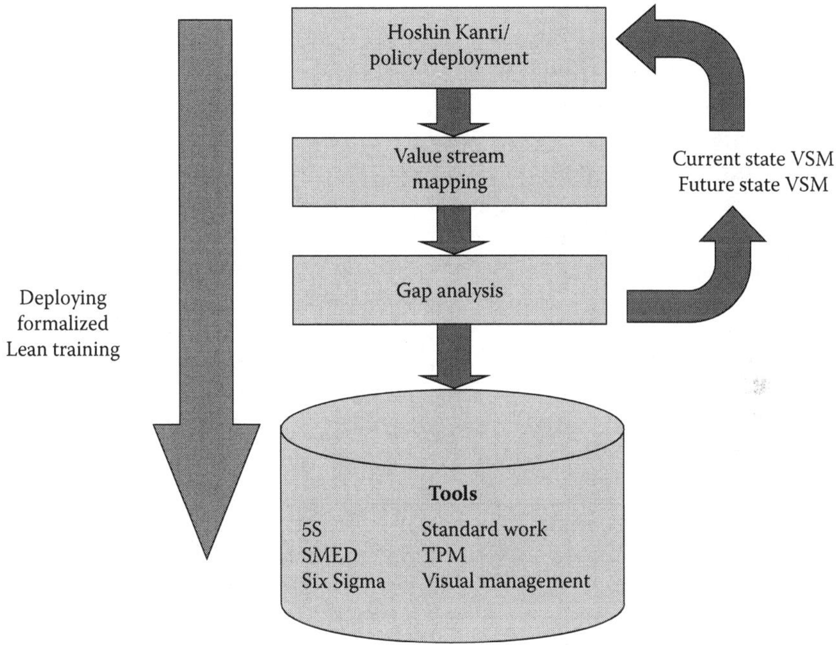

FIGURE 10.1
Improvement strategy. VSM, value stream mapping.

is currently operating. Kaizen bursts for areas of improvement are then identified. In phase 3, you develop the future state to design a Lean flow.

Next, in phase 4 you perform a gap analysis between the current state, future state, and strategic goals to prioritize the identified Kaizen bursts. This enables you to properly select the improvement project that will have the greatest impact on the entire organization rather than suboptimal improvements impacting only one area.

Finally, in phase 5 you focus on performing Kaizen events. Standard work and 5S must be the top priority as these techniques lay a foundation by improving consistency. Using the prioritized Kaizen bursts, you develop action plans or schedules to perform the Kaizen events or Six Sigma projects.

To promote strategic thinking, a corporation's strategic plan must be integrated with the macrolevel value stream map to identify optimal improvement opportunities. Often, improvement activities are identified with silo thinking without considering the effects on other systems or processes within the organization. Improvements in one area can have a negative impact on another business area.

Senior leadership including the chief executive officer (CEO) and directors should use Hoshin Kanri to develop long-term strategic objectives. Mid-level managers should then use macrolevel value stream mapping to identify areas of improvement to achieve strategic goals. Finally, department teams should use Lean and Six Sigma tools for process improvement.

To think strategically, the senior leaders of a corporation—in other words, you—should first determine where it is going, which is the vision. Then you need to identify your business' key processes. Next, you should perform a gap analysis between your organization's current state and vision. This will lead to a strategic approach to continuous improvement.

To become a Lean enterprise, you must integrate Lean throughout all levels of your organization. This means breaking down the silos and changing the focus of process improvement to a global perspective. What you need is a holistic approach to continuous improvement throughout your corporation. This will enable your corporation to make improvements that get the biggest bang for the buck, rather than suboptimal improvements. Lean strategy deployment breaks down these barriers and enables a holistic approach to continuous improvement that links to the long-term goals of your corporation, as shown in Figure 10.2.

Additional Lean and Six can be added to the base. A consolidated list is provided for graphical illustration. At the foundation of any Lean enterprise is people. People are the most important aspect of creating and sustaining a Lean initiative and culture.

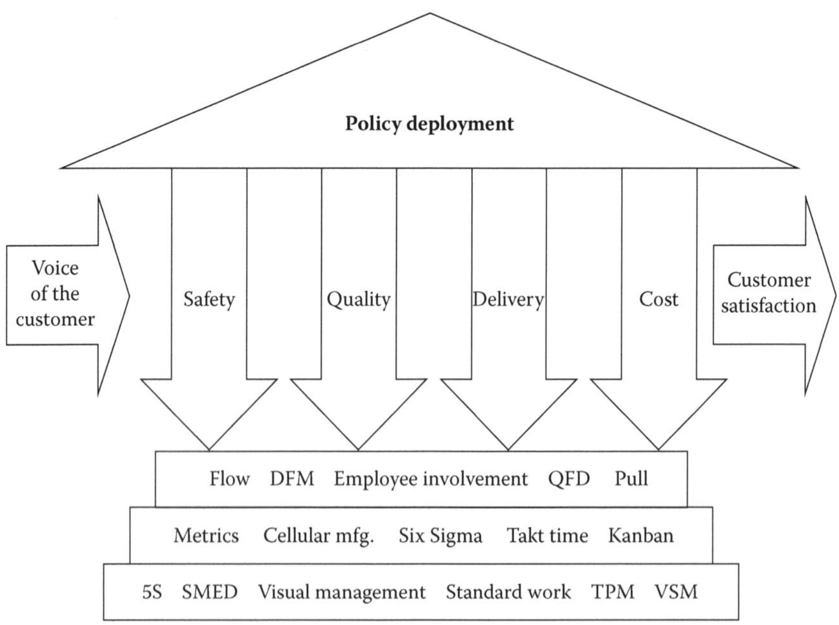

FIGURE 10.2
Strategic business system: DFM, design for manufacturing; QFD, quality function deployment.

Two Levels of Policy Deployment

Policy deployment is a methodology to capture strategic goals and integrate these goals with your entire organization's daily activities. The two levels of policy deployment include the following:

1. Management or strategic planning
2. Daily management

Planning for Policy Deployment

Effective planning is critical for the long-term success of a corporation. There are five main steps for effective planning:

- Step 1: Identify your critical objectives.
- Step 2: Evaluate the constraints.
- Step 3: Establish performance measures.
- Step 4: Develop an implementation plan.
- Step 5: Conduct regular reviews.

Daily Management of Policy Deployment

Daily management involves applying the plan–do–check–act (PDCA) cycle to daily incremental continuous improvement to identify broad system problems in your organization. Once you gain a breakthrough improvement in the system problem, the improvement becomes the focus of daily continuous improvement activities. Hoshin planning is the system that drives the continuous improvement and breakthroughs. Policy deployment involves both planning and deployment:

- Develop your targets.
- Develop your action plans to achieve your targets.
- Deploy both.

The concept of a hierarchy of needs was introduced by Maslow, and it outlines the basic needs that must be met before moving on to a higher need. Maslow's hierarchy of needs is illustrated in Figure 10.3.

In concurrence with the hierarchy of needs that must be met for an organization to move on to its next higher need, there are five levels of organizational needs:

- Level 1: Core vision
- Level 2: Alignment

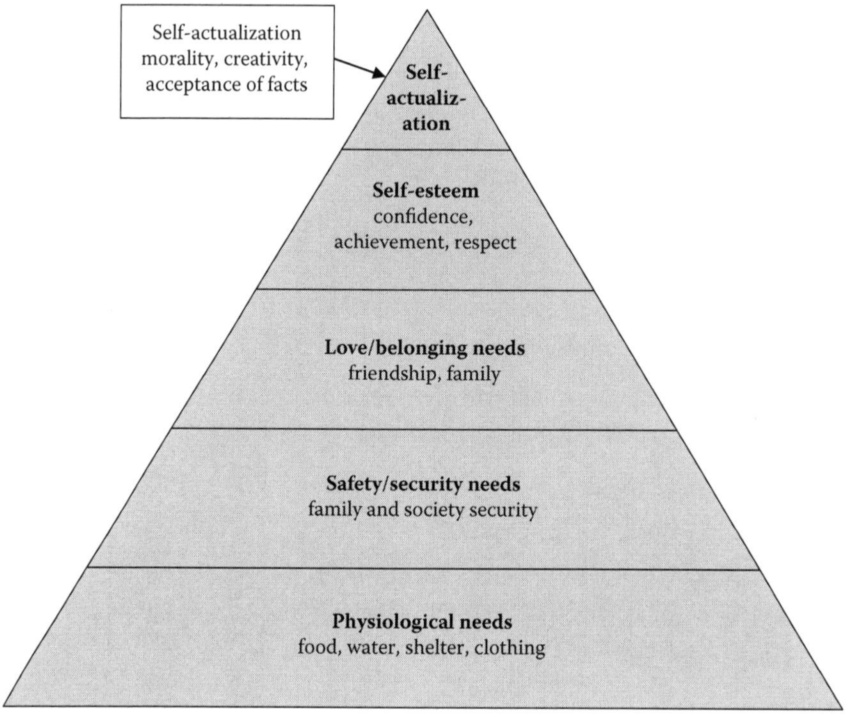

FIGURE 10.3
Maslow's hierarchy of needs.

- Level 3: Self-diagnosis
- Level 4: Process management
- Level 5: Target focus

Figure 10.4 illustrates the five levels of organizational needs.

The five levels of organizational needs are directly linked to the six Hoshin planning steps and five Hoshin methods. Figure 10.5 shows the linkage between the three.

Within the Hoshin planning process, there are six main steps. Each of these steps is discussed as follows:

Step 1: Develop a 5 year vision. Top management should develop a 5 year vision to define the strategic objectives of your organization based on the internal, external, and environmental challenges your organization faces. The 5 year vision is the future target for your organization. It is defined by your organization's goals, capabilities, and culture. It is a statement of where your organization wants to be in the future.

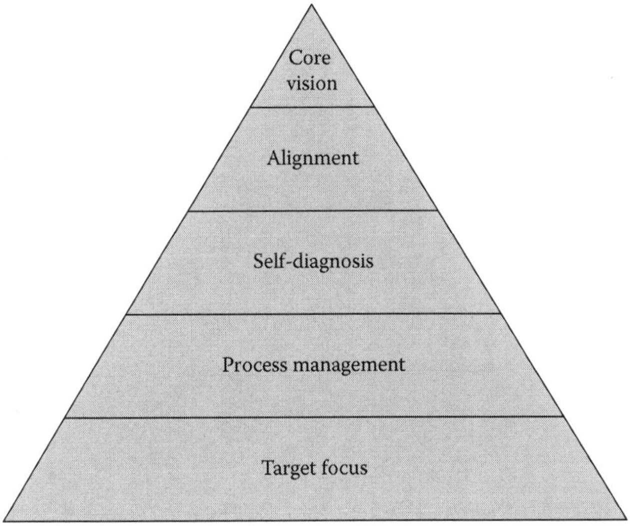

FIGURE 10.4
Hoshin planning phases.

Organizational needs	Hoshin planning steps	Hoshin methods
Core vision	5 year vision 1 year plan	Hoshin strategic plan summary
Alignment	Deployment	Hoshin plan summary
Self-diagnosis	Implementation	Hoshin action plan
Process management	Monthly reviews	Hoshin implementation plan
Target focus	Annual review	Hoshin implementation review

FIGURE 10.5
Linkage of organizational needs, Hoshin planning steps, and Hoshin methods.

The vision is a communication tool for senior leadership to relate to the ideal future for the organization. For example, a company may have a vision to be a top 10 company in its field in 5 years.

Step 2: Develop a 1 year plan. Based on your 5 year vision, you should develop a 1 year plan to outline continuous improvement activities that will enable your company to achieve its long-term strategy. The 1 year plan is linked to the 5 year vision by taking an incremental step and defining key targets to attain that step. The purpose of this step is to focus the activities throughout all levels of your organization on addressing your external issues and improving your internal problems. As part of this step, you should analyze external factors, including your competition and the economy as a whole. In addition, you should analyze past problems so that your organization does not repeat them. Top management must then prioritize the objectives based on safety, quality, delivery, and cost. Developing the 1 year plan blazes the path for the company to achieve its 5 year vision.

Step 3: Deploy your 1 year plan. The next step is to deploy your 1 year plan to all departments within your organization. Deploying your 1 year plan is where you begin to set measureable goals for each department. This is a planning step to determine specific opportunities for improvement within each department. At this point, a strategy is set to achieve the path set by the 1 year plan. This leads to step 4, which is implementation.

Step 4: Implement continuous improvement activities. Each department must drive continuous improvement activities that are aligned with your 1 year plan and 5 year vision. This step is where the improvement process begins. The prior steps (1–3) involved planning for improvement; in step 4, you are performing the improvement activities. For example, an improvement activity in this step could be a single-minute exchange of dies (SMED) event in a cell to reduce changeover time. The improvement activities in step 4 must tie directly back to the 1 year plan. This involves developing a master plan with appropriate measures and goals.

Step 5: Conduct monthly reviews. You should track the progress of your continuous improvement activities using quantitative metrics and communicate them to senior leadership (CEO and directors) in a monthly review. The monthly reviews should link directly to your deployment of the 1 year plan. You should monitor your actual improvements against your planned improvements as a monthly self-diagnosis and to ensure that the corrective actions are sustained. During a formal review to management, each department should present the tasks addressed, a problem analysis, and the problem-solving results. As the Hoshin planning implementation becomes more ingrained into your organization, the review may become more of a highlight or overview of the problems and corrective actions.

Step 6: Perform an annual review. Finally, you should conduct an annual review to monitor your progress and capture your organization's results. The annual review provides an opportunity to ensure that the implemented projects helped attain the 1 year plan and the 5 year vision. This is a check of how your implementation affected the organizational metrics set out in your 5 year vision. Based on the results of the previous year and the effectiveness of your implementations, a new 1 year plan is developed to set targets and goals for the upcoming year. In addition, the organization may redevelop their 5 year vision based on the current business environment.

Develop a Hoshin Strategic Plan Summary

The first step in policy deployment is to develop your strategic plan summary, also commonly known as the X Matrix. The Hoshin strategic plan summary links the strategic vision of the organization with measurable goals. The Hoshin strategic plan summary shown in Figure 10.6 illustrates the relationship between an organization's strategic goals, core objectives, metrics, and ownership. A value stream map gives you a picture of all the activities required to produce a product (Chapter 3 discusses the basics of value stream mapping). The Hoshin strategic plan summary also provides a picture of the overall strategy of an organization and how the strategy cascades throughout all levels of the organization. The linkage is clear on how each strategic goal is measured and who has the ultimate responsibility.

Hoshin Methods

How to Create a Hoshin Strategic Plan Summary for Your Organization

Let us look at each aspect of Figure 10.6 to illustrate how you can develop your own Hoshin strategic plan summary.

List your strategic goals: In Figure 10.6, the strategic vision is listed vertically on the left-hand side under the heading "Strategic Goals." These are the broad strategic goals of the organization for the next 5–10 years. The strategic goals should be what the organization needs to do to ensure long-term success.

List your core objectives: Next, we drive down a level to get more specific measurable goals. As you can see in Figure 10.6, these specific measurable long-term goals (for the next 3–5 years) are listed horizontally under the heading "Core Objectives."

Make sure your core objectives link to your strategic goals. To ensure that these goals are linked to the overall strategic goals in the plan summary (in Figure 10.6), a coding system is used to show the strength of the linkage:

- A filled-in circle (●) indicates a strong relationship between the strategic goal and the core objective.
- An open circle (O) indicates a direct relationship between the strategic goal and the core objective, but the core objective is not necessarily one of the key drivers for that strategic goal.

The key reason for showing the relationships in the strategic plan summary is to ensure that you are adequately addressing the strategic goals of your organization through your core objectives and that you are adequately measuring them by using the appropriate metrics.

HOSHIN STRATEGIC PLAN SUMMARY

Strategic goal 1	Strategic goal 2	Strategic goal 3	Strategic goal 4	*Core objectives* / *Measures*	Metric 1	Metric 2	Metric 3	Metric 4	Metric 5	Metric 6	Metric 7	Metric 8	Metric 9	Director of operations	Director of new business	Director of marketing	Director of engineering	Director of quality	Director of finance
		●		Core objective 1	O				●			●	●	●	O	●			●
●		O		Core objective 2					O			●	●		●	O	●		
	●		●	Core objective 3	●	●	O	●	●	O	●			●	●		●	O	●

Strategic goals

FIGURE 10.6
Hoshin strategic plan summary.

Identify who is responsible: Identify your metrics. Finally, you also want to ensure that proper ownership exists to drive the necessary improvements. You need to assign each improvement activity to a specific person for accountability.

Determine how you will measure improvements for meeting each goal: Now, you cascade the overall objectives into how you will manage the business and measure the progress of your process improvement activities. In this step, you now need to determine your metrics and short-term goals (for the next 1–2 years). These metrics will drive how you manage your business and how you prioritize your process improvement activities. List your metrics vertically on the right side of the matrix, as shown in Figure 10.6.

Make sure you tie your metrics to your core objectives. To ensure that your organization will meet your measurable core objectives, these metrics must meet several criteria:

- First, they must be "measurable" (quantitative not qualitative). These metrics assess the effectiveness of your process improvement efforts.
- The metrics must also be "baselined" to show your current performance or "benchmarked" against your competitors or your industry's standards.
- In addition, these metrics must be "achievable." When your metrics are unachievable (i.e., zero internal defects), your employees will become discouraged and your system will fail. In contrast, if you set realistic goals (i.e., zero external defects), your employees will team together to ensure your success.

Next, use the same coding scheme:

- A filled-in circle (●) to show strong relationships between your metrics and your core objectives.
- An open circle (O) to show direct relationships between your metrics and your core objectives.

Identify who is responsible for meeting your core objectives: Assign ownership. The final step in the strategic plan summary is to assign ownership of the core objectives. The ownership at this level falls on your organization's executive leadership because they own the responsibility of creating the strategic vision and driving it down through your organization. In Figure 10.6, the leadership team is listed vertically at the far right of the matrix. Here, again we use the coding scheme to illustrate ownership:

- Show ownership of a core objective with a filled-in circle (●).
- Show cursory responsibility with an open circle (O).

Because these responsibilities are tied into the strategic goals of your organization, there will be core objectives with several owners. For example, if one of your strategic goals is to drive new business through your product offerings, this will involve engineering and marketing (as well as possibly other owners). Therefore, in this example engineering and marketing would share ownership.

Drive Your Strategy Down to the Department Level: Develop a Hoshin Plan Summary

The Hoshin plan summary details the strategic goals and cascades them down to the department level. Whereas the Hoshin strategic plan summary is at the highest organization level, the Hoshin plan summary is the tactical plan for each department. Now, you need to drive the strategy down to each department. Figure 10.7 is a blank template of the Hoshin plan summary.

The first column in Figure 10.7, labeled "Core objectives," should correspond to the core objectives listed in your Hoshin strategic plan summary (in Figure 10.6). Now that you are moving down to the department level, the management owner will be the person in charge of that department. Each department will have its own Hoshin plan summary. In some cases, depending on the management structure of your organization the management owner will be the same here as the core objective owner in your Hoshin strategic plan summary.

Hoshin plan summary								
Core objectives	Management owner	Goals		Implementation strategies	Improvement focus			
		Short term	Long term		Safety	Quality	Delivery	Cost

FIGURE 10.7
Hoshin plan summary.

The next two columns are for your short-term and long-term goals. Your long-term goals may, in some cases, correspond to the strategic goals from the left side of your Hoshin strategic plan summary. However, your goals should definitely correspond to the metrics you outlined in the Hoshin strategic plan summary. This ensures that you align the proper activities with your overall strategic vision. Based on the metrics you previously outlined, your short-term and long-term goals should already be developed. These may not necessarily be the same for each department.

For example, let us look at external defects (measured as external parts per million [PPM]): a manufacturing department's short-term and long-term goals should be very aggressive. On the other hand, the engineering department should also be focusing on improving the product design using tools, such as design for manufacture and assembly, which will in turn reduce external defects. Therefore, the engineering department's short-term and long-term goals for reducing external defects will not be as aggressive as those of the manufacturing department. Also, your organization as a whole must be aligned and managed by your senior leadership team to ensure that your various departments come together to provide the overall necessary reduction in external defects that your organization is seeking. This provides a common goal for multiple departments to work together to achieve and eliminate silos.

In addition, with respect to goals there may be several metrics that relate to a core objective. As noted in Figure 10.7, you may need to list your core objective in multiple rows to correspond to the appropriate metrics. The various metrics for a core objective may then call for a different implementation strategy. For example, if your core objective is to improve product quality, this can be measured with internal PPM and external PPM. Internal PPM may be handled with an internal Six Sigma project as the implementation strategy. On the other hand, the implementation strategy to reduce external PPM may be to implement a *poka-yoke* device. So you would want to list these in two different rows to highlight that they are two different metrics with different implementation strategies.

Develop Implementation Strategies for Your Hoshin Plan Summary

The next step in developing your Hoshin plan summary is to develop your implementation strategies. This is critical in how your organization makes process improvements appropriately, using the most efficient and effective technique. Each department must develop a strategy on how it will achieve its short-term and long-term goals. The team members developing the strategy should revisit their current state maps to understand all the activities involved. This will enable them to select the most effective technique—whether it is a Six Sigma project, 5S, standard work, SMED, total productive maintenance (TPM), and so on.

Decide Where You Want to Focus Your Improvement Efforts

The final step in completing your Hoshin plan summary is to determine your improvement focus. Typical focus areas for organizations include safety, quality, delivery, and cost. Here, we again use the same coding scheme:

- Use the filled-in circle (●) to show strong relationships between the implementation strategy and its impact on safety, quality, delivery, and cost.
- Use the open circle (O) to show cursory relationships.

The purpose of showing the relationships in this matrix is slightly different. Here, you want to balance your improvement efforts. However, you still need to ensure that you link your implementation strategies to your improvement goals, which are linked to your core objectives. This common thread of linkage must be well defined and visible. But you also want to make sure that the implementation strategies you develop will have an impact on your improvement focus areas. For example, if you develop an implementation strategy that impacts quality but does not impact safety, delivery, or cost, then this may be a signal that it is not the most effective strategy. For balance, you want an implementation strategy that impacts more than one focus area. If you have a critical safety issue, then it should probably take precedence and may not impact any of the other improvement focus areas. But, in general, because you will be expending time and money for process improvement, you would want to impact multiple focus areas.

Develop Your Hoshin Action Plan

Next, you need to develop your Hoshin action plan. This further drives down your core objectives into the daily activities of your organization for process improvement by creating a detailed action plan. You should present this action plan to your leadership at a set frequency (typically a weekly walk-through or monthly management review).

Figure 10.8 shows a blank template of a Hoshin action plan. The top portion of the Hoshin action plan provides the necessary information to show the linkage between your action plan and each of your strategic core objectives. The following information is necessary:

- Core objective
- Management owner
- Department
- Team

Hoshin action plan		
Core objective: Management owner: Department:		Team: Date: Next review:
Situation summary:		
Relationship to Core objective:		
Short-term goal Long-term goal	Strategy:	Targets and milestones:

FIGURE 10.8
Hoshin action plan.

- Date
- Next review

To illustrate the Hoshin action plan, let us continue with the example of improving product quality.

The next section is situation summary. The situation summary provides a problem statement of the current status. It should clearly state why an improvement is necessary. An appropriate situation summary might be as follows:

> *Product quality is a key market driver in our industry. The external PPM for Product A has increased from 3,861 PPM to 4,725 over the past six months. As a result, our supplier rating has dropped from an A in the first quarter to a B in the second quarter.*

Next, define your overall objective. This objective should relate to one of your core objectives. The objective statement might then be as follows:

> *To improve product quality by decreasing external defects by 50%.*

The next step is to complete the short-term and long-term goals using the metrics you previously detailed in your Hoshin plan summary (shown in Figure 10.7). You should identify your short-term goals for a period of the next 3–6 months, and your long-term goals should focus on improvements for the next 12 months.

The next step is to discuss the implementation strategy. This should flow down from your Hoshin plan summary (shown in Figure 10.7). At this point, however, it should be more detailed. For example, with the core objective of "improving product quality" your implementation strategy in the Hoshin plan summary might have simply stated "Six Sigma project." In the Hoshin action plan, you would want to clarify this in more detail, for example, you might explain this as follows: "Six Sigma project on oversize cylinder bore."

The final step in the Hoshin action plan is to outline the targets and milestones of your strategy. Continuing with the example of the Six Sigma project on the oversize cylinder bore, your targets might be "perform a hypothesis test" or "run a design of experiments." The milestone would be the anticipated completion date.

Develop Your Hoshin Implementation Plan

Now you should develop your Hoshin implementation plan, which records your progress and lists the implementation activities. Figure 10.9 is a template you can use for this. Your implementation plan compares the current status of milestones to your initial projections. It is typically shown in a Gantt chart format.

Review your Hoshin implementation plan with your organization's senior leadership at a set frequency, typically monthly. This requires each department to outline its expected improvement gains by month for the following year.

The top portion of the Hoshin implementation plan details each core objective, its management owner, and the date you are targeting to achieve that objective. Your core objectives on the Hoshin implementation plan should link back to the high-level Hoshin strategic plan summary (shown in Figure 10.6), and the management owner should link back to the Hoshin plan summary (shown in Figure 10.7).

In the first column on the left of Figure 10.9, list your implementation strategies, as outlined by each department. Each department should have a Hoshin action plan for each implementation strategy.

In the next column of Figure 10.9, define your target and the actual performance for each implementation strategy. The performance should be measured using the metrics defined in your Hoshin strategic plan summary (Figure 10.6). Then, break down the performance by month to monitor the performance improvement trends for the year. One way to visually show which metrics are on track by month is to color the background of that

Hoshin implementation plan													
Hoshin objective: **Strategy owner:** **Date:**													
Strategy	**Performance**	**Schedule and milestones**											
		Jan	**Feb**	**Mar**	**Apr**	**May**	**June**	**July**	**Aug**	**Sept**	**Oct**	**Nov**	**Dec**
	Target Actual												
	Target Actual												
	Target Actual												
	Target Actual												
	Target Actual												

FIGURE 10.9
Hoshin implementation plan.

month's performance in green; months that do not meet the target performance can then be colored red. This makes it easy for you to hone in on those implementation strategies that are not meeting their target performance.

Conduct a Hoshin Implementation Review

Finally, you should conduct a Hoshin implementation review, which records the progress of your performance. Figure 10.10 provides a blank template. The Hoshin implementation review also records your company's performance relative to your industry's overall performance. The implementation plan also lists your highest-priority implementation issues.

During your presentation to your senior leadership team, use your Hoshin action plans as backup information to show what targets and milestones you have met as well as a recovery plan to get performance back on track.

Hoshin implementation review	
Core objective: **Strategy owner:** **Date:**	
Performance status	**Implementation issues**

FIGURE 10.10
Hoshin implementation review.

There are five key steps for implementing an effective strategy, which are described in the following subsections.

Step 1: Measure Your Organization's System Performance

In measuring your organization's system performance, it is critical to develop a plan to manage the strategic change objectives. The initial direction must be adaptable. The planning process must also be adaptive to respond to business changes. Then, regular assessments of planning and implementation are necessary.

Step 2: Set Your Core Business Objectives

To set your core business objectives, a technique called "catchball" is effective to incorporate group dialog. Catchball is equivalent to tossing an idea around, which provides the holistic and balanced organizational objectives for the overall business system.

Step 3: Evaluate Your Business Environment

The business environment must be evaluated to understand the needs of the organization's customers. These customers include stockholders, employees, external customers, and so on. Environmental analysis includes the technical, economic, social, and political aspects of the business. The purpose is to answer the following question: How does the business perform relative to its competitors?

Step 4: Provide the Necessary Resources

For strategic alignment to be successful, management must also provide the necessary resources to lead the efforts for both strategic objectives and daily

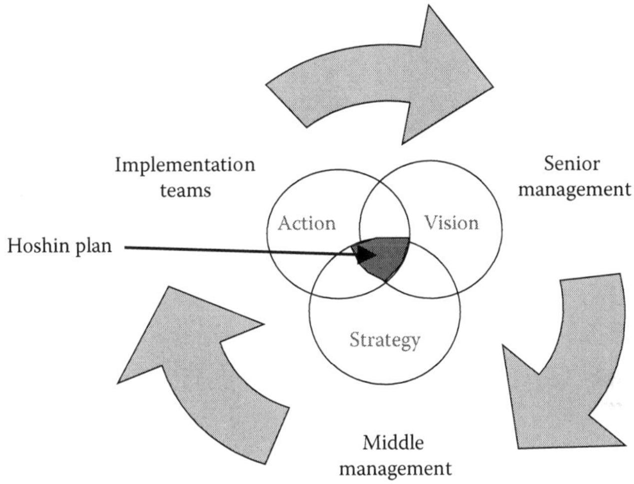

FIGURE 10.11
Hoshin plan alignment.

management. Remember that the purpose of Hoshin is to align the system to strategic change initiatives. This requires resource commitment.

Step 5: Define Your System Processes

Another key aspect is to define the system processes. Hoshin enables consensus planning and execution between all levels of an organization, as shown in Figure 10.11. The Hoshin plan aligns the strategic vision, strategy, and actions of the organization. The actions of senior management, middle management, and implementation teams (all levels of the organization) are aligned around the common Hoshin plan.

Three Main Tools of Policy Deployment

Policy deployment is one of the pillars of total quality management (TQM). TQM is based on Deming's PDCA cycle. The three main tools of policy deployment are as follows:

1. PDCA cycle
2. Cross-functional management (CFM)
3. Catchball

Let us look at each one in more detail to show how you can use these three tools for consensus planning and execution.

Deming's Plan–Do–Check–Act Cycle

Deming developed the PDCA cycle (shown in Figure 10.12) as an iterative four-step problem-solving process:

1. Plan consists of establishing objectives and processes to achieve specific results.
2. The do step involves implementing the processes.
3. In check, processes are monitored and evaluated against specifications.
4. In the fourth step, act, actions are taken to improve the outcome to meet or exceed the specifications.

One of the key differences between PDCA and Hoshin Kanri (aka policy deployment), though, is that policy deployment begins with the check step, which is step 3 of Deming's cycle. Therefore, the cycle is really CAPD—check–act–plan–do. You start by checking the current status. This propels the Hoshin Kanri process. Each company-wide check begins the deployment of a new target and an action plan for achievement.

Cross-Functional Management

CFM enables the continuous checking of targets throughout the product development and production processes. It is critical when developing the Hoshin plan to include a cross-functional group from marketing, design engineering, quality, manufacturing, finance, operations, and sales depending on your organization's structure. This diverse group can address the needs of all the stakeholders (internal customers, external customers, stockholders, etc.) in the Hoshin plan. By involving a cross-functional group, you can ensure a balanced representation of the needs of your organization's customers.

FIGURE 10.12
Deming's Plan–Do–Check–Act cycle.

Catchball

Catchball involves continuous communication and refers to communication and ideas starting at the worker level and disseminating up the hierarchical ranks to upper management, much like a ball being tossed in the air. Like a ball falling, upper management evaluates the plan or idea and communicates this back down to the workers who originally thought of the idea. In addition, management communicates the idea to other hierarchical ranks and to departments and areas, disseminating the idea to others through transparent communication. It is essential for the development of targets and action plans. Catchball is also essential for the deployment throughout the organization. You must create feedback systems to allow bottom-up, top-down, horizontal, and multidirectional communication. There must be commitment to total employee involvement for transparency and communication to work well.

Conclusion

Using an approach to link Lean to the strategic vision of an organization enables the organization to realize the full benefits of Lean. Lean projects that are selected based on their impact on the entire organization have the most effective results. For Lean efforts to be successful, employees must be adequately trained and coached to develop their skills. Effective mentoring is essential for employees to understand and implement new techniques. Therefore, the approach linking Hoshin Kanri and Lean must address training and coaching to develop the skills of employees at all levels of the organization.

It is critical to ensure that your strategic vision cascades down throughout your organization into the daily activities of all employees. This clear linkage enables the organization to move in a common direction with common goals. When employees understand the direction of the organization, they can make appropriate improvements that will enable long-term success. By using the strategic vision, an organization can employ Lean techniques to eliminate waste and improve flow.

Homework Problems

1. Define Hoshin Kanri.
2. What does Hoshin mean?
3. What are the two levels of Hoshin Kanri? Explain each level.
4. What are the five main steps for effective planning?
5. What are the five levels of organizational needs? How do these relate to Maslow's hierarchy of needs?

6. Describe the six main steps of the Hoshin planning process.
7. What is the purpose of the 1 year plan? What should be analyzed and considered when developing the 1 year plan?
8. What is the purpose of the annual review?
9. What does the Hoshin strategic plan summary link? How are these linked? What information is provided in the Hoshin strategic plan summary?
10. What is the key reason for showing the relationships in the Hoshin strategic plan summary?
11. What are the criteria for the metrics on your Hoshin strategic plan summary?
12. What is the purpose of the Hoshin plan summary? At what level in the organization should the Hoshin plan summary be developed?
13. How is the purpose of showing the relationships in the Hoshin strategic plan summary matrix different from the Hoshin plan summary?
14. What does the Hoshin implementation plan record, and how often should it be reviewed?
15. Describe the key steps for implementing an effective strategy.
16. In Hoshin Kanri, a key step is to conduct an environmental analysis. What is an environmental analysis?
17. How is Deming's cycle different in Hoshin Kanri? Explain why it is different.
18. What is CFM? Why is it important in Hoshin Kanri?
19. Define catchball. Give a specific example.
20. You are the manager of your local campus bookstore. After studying Hoshin Kanri, you have determined that you need a more transparent, clear strategic focus for the bookstore. You have identified four strategic goals for the bookstore in the near term: (1) grow the customer base, (2) lower administrative costs, (3) adopt new technology, and (4) expand the size of the store. With these strategic goals in mind, fill out a strategic plan summary table, refer to Figure 10.6. Be sure to identify objectives, metrics, and ownership.

Reference

Cudney, E. (2009). *Using Hoshin Kanri to Improve the Value Stream*, Productivity Press, New York.

Section III

Manufacturing, Service, and Healthcare Case Studies

11

Lean Restaurant

Corbin LeGrand, Neha Pawar, Snehal Digraskar, Sneha Mahajan, Sukhada Mishra, and Susan Polson

CONTENTS

Executive Summary

The current state of XYZ Restaurant provided the opportunity for significant improvements by utilizing Lean methodologies. Such improvements would increase overall customer satisfaction and, in turn, enhance sales. XYZ Restaurant is a relatively new business. According to the U.S. Department of Commerce, only 7 out of 10 businesses survive the first 2 years, and only half survive the first 5 years in business.* It is imperative that new businesses understand and respond to the voice of the customer (VOC) to ensure viable longevity. The implementation of Lean tools at XYZ Restaurant will assist in minimizing costs and developing a culture of sustainability through Lean.

A customer survey was developed and administered to correctly identify the issues with the kitchen. The survey was ranked on a scale of 1 through 5 with a score of 5 being best. On the survey were six categories: (1) food quality, (2) menu variety, (3) service & atmosphere, (4) delivery time, (5) cleanliness, and (6) overall experience. In addition, the survey respondent was asked to describe how frequently they dined at the restaurant and describe their overall experience. The survey revealed that customers' wait time on food delivery, the accuracy of the order, as well as lengthy delays in settling payment were frequent complaints. The team decided that by implementing various Lean strategies, XYZ Restaurant could reduce customer wait for food delivery to an average of 15 minutes, improve the accuracy of the orders placed by 60%, and reduce the time to pay the bill by 100%.

The outline of the project followed the DMAIC form: Define, Measure, Analyze, Improve, and Control. The project charter was developed during the Define phase by detailing the problem statement, stakeholders, and scope of the project. In the Measure phase, the team utilized a current state value stream map of the overall process to highlight potential areas for improvement.

A spaghetti diagram was also used to capture the motion of the restaurant workers during the order fulfillment process. After reviewing these areas, during the Analyze phase, the team decided on tools to use to improve process efficiency.

Implementation of 5S activities, a kanban system, Kaizen, and standard work procedures into XYZ Restaurant's procedures produced positive results. The cost and benefit analysis and time improvement calculations utilized in the Improve phase are explained in this chapter.

Once the implementations were complete, it was crucial that a control was developed to ensure the continued success and use of Lean tools at XYZ Restaurant.

* U.S. Department of Commerce, Census Bureau, Business Dynamics Statistics; U.S. Department of Labor, Bureau of Labor Statistics, BED.

Introduction

Lean Thinking Overview

Lean principles were not implemented in the food service industry until the 1970s. Quick and accurate food service became a necessity, and techniques to offer this to the customer at the lowest cost and effort have been and continue to be investigated thoroughly. This includes reducing customer wait times and increasing profits, all of which lead to the central idea of Lean, that is, maximizing production with the bare minimum resources. The goal of this project is to provide the customer what they want, when they want it, and at the desired cost. The main idea is to have the means to quantify the operation, which provides a way to track the effect of decisions on customer satisfaction and the financials.

Lean thinking is an innovative view of processes that allows for identification and elimination of waste, or muda. Through Lean tools, it is possible to significantly improve operations by reducing muda. These tools are used to identify and eliminate areas of waste, create flow and pull, and refocus the organizational culture. Done correctly, Lean thinking will result in reductions across the board, while the gains can be reallocated to produce more value for the organization.

Project Overview

XYZ Restaurant is a breakfast spot that is famous for its homemade bread and delicious omelets. This restaurant has seen a fair share of change in its management; thus, the operations have been constantly transforming. Recently, new management took over and a need was identified for introducing Lean to the management of this restaurant. The inspiration came from a visit to this restaurant and personal experience. A thorough investigation and analysis of their work processes was performed. This investigation was aimed at optimizing their processes by improving quality and reducing the waste in their operations. These improvements promise significant savings for the restaurant owners, which would, in turn, be passed on to the customers, both quantitatively and qualitatively. The current design will be the starting point for the proposed improvements. Further improvement will also be suggested.

The project followed the DMAIC outline: Define, Measure, Analyze, Improve, and Control:

1. *Define*: The current status of the operations is documented here.
2. *Measure*: Data collection.
3. *Analyze*: Root cause analysis of defects are synthesized from the data collected.

4. *Improve*: Improve the current situation by implementing suggested solutions.
5. *Control*: All the identified areas of improvement are monitored constantly to sustain the improvements.

DMAIC provided structure and organization to the project that details areas of waste and tools to reduce waste and sustain the results. This project is classified as Lean food service.

Define

Problem Statement

As a way to capture the VOC, a customer survey was developed and implemented to provide the baseline for customer satisfaction. The survey illuminated three issues that needed to be addressed. First, the time it takes to deliver food to the table is too long. Secondly, the accuracy of the orders needs to be improved. Finally, the time it takes to pay the bill needs to be reduced.

Project Objectives

The first objective of this project is to reduce food delivery time to a maximum of 15 minutes. In some instances, this would be an improvement of 40%. The next objective would be to eliminate the time required to pay the bill at the end of a meal from an average of 3 minutes. The third and final objective should seek to increase the accuracy of the orders delivered to the customers by 100%. Various methods will be analyzed to reduce the lead time by reducing worker movement and the steps required to prepare a dish in the kitchen, improve inventory storage levels, and increase overall customer satisfaction.

Project Boundaries

The project boundaries of XYZ Restaurant were limited for this project to the kitchen, dining area, pantry, and cashier area. The Lean tools utilized for this project should be expanded to other areas of the restaurant at a later date by restaurant management after proven sustainability of current Lean improvements.

Measure

Baseline

Baseline information must be collected to understand and define a starting point to determine the overall effectiveness of Lean tool improvements. Baseline data captured will help identify the areas where improvements can be made. The peak rush hours of lunch and dinner were chosen to capture the baseline data.

Data Collection Plan

Discussions with the restaurant manager prompted a data collection plan. Two methods were decided upon for data collection. The first was the customer survey. The customer survey was selected to define what exactly is important to the customer. The second method selected was to visit the restaurant during various times of the day. This would enable the team to capture the activities and processes utilized in the dining and kitchen areas. The details of the activities and processes would be highlighted during high-demand periods as well as low-demand periods.

Current State Value Stream Map

The initial visit to the restaurant was quite in-depth. Information regarding the current state of a customer's experience in the restaurant was captured (Figure 11.1). Each process in the customer's experience was identified. The processes were separated into seven steps to develop a process flow map. These seven processes are as follows to create the current state value stream map illustrated in Figure 11.2.

1. Customer arrives and is seated by the hostess.
2. Customer receives a menu.
3. Customer receives water and their order is taken.
4. Order is prepared.
5. Order is delivered to customer and consumed.
6. Customer pays bill.
7. Customer departs.

The current state value stream map (Figure 11.1) provides an easy way to identify the nature of the time required to complete activities. Time can be broken down into two components: value-added, which is necessary to transform the product, or non-value-added, which does not contribute to transforming the product. Value-added time (VAT) for an order cycle is 50 minutes. It has been measured that the non-VAT (NVAT) is 17 minutes.

Statistical Analysis of Baseline Data

From the baseline data, we were able to determine that the average time of customer delivery for the current processes is 17.166 minutes, with a standard deviation of 6.14 minutes. A high standard deviation value (6.14) shows

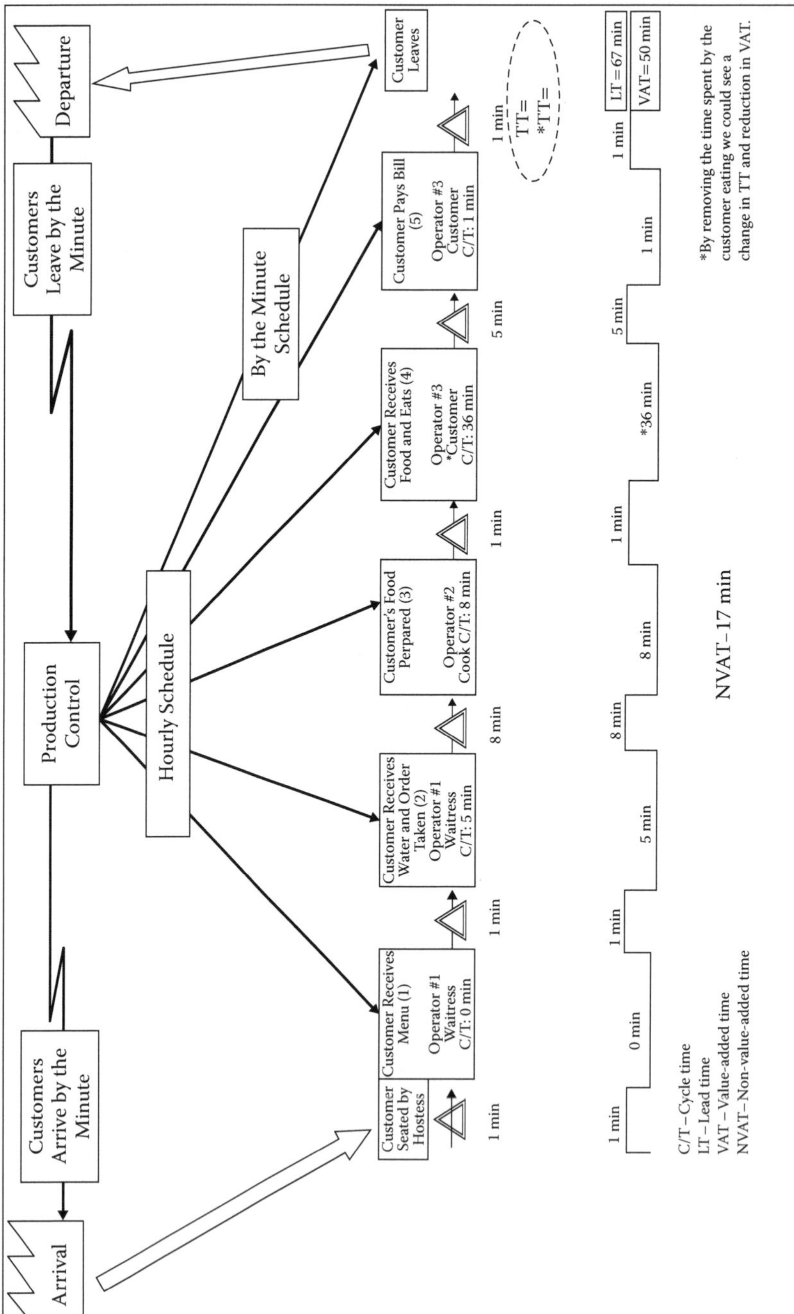

FIGURE 11.1
Previous state value stream map.

Table No.	Arrival (AM)	Menu Card and Water (AM)	Order Placed (AM)	Order Received (AM)	Time Delivery	Left Table (AM)	Table Cleaning (AM)
1	10:06	10:07	10:12	10:33	21 minutes	11:04	11:08
13	9:28	9:29	9:32	9:52	20 minutes	10:20	10:22
13	10:20	10:24	10:29	10:45	6 minutes	10:53	10:55
14	10:08	10:09	10:12	10:25	13 minutes	10:55	11:00
10	10:00	10:01	10:05	10:26	25 minutes	10:40	10:44
17	9:51	9:52	9:58	10:16	18 minutes	10:40	10:43

FIGURE 11.2
Baseline data of order fulfillment process.

Mean	17.166
Median	20.5
Min	6
Max	25
Range	19
Variance	37.80
Standard Deviation	6.14

FIGURE 11.3
Baseline statistical data.

that the data is widely spread. This means that there is a lack of standardization, which leads to variable customer delivery time. The principal goal of implementing Lean tools should be reduction in the average customer waiting time as well as reduction in standard deviation (see Figure 11.3).

Analyze

Customer Expectations

Listening to the VOC is important. This ensures that the resulting output is actually seen as a benefit to the customer. The customer survey was used to determine what was of value to their visit. Customers rated the restaurant quantitatively using a scale that ranged from 1 to 5. Poor was rated (1) while excellent was rated (5). There were six different categories including food quality, menu variety, service and atmosphere, delivery time, cleanliness, and overall experience. The quantitative results of the surveys collected have been summarized in Figure 11.4.

Category	Results
Food Quality	Good
Menu Variety	Good to Excellent
Service & Atmosphere	Moderate
Delivery Time	Moderate
Cleanliness	Moderate to Good
Overall Experience	Moderate to Good

FIGURE 11.4
Customer survey result summary.

In addition to the ranked categories, three subjective questions were asked in an effort to acquire additional information from the customer.

The first qualitative question was to determine how often a customer visits XYZ Restaurant. The regularity of customer visits is scattered. Some customers visit rarely, other customers visit weekly, or some visit monthly. The second question solicited a description of their visit. This item had few responses. Third, the customer was asked to make suggestions for improvement. The results of the subjective questions have been compiled into the following two groups:

Recommendations for Improvement:

1. Shorter delivery times
2. Extended service hours
3. Eliminate mistakes in orders
4. Bar stools in dining area
5. Better ambiance

Appreciative of:

1. Homemade bread
2. Coffee
3. Omelet

Improvement Opportunities

The customer survey resulted in only one category, menu selection, with a good to excellent rating. The results of the survey clearly indicated that customers view the restaurant as having room for improvement in all six categories. Specifically in the subjective portion of the survey, customers stated that food delivery in a shorter time frame is desired. On the current state value stream map of XYZ Restaurant (Figure 11.1), the amount of NVAT is highlighted.

Several tools are used to further determine the non-value-added activities such as a Pareto analysis, a cause and effect diagram, and a spaghetti diagram.

Visual observation of the kitchen, dining, and billing areas was also compiled into detailed lists of potential items for improvement (see Figure 11.5). In the dining, kitchen, and billing areas, there was a lack of continuity between workers. This would impact the customers by providing inconsistent service. The organization of supplies was also inconsistent. This caused excessive time to be used walking and searching for items. Figure 11.6 shows the data analysis captured from the current state.

Dining Area	Kitchen Area	Billing Area
Lengthy food delivery times	Wasted space	Cashier unfamiliar with pricing of some items
Waiter not acquainted with menu	Unnecessary items in kitchen (empty milk containers)	Long queue for billing
No standardized work pattern of order taking and delivery	Unorganized kitchen items	Average billing time ~ 3 minutes
Server discontinuity	Supplies sitting on the floor and on top of the appliances	
Error on food ordered versus food delivered	Unsorted inventory	
Delay in cleaning tables	No visual indication for diminishing inventories	
Absence of adequate supplies on table (beverage, napkins, condiments)	No labeling system	
No consistency in taste		

FIGURE 11.5
Potential areas of improvement.

Reduced Serving Time at Restaurant				
Problem	**Count**	**Percent of Total**	**Cumulative Percent**	**Horizontal Line Value**
Lack of standardized processes	23	30.67	30.67	80
Time waste—motion	15	20.00	50.67	80
Rush hours	11	14.67	65.34	80
Organization in kitchen	10	13.33	78.67	80
Availability of tools	7	9.33	88.00	80
Availability of cook	5	6.67	94.67	80
Food waste	4	5.33	100	80
Total	**75**	**100**	**100**	**80**

FIGURE 11.6
Data analysis of current state.

Pareto Analysis

Data were gathered to identify areas for improvement. A Pareto analysis was completed to identify the top issues contributing to long food delivery time. The Pareto provides a map of where to gain the largest improvements and allows businesses to move forward with justification for addressing the issues based on time and cost to implement.

Charting this data (see Figure 11.7) provides an easy graphical representation of the magnitude of each issue and its contribution to the overall delay in food delivery.

Cause and Effect Diagram

A cause and effect diagram was constructed to identify contributing factors leading to customer satisfaction or dissatisfaction. The chart graphically shows the first tier issues that affect customer satisfaction, food quality, poor service, and high prices. Second tier issues, such as slow response time, long hold time, or wrong orders, combine to cumulatively be defined as poor service (see Figure 11.8).

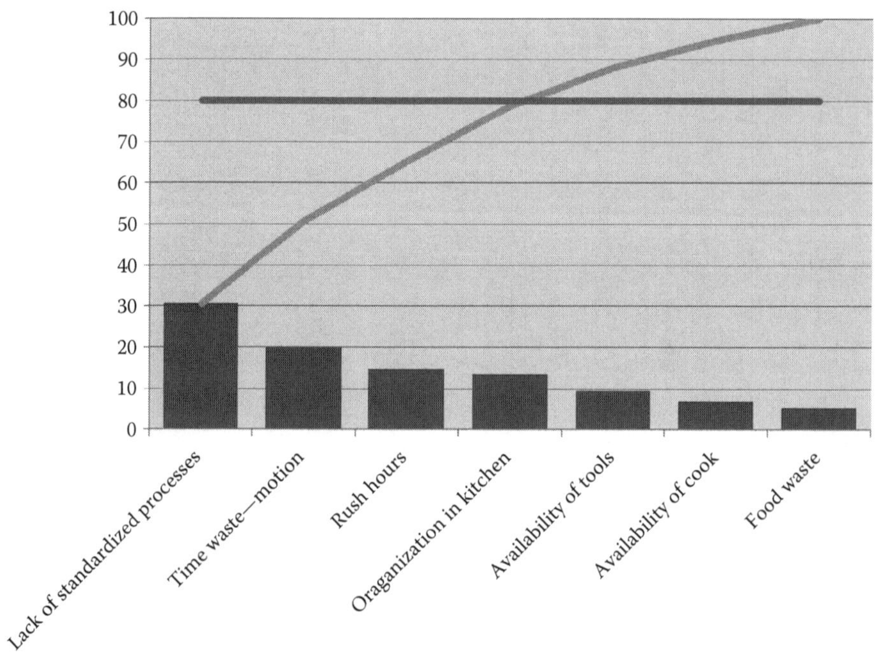

FIGURE 11.7
Pareto chart for serving time at XYZ Restaurant.

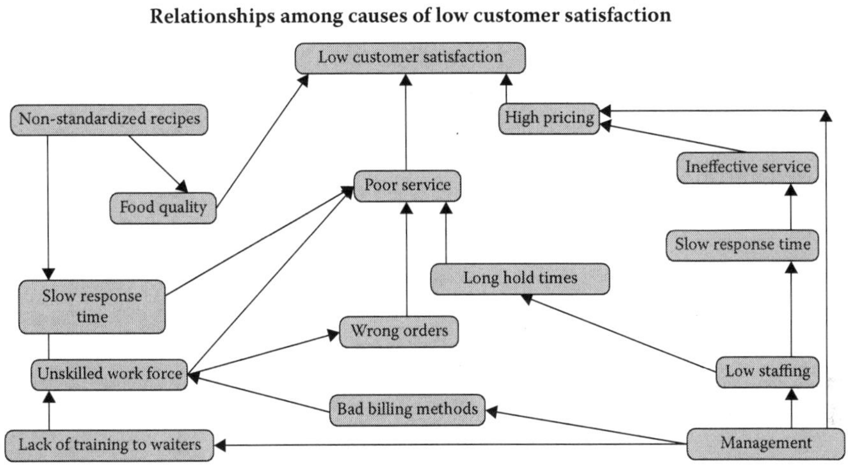

FIGURE 11.8
Cause and effect diagram for customer satisfaction.

Improve

5S Implementation

The 5S (Sort, Straighten, Shine, Standardize, and Sustain) activity was one of the Lean tools chosen to improve the delivery time of food orders and order accuracy. The boundaries of the 5S activity were limited to the kitchen, pantry, and dining areas.

Sort. Items in the refrigerators, and in the pantry area, were unorganized before starting the Lean activity. The items in the refrigerator were organized in different sections as per their use and their expiration dates. The inventory was organized by type of item such as salad dressings, flour, and spices. Each of these items was then labeled. Some of these frequently used items were moved to a more easily accessible space in the kitchen.

Straighten. This part of the activity seeks to provide a clear and organized workspace to eliminate waste in movement and time. We improved this problem by providing clear labeling on all food products and tools used in the workplace as seen in Figure 11.9. These labels mention clearly the designated areas for particular items at all times. This will reduce waste in motion and time associated with searching for desired items in the workplace. It will also provide a more functional workspace as the labels were grouped together according to use and appropriate designations.

Shine. The goal here is to provide a clean workspace. The team created standard checklists for opening and closing procedures for the restaurant (Figure 11.10). These will provide a set of tasks for staff to follow and ensure

Item Name	__________
Date Prep	__________
Exp. Date	__________
Shelf	__________

FIGURE 11.9
Label example.

Closing Sheet (for Kitchen)		**Completed by: Date:**	
No.	**Check Item**	**Description**	**Check**
1	Floors	Are floors shiny, clean, and free of waste and water?	
2	Appliances	Are the appliances wiped clean?	
3	Dining area	Are beverage station, counters, tabletops, and chairs clean?	
4	Replenish dining area supplies	Does each table have enough supplies of sugars, salt, peppers, jams, tissue, etc.?	
5	Utensils	Are all utensils cleaned and placed at respective positions?	
6	Garbage	Are all garbage bags disposed of?	
7	Fryer oil	Is fryer oil replaced with new oil?	
8	Left over items	Are all leftovers labeled and refrigerated?	
9	Today's special	Is the special item for tomorrow decided?	

FIGURE 11.10
Closing activity checklist.

that everything is in the correct place to develop a smooth flow throughout the rest of the day.

Standardize. This part of the 5S activity seeks to provide a set of direct procedures for frequent tasks to create a reproducible, consistent outcome. The team also noticed that the staff would benefit from a displayed standardized ingredients list for certain menu items instead of trying to remember what the ingredients were or having to go look them up. These improvements will save time in the ordering process. These standardized ingredients lists are similar to what you would find in a common cookbook.

Sustain. Here the goal is to provide ways to develop a culture of 5S on the above-mentioned 4S. The team developed assigned weekly task sheets and signs to provide a means of organization and culture paradigm shift, respectively. The signs will serve as a constant reminder of Lean thinking to all staff. The assigned weekly task sheets will effectively help workers fall into certain Lean work patterns. The assigned weekly task sheets will also serve as a means to standardize the workplace. Additional signs and the assigned weekly task sheet template can be found in Figure 11.11.

Week No.	Beverage Station	Kitchen Cleaning	Dining Area Cleaning	Floor Cleaning	Inventory Check	Food Prep	Food Serving	Dish Washing	Cashier
1									
2									
3									
4									

FIGURE 11.11
Weekly task assignment.

Inventory Card			
Items	Qty.	Date	Check

FIGURE 11.12
Kanban card example.

Kanban System

Inventory management is an important issue that was unaddressed previously. Inventory kanban cards were developed and assigned a proper place or wooden container for storing these cards. The cards contain information about the item required, quantity required, and date when the supply ended. These cards can be marked when that particular item is replenished and will be placed in the adjacent container again. A sample kanban card can be seen in Figure 11.12.

Future State Value Stream Map

The future state map was then developed to offer continuous improvement. It is provided in Figure 11.13.

Standard Work and Spaghetti Diagrams

As mentioned in the "Project Overview" section, the restaurant is famous for homemade bread, which was appreciated by many customers. This advertisement is aimed at attracting more customers and will in turn lead to financial profits. A standard set of questionnaires or procedures to be followed by the waiter to address the customer was prepared. Standard work is one of the Lean tools that provides a guideline for how a process should be performed. The tasks are identified and broken down into smaller well-defined steps. Standard work is then utilized to maintain consistent quality, costs,

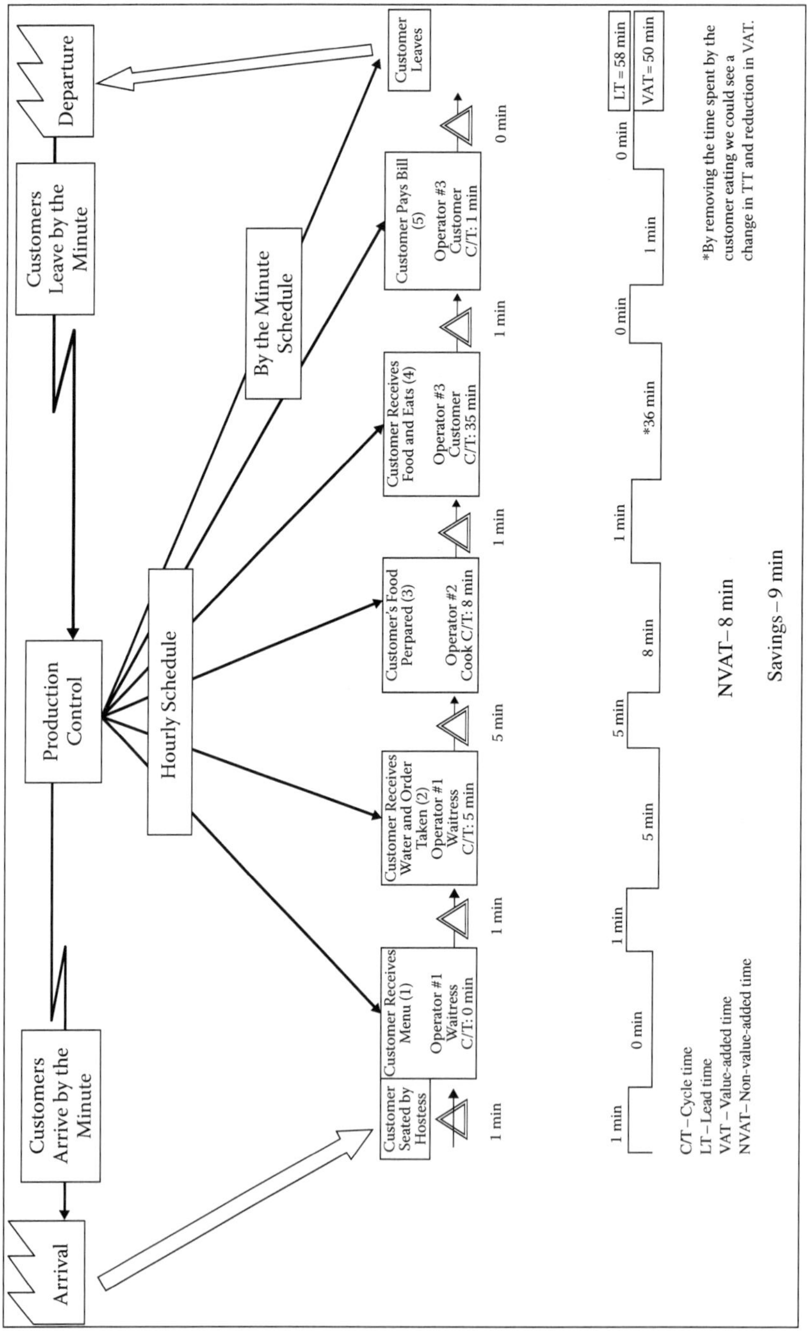

FIGURE 11.13
Future state map.

Standard Work Instruction Sheet			Document#: 1000WI	Sheet: 1
		Prepared By:	Approved By:	Revision:
Order Taking		Mst Team 1	XYZ Manager	AA
		28-Nov	30-Nov	
No.	Operation Description			Quality Key and Safety Point
1	Greet the guests and escort them to the table			
2	Present menus			
3	Take order for beverage			
4	Describe the special of the day			
5	Take food order			
6	Suggest customer order homemade bread			
7	Ensure each dish is complete and according to the guest's request			Check order
8	Serve everyone at the table at the same time			
9	Check back in the first few minutes			
10	Clean and clear constantly			
11	Resupply beverages and condiments			

FIGURE 11.14
Standard work instruction sheet.

Table No	Arrival (AM)	Menu Card and Water (AM)	Order Placed (AM)	Order Received (AM)	Time Delivery	Left Table (AM)	Table Cleaning (AM)
1	9:08	9:09	9:12	9:22	10 minutes	10:04	10:05
3	9:40	9:41	9:48	10:03	15 minutes	10:30	10:31
9	10:02	10:03	10:08	10:25	17 minutes	10:58	10:59
14	8:08	8:09	8:12	8:24	12 minutes	8:55	8:56
10	10:12	10:13	10:20	10:37	17 minutes	11:00	11:05
17	8:40	8:41	8:50	9:01	11 minutes	9:40	9:43

FIGURE 11.15
Time monitoring.

and cycle time. Once established, it was then integrated into the order-taking process. These procedures are described in Figure 11.14.

Biweekly meetings are arranged targeting the customer survey cards for those 2 weeks, necessary notifications, communication, and sharing of ideas. The best reviews from customers will be displayed for employee motivation. Employment of proper tray placement procedures considerably reduced the delivery time.

Figure 11.15 shows the time monitoring results after Lean implementation.

Statistical Analysis

Based on the statistical analysis performed on the baseline data, it was necessary to reduce the average delivery time as well as the standard deviation. To improve customer satisfaction, achieving an average delivery time of 10 minutes was set as an objective. It should have a lower control limit of 10 minutes and an upper control limit of 15 minutes. The statistical control chart shown in Figure 11.16 provides the statistical analysis. It demonstrates that the average delivery time was reduced from 17.17 minutes to 13.66 minutes after implementation of Lean tools. There is also a reduction of standard deviation, from 6.14 to 2.08, which signifies that the variation in the processes has been reduced through standardization, as shown in Figure 11.16.

Cost Benefit Analysis

In addition to the statistical results, a cost benefit analysis was performed to show the cost savings. The data provided in Figure 11.17 is based on information from the manager and the savings are calculated based on the profit earned in 1 month. The profit is expected to increase, after a successful utilization of Lean tools, over the period of 1 year.

Mean	13.66
Median	20.5
Min	6
Max	25
Range	19
Variance	7.888
Standard Deviation	2.808

FIGURE 11.16
Statistical analysis chart.

Number of customers (per day)	40
Average expenditure by individual customer	$20
Monthly revenue	$19,200
Monthly expenditure by owner	$12,000
Profit	$8,000
After implementation phase:	
Profit achieved	$8,300 ($300 profit)
Increase in profit	5.556%
Projected profit (yearly)	$9,600 (10%)

FIGURE 11.17
Summary of results.

Control

Implementation and Control Plan

Hoshin Planning

Hoshin planning focuses on achieving a vital annual stretch goal and it has been used successfully by Toyota and other companies in Japan since the 1960s. Hoshin is a cyclic process, consisting of the review of the previous year's performance, which creates the basis for the next year's plan. The two levels of hoshin kanri—strategic planning and daily management—take care of the development of action plans as well as its implementation. Hoshin Kanri focuses on safety, quality, delivery, and cost. Analogous to the DMAIC approach, it also consists of a few steps. It measures the previous performance, based on the analysis objectives, to improve conditions. To continuously monitor the action plan, a hoshin plan summary is created and documented. Hoshin action plans take long-term and short-term goals into account, the strategy applied, and the successes achieved are documented in terms of the summary. The hoshin implementation review keeps the plan up to date and records performance measurements in terms of data. As this is the first year of establishment of XYZ Restaurant, the current hoshin planning is based on performance from the past 6 months. The hoshin strategies using hoshin kanri is developed as shown in Figure 11.18.

Conclusion and Future Scope

This project details the application of Lean management tools in the food service industry. A small restaurant establishment was chosen for Lean implementation. Customer survey and value stream mapping were the most important tools to differentiate the value-added activities from the non-value-added activities. Various wastes such as an unorganized work area, movement of servers, and long billing queues were identified. Lean tools such as 5S, hoshin kanri, spaghetti charts, and kanban cards were used to improve operations in the processes. The NVAT was improved by approximately 50% (initially 17 minutes improved to 8 minutes). The projected financial savings are 5.56%.

Future work in this project includes attaining a takt time of 50 minutes by further reducing the waste. Also, a reduction in the refrigeration costs can be achieved by use of proper inventory management.

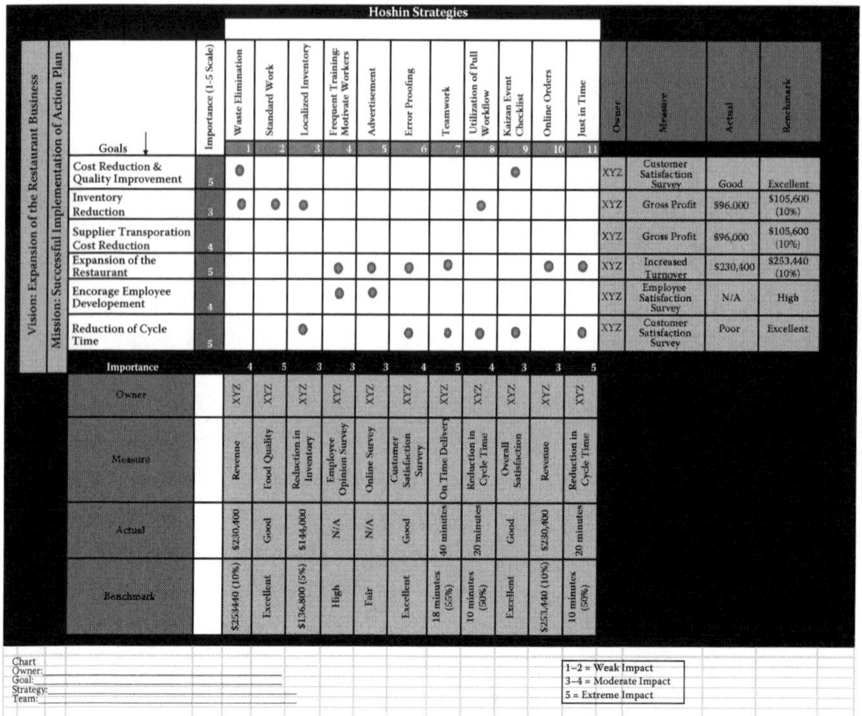

FIGURE 11.18
Hoshin planning.

Bibliography

Abdi F., Shavarini S.K., Hoseini, S.M. July 2006. "Glean Lean: How to Use Lean Approach in Service Industry?" *Journal of Services Research*, 6.

Seddon J., O'Donovan B. 2010. "Rethinking Lean Service." *Management Services* 54 (1): 34–37.

Sullivan W.G., McDonald T.N., Van Aken E.M. 2002. "Equipment Replacement Decisions and Lean Manufacturing." *Robotics and Computer Integrated Manufacturing* 18: 255–265.

Womack J. 2003. *Lean Thinking*, 2nd edition, New York, NY: Free Press, pp. 348–352.

12

Achieving Flow in a Rapid Prototyping Laboratory

Shirish Sreedharan, Elizabeth A. Cudney, and Frank Liou

CONTENTS

Executive Summary

The goal of this project was to apply Lean manufacturing techniques in a rapid prototyping laboratory. Lean manufacturing focuses on eliminating waste and improving flow using techniques such as value stream mapping (VSM), standard work, 5S housekeeping, single-minute exchange of dies (SMED), and visual management. The ultimate goal of this project was to achieve smooth flow in the rapid prototyping laboratory through the implementation of 5S housekeeping principles and implementation of the future state created in the VSM exercise. VSM allowed the identification of gaps between where the laboratory was and where it was expected it to be. The current and future state maps developed in the VSM exercise allowed the creation of a roadmap for implementing process improvements. To sustain the program, a 5S checklist was developed and commissioned, which is being presently utilized to maintain the future state achieved.

Introduction

Lean manufacturing emphasizes the elimination of waste and creation of flow within an enterprise. Lean manufacturing's primary focus is on the customer, to address value-added and non-value-added tasks. Value-added tasks are the only operations for which the customer is willing to pay for. The idea in creating flow in Lean manufacturing is to deliver products and services just-in-time, in the right amounts, and at the right quality levels at the right place. This necessitates that products and services are produced and delivered only when a pull is exerted by the customer through a signal in the form of a purchase. A well-designed Lean system allows for an immediate and effective response to fluctuating customer demands and requirements. The Lean manufacturing tools that are most commonly used to eliminate waste and achieve flow are VSM, standard work, 5S housekeeping, SMED, and visual management. In this project, two of the primary tools, VSM and 5S housekeeping, are demonstrated.

The Approach

The purpose of this project was to demonstrate the usefulness of VSM and 5S housekeeping in starting any Lean project and also to configure and enhance a laser-aided manufacturing process laboratory. The lab is a rapid prototyping system that is composed of a central chamber that houses the CNC machine and laser input, a powder feeder, a PC setup for executing machine codes, gas outlets, and a diode laser as shown in Figure 12.1. The lab is used by students for conducting research experiments.

The approach taken was to invite all students and faculty who were involved with the lab to an initial kickoff meeting for the project. During this meeting, a team to execute the VSM exercise and the 5S housekeeping project was created. Based on brainstorming, the specific tasks that were identified to be completed were as follows:

1. Formation of a team.
2. Creation of a team charter to execute the project.
3. Walk the flow in the lab and collect data for the current state.
4. Create a current state value stream map for the lab.
5. Brainstorm solutions to incorporate in the future state.
6. Create the future state value stream map for the lab.
7. Identify Kaizen blitz activities to be undertaken to achieve the envisioned future state.

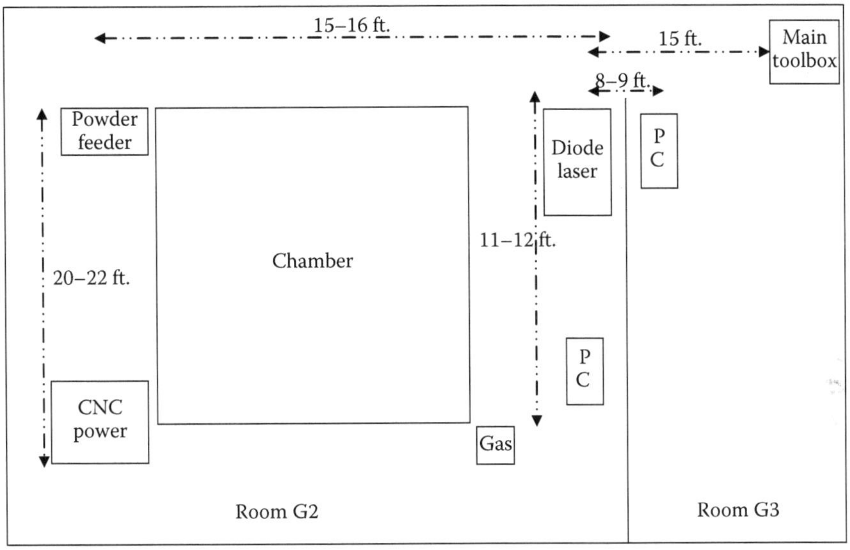

FIGURE 12.1
Prototyping laboratory layout.

8. Hold the Kaizen event with team members.
9. Achieve the future state.
10. Put controls in place to maintain the future state.

It was believed by those participating in the kickoff meeting that conducting 5S housekeeping and incorporating visual management into the lab are self-evident and should be tackled as a Kaizen blitz during the implementation phase. A team was formed and the team began data collection for the VSM exercise by walking through an experimental procedure in progress by several students.

Value Stream Mapping

VSM is a flowcharting method originally created in the Toyota Production System to document the entire process on a single sheet of paper to encourage dialog and understand the process better. A current state map of how the value presently flows is first created and then, utilizing the principles of Lean manufacturing, an envisioned better state as to how value should flow is created. Kaizen activities, which are events to overcome deficiencies in the current state that will allow the company to reach the future state, are identified and implemented.

In VSM, the flow of information and material is simultaneously analyzed to enable the elimination of waste in both. By considering material flow and information flow concurrently, whether one is hampering the flow of the other is easy to identify and correct.

VSM captures the flow of a product from the point raw material enters the premises to the point where a final product is delivered to the customer. This exercise includes all value-added and non-value-added activities that are currently taken to produce the product. VSM can be applied to any product or service, and therefore, it is finding widespread application to nonmanufacturing processes such as designing and commissioning a product and to business processes such as purchasing, billing, selling, and so on.

Also, since in VSM the entire process is visualized as a holistic set of operations rather than as a disjointed movement of material through discrete operation steps, it enables interaction between the operators to help see how the entire process can be optimized rather than seeking out only local optimization for the specific operations. In doing so, the seven classical wastes of transportation, inventory, motion, overproduction, overprocessing, and defects that are hidden in the process become evident, which allows the team to undertake projects to eliminate these wastes.

VSM also provides a common platform to apply various Lean principles and tools and allows the creation of an integrated plan to follow for implementation. Furthermore, since it captures the delay caused by material and information flow on the same page, it is possible to tackle the two together to create an optimized process such that the lack of one is not hampering the flow of the other.

Current State Map

After acquiring an initial understanding of the VSM process, the team set out to do an initial walkthrough of the lab. "Walking the flow," as this exercise is called, allows the team members to identify where problems exist in the present state of affairs. This can then lead into being able to see opportunities where principles of cellular manufacturing, one-piece flow, and SMED can be implemented. Incorporating the new ideas, the envisioned improved process can then be mapped as the future state map.

The team also captured how information flow and material flow interact within the facility to enable the processing of the product. To define value as perceived by the customer, team members interviewed faculty, staff, and students who used the lab. Next, all the steps were documented and each step was classified as either value-added or non-value-added based on the input from the customers.

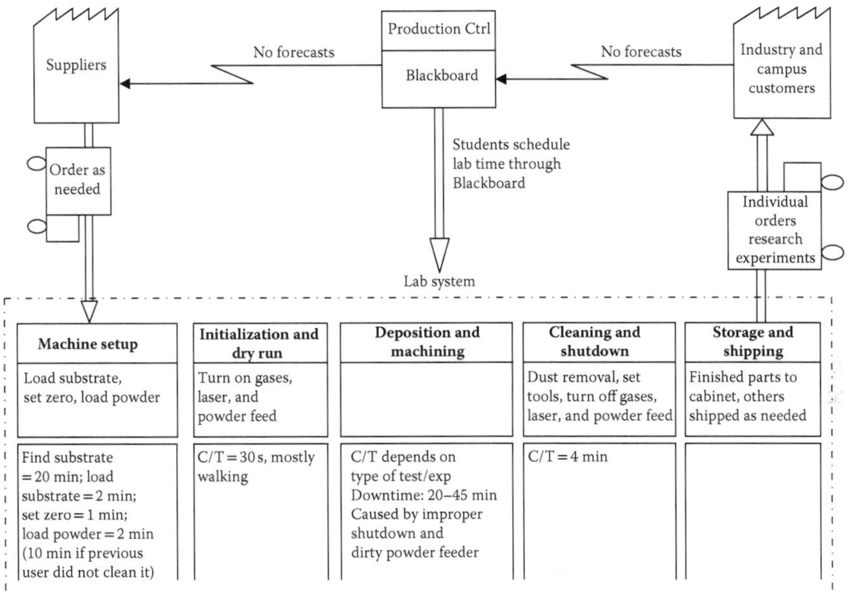

FIGURE 12.2
Current state map.

The current state map, shown in Figure 12.2, captures the lab system as it was before implementing Lean principles. It shows the flow of information (top, right to left) and the flow of operations (bottom, left to right). Being an experimental laboratory, the activities that constitute a typical experimental setup are grouped and tagged under various process boxes. Since each experiment is unique, it is difficult to tabulate an overall lead time for executing a particular experiment. The operation times necessary for the various activities in the lab have been also shown in the process boxes in Figure 12.2.

Future State Map

After creating the current state map, the team again "walked the flow" to identify ideas to incorporate in the future state map. The central theme that emerged for the future state map was to merge operations so that there is no wasted time built in between the processes. To achieve the ideal state, improvement efforts that will have to be undertaken in the form of a Kaizen blitz events were then identified. The future state map that was developed is shown in Figure 12.3.

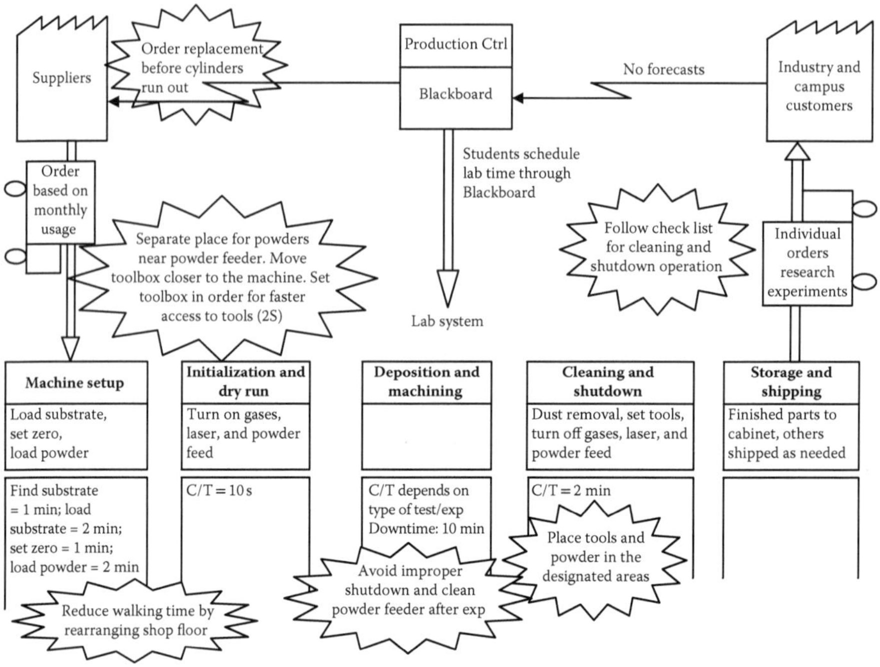

FIGURE 12.3
Future state map.

It was estimated that the improvements envisioned should reduce the lead time for conducting an experiment to half what it was before. The deficiencies identified and improvements recommended for the lab were as follows:

1. Provide a separate place for powders near the powder feeder: The powder for the rapid prototyping machine was being stored very close to the machining cell and resulted in contamination and wastage of the powder. To remedy this, the team decided to find a permanent safe place for the powder, which was out of the span of the working area.
2. Move toolbox closer to the machine: Tools needed for conducting the setup before starting the rapid prototyping machine were stored in a toolbox in an entirely separate area. This resulted in students having to walk to the toolbox and created a drawer full of mixed and mingled tools (Figure 12.4). To alleviate the problem, it was decided to implement principles of 5S in the lab and move the toolbox closer to the machine with only tools that were needed for the setup operation.
3. Order replacement gas cylinders before the old ones run out: Due to lack of monitoring of the gas cylinders, it was common to run out of gas in between a rapid prototyping run. A set of total productive maintenance rules that included a rule to check the amount of gas

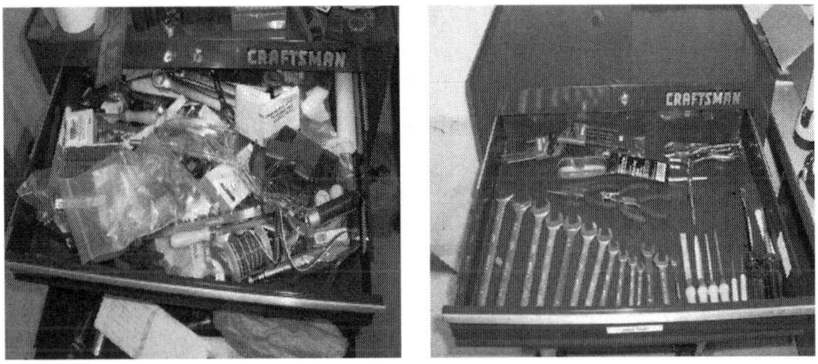

FIGURE 12.4
The toolbox before and after 5S.

1	Turn off the laser
2	Turn off the powder feeder
3	Shut off the value on the gas cylinders
4	Clean the powder feeder system
5	Do not leave any work piece in the machine
6	Use vacuum cleaner to clean chamber of all dust
7	Using a dusting cloth, wipe all machine areas
8	Check gas cylinder; if only quarter tank is left, release an order
9	Place powder bottles back in designated area
10	Return all tools to the toolbox

FIGURE 12.5
Rules developed and implemented in the lab.

left in the cylinder at the end of an experimental run were developed and placed in a prominent location to be followed by whoever utilized the lab as shown in Figure 12.5.

4. Reduce walking time by rearranging the shop floor: The layout of the lab was never designed and had evolved with each student, staff, and faculty member making changes as they saw fit. This had resulted in a layout ripe with excessive walking and backtracking. To remedy this, it was decided to create a new layout based on world-class plant layout and Lean manufacturing principles.
5. Avoid improper shutdown of the machine and clean powder feeder after experiments: Without proper instructions and the appreciation for the need for proper shutdown of the machine to enable easy restarting, students were in the habit of shutting off the machine abruptly in a cycle, which resulted in powder being thrown all around the work area. This not only resulted in a messy workplace but also allowed the powder to fall into the tracks and slides of the machine, resulting in maintenance shutdown. To prevent this from

happening, a procedure for the proper shutdown of the machine was developed and a policy of cleaning up the machine area before leaving the work area was instituted. The list of rules shown in Figure 12.5 contributed to remedying this.

6. Follow checklist for cleaning and shutdown operation: To enable the proper implementation of the above fix, a checklist as shown in Figure 12.6 was developed and implemented.
7. Clean up excess mess: A 5S sorting program was instituted wherein everything that was in the area was scrutinized and whatever did not seem to have use in the lab was red-tagged as shown in Figure 12.7. After an allocated time, if no students, staff, or faculty found use for the red-tagged items, they were disposed of.

5S Evaluation Sheet

Engineering Lab	Nov		Dec		Jan	
	Yes	No	Yes	No	Yes	No
1S Simplify (Seiri) Clearly distinguish what is needed and what is not needed.						
Do work areas only have the needed materials and items?						
Are only actively utilized equipment and tools present?						
Are unneeded items marked with a red tag and place in the red tag area?						
Are test items marked for identification or disposition?						
Are work standards and operating instructions posted and current?						
2S Straighten (Seiton) Standardize how necessary items are kept.						
Are equipment and other stationary items marked?						
Are materials, equipment, and tools visually controlled and at the point of use?						
Are materials, equipment, and tools in their designated location when not in use?						
Are walkways and storage areas clearly marked?						
Are materials and tools clearly identified?						
3S Scrub (Seiso) Sweep floors and clean office equipment.						
Are floors free of trash and fluids?						
Are equipment and materials free of dirt and fluids?						
Is the area bright, well ventilated, and free of dust?						
Are cleaning plans documented using visual management?						
Are cleaning supplies easily available?						
4S Stabilize (Seiketsu) Maintain the first three S's.						
Does machinery and equipment have systems to remove waste?						
Are TPM procedures in place and current?						
Is there a regular cleaning schedule?						
Are waste receptacles emptied regularly?						
Are disposition materials removed regularly?						
5S Sustain (Shitsuke) Adopting a discipline of 5S.						
The 5th S is awarded when three consecutive months of 1S, 2S, 3S, and 4S have been sustained.						

FIGURE 12.6
Engineering lab 5S evaluation sheet.

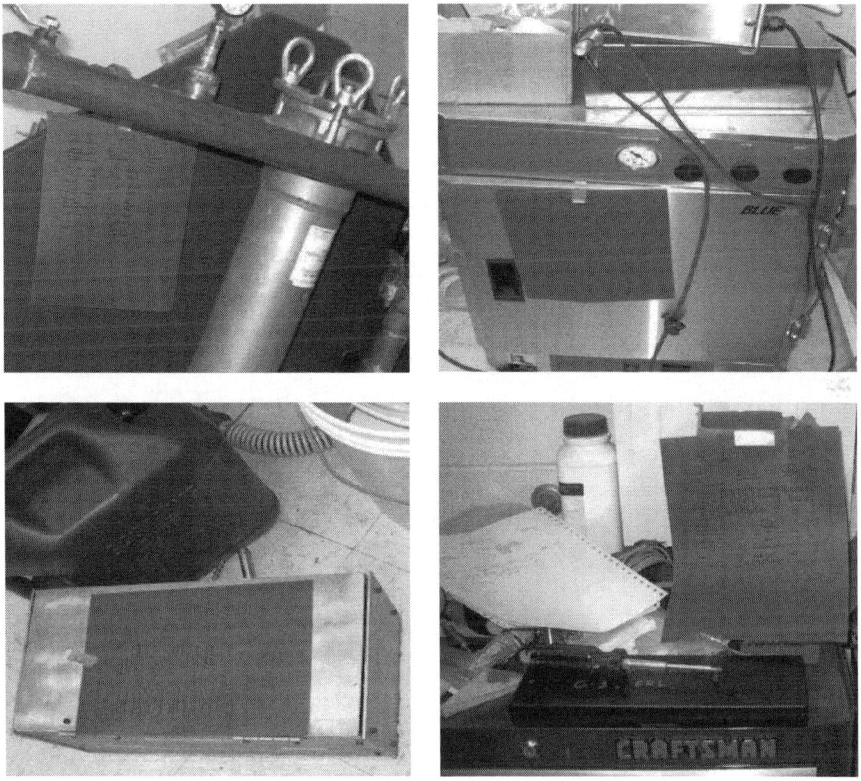

FIGURE 12.7
Red-tagging operation in the lab.

Results and Conclusion

Based on the above-mentioned improvements, a new layout for the lab as shown in Figure 12.8 was developed and is in the process of implementation. The proposed future shop floor layout that was developed focuses on creating a streamlined layout to enhance flow within the rapid prototyping laboratory. The new layout based on Lean principles will also substantially reduce the walking that the student/operator will have to do to carry out the operation. The 5S workplace reorganization efforts will create an ergonomic workplace in which the student/operator will be able to achieve significant improvement in efficiency. One of the major changes in achieving this was to move the powder feeder and main toolbox closer to the machine chamber where the users end up spending most of their time. A secondary change was also proposed to rewire the CNC power outlet close to the machine so that the user does not have to walk across the room after switching the machine on.

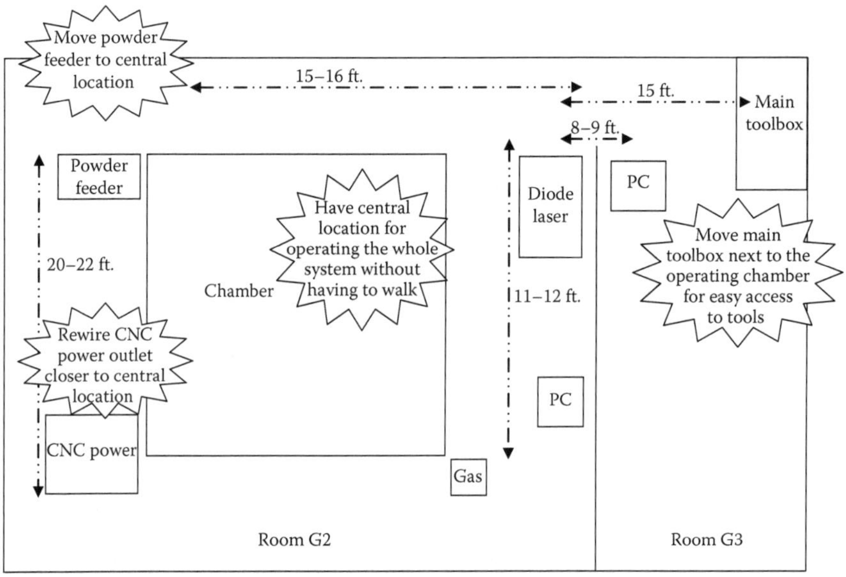

FIGURE 12.8
New proposed layout for the lab.

At the conclusion of this project, a basic level of workplace reorganization beyond the 2S stage of 5S was achieved. For the system to stabilize and to derive the maximum benefit from the 5S effort, it is proposed to go ahead and implement the remaining 3S objectives.

For sustaining the achieved improvements, internal audits on a monthly basis were instituted. Our hope is that the implemented visual control tools will play an important role in reminding users of the need to maintain a Lean and efficient work area.

In articulating lessons learned from this project, it can be said that applying Lean principles in a lab environment is especially challenging because of the stand-alone job shop type nature of the setup. However, flow is a universal concept that can improve any process, manufacturing or nonmanufacturing, and through the use of Lean tools such as VSM and 5S housekeeping, we showed that these tools can even be used to reduce waste and improve efficiency for a rapid prototyping lab.

13

Implementing Lean Manufacturing Techniques to Achieve Six Sigma

Elizabeth A. Cudney

CONTENTS

Executive Summary

This case study presents how Lean manufacturing techniques such as one-piece flow, kanban, and U-shaped cells were used to achieve Six Sigma through continuous improvement efforts and reducing non-value-adding processes in a case study. The objective of the project implementation was to improve quality to a Six Sigma level by using the Define–Measure–Analyze–Improve–Control (DMAIC) methodology and implementing Lean manufacturing techniques. Six Sigma is a continuous improvement technique with a goal of 3.4 defects per million. In this case study, Six Sigma tools are applied to a process that is incurring an unacceptable level of defects and not meeting production requirements. The initial process was mapped and analyzed to collect data on defect levels, work in process (WIP), cycle times, product throughput, machine utilization, and labor utilization. Next, an alternative process layout utilizing several Lean manufacturing techniques including kanban, cellular manufacturing, line balancing, one-piece flow, and a pull system was implemented. Finally, after applying effective poka-yokes, standard operating procedures, and training, the revised process was compared to the initial process to quantify the results of the improvements. The results included an improvement from 2.6 to 2.8 sigma level, a 43% reduction in defects, a 72% increase in production, a machine utilization increase of 50%, a labor utilization increase of 25%, a labor cost reduction of 33%, and a work-in-process reduction of 97%.

Introduction

Six Sigma is a customer-focused continuous improvement strategy and discipline that minimizes defects and variation toward an achievement of 3.4 defects per million opportunities in product design, production, and administrative processes. It is focused on customer satisfaction and cost reduction by reducing variation in processes. Six Sigma is also a methodology using a metric based on standard deviation. Traditionally, Six Sigma targets aggressive goals that include developing a world-class culture and leaders and supporting long-range objectives. There are numerous benefits of Six Sigma including a stronger knowledge of products and processes, a reduction in defects, an increased customer satisfaction level that generates business growth and improves profitability, increased communication and teamwork, and a common set of tools.

The objective of this case study was to utilize the Six Sigma DMAIC methodology in conjunction with Lean manufacturing implementation to meet customer requirements in terms of both the level of quality performance and production requirements. Lean manufacturing techniques were implemented including total employee involvement (TEI), smaller lot sizes, U-shaped cells, kanbans, and one-piece flow.

In considering how to develop the most effective process, the focus was on reducing product cost by improving quality, decreasing WIP, and increasing production throughput. The most obvious approach was a U-shaped cell to develop line flexibility. The following steps were taken to determine if this focus met our needs:

- Obtained and analyzed information on the current process
- Analyzed production and quality requirements
- Evaluated material handling

Define

The objective of the project was to reduce quality defects to a Six Sigma level by developing a more efficient process that also reduced product cost through reduction of not only quality defects but also by a reduction in WIP and lot sizes and an increase in production throughput. The tube line was also used as a pilot cell for other layout designs throughout the plant.

Based on the current method of production, the performance measure used to determine the achievement of target goals and the design effectiveness is sigma level. Supplemental performance measures include

- Scrap
- WIP level
- Production throughput
- Machine utilization
- Output quality
- Labor utilization
- Labor costs

The current process design does not meet an acceptable sigma level and current production requirements. The design of the new process needed to meet the following department objectives:

- Improve quality
- Decrease scrap
- Delivery to the point of use
- Smaller lot sizes
- Pull system
- Better feedback
- Increase production
- Individual responsibility
- Decrease WIP
- Line flexibility

The first step in understanding the process was to develop a process flow diagram (PFD) with the cross-functional Six Sigma team. PFDs allow a team to identify the actual flow or sequence of events. It also shows problem areas, redundancies, unnecessary steps, and areas for simplification and standardization. In addition, PFDs may be used to compare the actual versus the ideal flow and allows a team to identify activities that may impact performance. Also, PFDs show locations where additional data can be collected and identified. The existing tube line consisted of four operations (Figure 13.1). First, an aluminum tube is machined to face and bore three inner diameters and an outer diameter. Next, the tube is washed to remove dirt, turnings, and oils in a batch size of five parts. A bearing is then pressed into the bearing bore. Finally, a bushing is pressed into the bushing bore creating a final product. The tube is sold to an assembly plant as a component for an axle assembly.

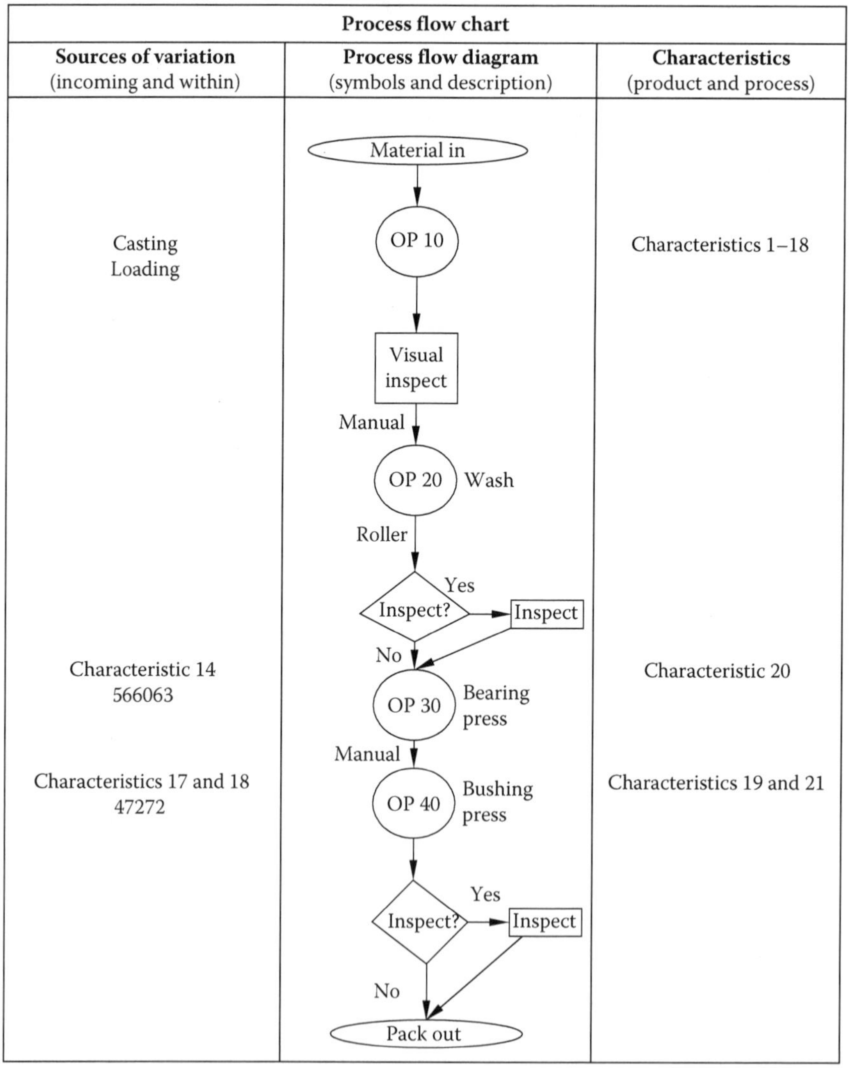

FIGURE 13.1
Process flow diagram.

Measure

Once the process was mapped, data were collected from the material review board (MRB) bench to measure and analyze the major types of defects using a Pareto diagram (Figure 13.2). A Pareto diagram is a bar chart where the bars are placed in descending order relative to quantities and cumulative percentages are listed concurrently. Although the Pareto diagram is traditionally used by statisticians, in Six Sigma, it focuses efforts on the problems that offer the greatest potential for improvement by showing their relative frequency or size. It is based on the Pareto principle: 20% of the sources cause 80% of the problem. A Pareto diagram displays the relative importance of a problems in an easily interpreted, visual format.

To better understand the process, a time study was performed. Initially, two operating stations were used to perform the necessary procedure for the final product. The first problem discovered was the unbalanced line. The first station was utilized approximately 70% of the time and the second station was only utilized approximately 30% of the time. The operators of the second station were spending a considerable amount of their time waiting between cycle times. By combining stations 1 and 2, room for improvement is evident with respect to individual responsibility, control of inventory by the operator, and immediate feedback when a problem occurs. A time study of the existing operation (Figure 13.3) and the department layout (Figure 13.4) reflect these findings.

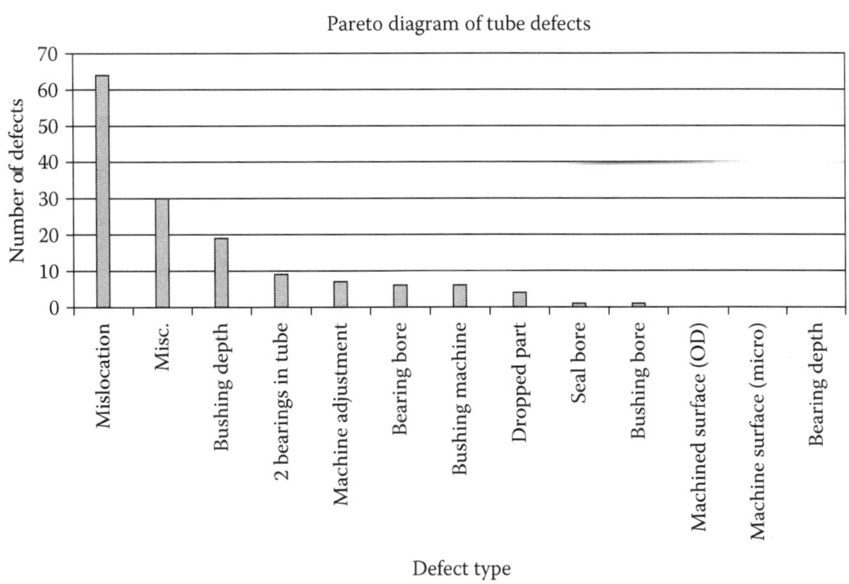

FIGURE 13.2
Pareto diagram.

Machining	Cycle and Pcs/Cycle	0.417	1	
Washer	Cycle and Pcs/Cycle	0.383	1	By: E. Cudney
Bearing Press	Cycle and Pcs/Cycle	0.217	1	Page 1 of 1
Bushing Press	Cycle and Pcs/Cycle	0.233	1	
	Total Pcs/Cycle		1	
Operation				
Face and bore IDs and OD				
Wash to remove chips and coolant				
Press needle bearing				
Press bushing into bore				

Element Description	**Min Occur**	**Frequency**	**Internal**	**External**
Load machine from box and cycle	0.167	1/1	0.167	
Unload part from machine into washer and cycle washer	0.083	1/1	0.083	
Unload part from washer to table	0.267	1/5		0.053
Load part and bearing onto press and cycle press	0.217	1/1	0.217	
Unload bearing press, load bushing, load part, and cycle press	0.167	1/1	0.167	
Unload part to table	0.117	1/1		0.117
Packout from table to box	0.133	1/2		0.067
Total internal	(cycle time × 20% if greater)		0.634	0.634
Total external				0.237
Total floor to floor time				0.871
Tool change and adjust				
Gaging	1.783	1/40		0.045
Gaging	0.083	1/1		0.083
Misc. and chip pulling				
SPC charting	0.45	1/40		0.011
Personal allowance	28.00 min/day			
Delay allowance	26.00 min/day			
Fatigue	8.00 min/day	Total cycle		1.01
Total allowance	62.00 min/day	14.80%		0.14948
		STD min/part		1.15948
		STD hours/100		1.93245
		STD parts/hour		51.7473

FIGURE 13.3
Initial time study.

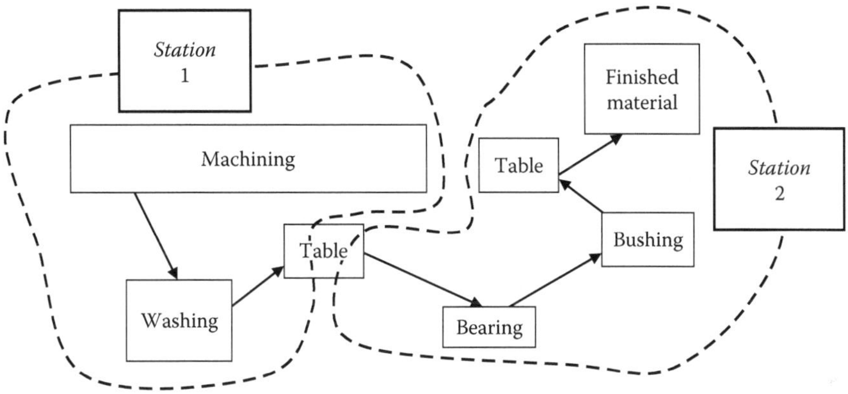

FIGURE 13.4
Initial layout based on operator stations.

The existing tube line consisted of one operator and four operations separated into two stations by a large table using a push system. The table acted as a separator between the second and third operations creating an inventory of WIP of approximately 172 parts. The first two operations consisted of boring the three inner diameters and one outer diameter followed by washing. Operators continued to run parts until they could not stack them any higher. The operator would then move to the other side of the table and run the bushing press and the bearing press. The finished parts were gaged 100% for assembly dimensions and stacked on a cart. After approximately 30 parts were finished, the operators packed out the parts into a cardboard box two at a time. Double handling occurred between each grouping of operations. Also, due to lack of floor space with the large table, the inefficient process flow and large lot size caused a higher probability of an increased number of defects. The process flow also allowed for the possibility of missed operations. Missing the bearing press was a common defect that caused sorts of not only the in-house inventory but also the product that was shipped to the customer. The number was not accounted for in the defect data because the operators were able to rework the parts and did not enter it into the defect log. After training, all defects were entered into the defect log before rework. Four sorts of the product had been performed over an 8-month period just before implementation of this project for missing bearings.

A second problem existed as the production rate did not allow the production schedule to be met with the two stations because of the current process flow. The machines were waiting for operator attention because operators lost track of machine cycles. The operators also tried to push parts through the first station because it was the bottleneck operation in the process and then continued to manufacture the parts at the last two operations. Long runs of WIP built up in which quality problems were not caught until typically a substantial number of defectives were produced.

A failure mode and effects analysis (FMEA) was created by the cross-functional Six Sigma team. An FMEA allows a team to understand what problems can occur. The FMEA facilitated the team in identifying where problems were likely to occur using the Risk Priority Number and subsequently led to the development of a more robust process.

A control plan was also developed. A control plan enables a team to understand how a defect is controlled. It also includes the measurement systems and frequencies of inspection. This was a critical step because we discovered inadequate measurement systems using Gage Repeatability & Reproducibility and features that were not being controlled.

Analyze

The results of the initial time study showed extremely unbalanced time utilization between the operations. The first station was utilized approximately 70% of the time and the second station was only utilized approximately 30% of the time. The operators of the second station were spending a considerable amount of their time waiting between cycle times. By combining stations 1 and 2, room for improvement is evident with respect to individual responsibility, control of inventory by the operator, and immediate feedback when a problem occurs.

The next step in analyzing the problem was to chart the process flow. This identified the critical opportunities for time-saving steps and showed potential problems. The process flowchart sequenced the most effective process of operating the line to reduce the amount of machine idle time and optimize throughput. The initial layout was also traced to show operator travel (Figure 13.5).

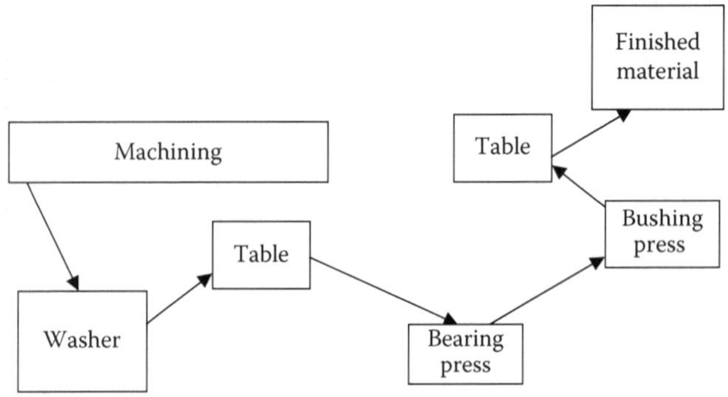

FIGURE 13.5
Initial layout based on operator travel.

Analyze Production and Quality Requirements

The capacity for the upcoming year was not determined. Therefore, a U-shaped cell, which provided no limitation to the number of operators and reduced material handling within the process, was selected. Capacity for our facility with the initial line was 330 pieces per shift, or 990 pieces per day. The average actual production rate was 300 pieces per shift or 900 pieces per day. However, the current total usage by all customers was 1528 pieces per day based on a 5-day schedule per week. Therefore, this line was required to run overtime to meet current demands. If this project meets production expectations, the process should be able to meet all of the customer requirements for this part. Also, due to the number and severity of recent quality problems, the quality must surpass the initial process quality.

Evaluate Material Handling Requirements

Upon day-to-day inspection of the current process, it was noted that workers spent a considerable amount of time either waiting for material handling personnel or performing the material handling themselves. It became evident that to the point of use would be a better system for receiving raw material.

Improve

The project involved the implementation of just-in-time (JIT) with kanban control and theory of constraints (TOC) to continuously improve and reduce non-value-adding processes of the tube line. The existing process wasted labor and incurred a high number of quality defects. A time study of the existing line was performed. The findings of this time study provided the basis for reducing the cycle time, balancing the line, designing the flow using JIT kanbans, and scheduling with finite capacity using TOC to reduce material handling, improve quality, decrease lot sizes and WIP, and improve flow.

JIT and TOC methodologies were used to successfully achieve the predetermined goals. Some of the concepts used included TEI, smaller lot sizes, finite capacity, scheduling, point of use inventory, and improved layout. All employees and supervisors in the department were involved in all phases of the project. Their ideas and suggestions were incorporated into the planning and implementation process to gain wider acceptance. Smaller lot sizes were introduced to minimize the number of parts produced before defects were detected. Kanbans, in the form of material handling racks, were introduced to control WIP and to implement a pull system. The department was also changed to form a cellular layout to decrease travel between operations. A simulation model was used to illustrate the difference in performance measures before and after the implementation.

Total Employee Involvement

The first key element in the success of any project is to have TEI and management commitment. This project was proposed 9 months earlier, but with the lack of support from key members of management, the project sat idle. By illustrating the cellular layout and walking through the production flow with each operator and the production manager, we received approval to rearrange the machinery. With the use of a plant-wide idea implementation system, the operators contributed suggestions to improve the process and eliminate waste on their line.

Kanban Control System

The cellular layout reduced WIP and increased production immediately. The cell only allowed one part in each machine and one in route at any given time. The table in the initial layout was removed, almost eliminating WIP. The pull system goal of increased production was accepted under the idea that WIP does not improve production numbers; instead it is the actual number of finished products. Kanbans were implemented in the form of racks, which hold only one in-process part, reducing WIP and increasing production.

Elimination of Defects

With the reduction of WIP and a pull system implemented, defects were detected immediately, therefore allowing operators to adjust the machinery back into tolerance or notify maintenance for timely repair. This dramatically reduced the amount of scrap generated.

Individual Responsibility/Immediate Feedback

The operators were authorized to stop the line when problems arose. In the initial arrangement, the operators in some instances were still able to continue running parts when a subsequent operation was down. The layout with kanban control eliminated the possibility of storing WIP requiring the operator to shut down the entire line. The cellular layout provides excellent opportunities for improving communication between operators about problems and adjustments to achieve better quality.

Flexibility/Delivery to the Point of Use

With the cellular layout, the number of operators is now flexible. Previously, the layout was set up where a maximum of two people could run the line. Now, the number of operators can vary depending on demand. With the U-shaped cell, delivery to the point of use is more convenient for the operator. The operator places boxes of raw material on six moveable roller carts, easily accessible. The six boxes are enough to last a 24-hour period.

Setup Time Reduction

To reduce setup times, the tools needed for machine repair and adjustments are located in the cell. The screws are not standardized; therefore, tools are set up in order of increasing size to quickly identify the proper tool. At this point, the tools only consist of various sizes of Allen wrenches due to the fixture design.

Control

An evaluation of the improvements was conducted to quantify the actual benefits achieved. The process was monitored for 3 months to verify the process was in control.

The time studies from the initial arrangement (Figure 13.3) and the implemented layout (Figure 13.6) were compared to show the increase in production from 300 to 514 finished products per shift. The new layout eliminated double handling between the second and third operations as well as at the pack out (Figure 13.7). The new layout also reduced throughput time by making it easier to cycle all four operations in a pull system order.

Benefits and Costs

The benefits and costs were evaluated to determine the success of the new layout (Figure 13.8). Meeting our current customer demand with two shifts reduced the labor cost. The benefits achieved by a cellular layout include

- Quality level increased
- Decreased scrap
- Decreased material handling
- Better feedback
- Increased production
- Line flexibility
- Decreased WIP

Machining	Cycle and Pcs/Cycle	0.417	1	
Washer	Cycle and Pcs/Cycle	0.383	1	By: E. Cudney
Bearing Press	Cycle and Pcs/Cycle	0.217	1	Page 1 of 1
Bushing Press	Cycle and Pcs/Cycle	0.233	1	
	Total Pcs/Cycle		1	
Operation				
Face and bore IDs and OD				
Wash to remove chips and coolant				
Press needle bearing				
Press bushing into bore				
Element Description	**Min Occur**	**Frequency**	**Internal**	**External**
Remove raw material from box, unload machined tube from machine, load raw material into machine, and cycle	0.183	1/1	0.183	
Unload part from washer, load unwashed part, and cycle washer	0.117	1/1	0.117	
Unload part from bearing press, load new part and bearing onto press, and cycle press	0.200	1/1	0.200	
Unload bushing press, load bushing, load part, and cycle press	0.133	1/1	0.133	
Packout to box	0.117	1/1		0.117
Total internal	(Cycle time × 20% if greater)		0.633	0.633
Total external				0.117
Total floor to floor time				0.750
Tool change and adjust				
Gaging	1.783	1/40		0.045
Gaging	0.083	1/1		0.083
Misc. and chip pulling				
SPC charting	0.45	1/40		0.011
Personal allowance	28.00 min/day			
Delay allowance	26.00 min/day			
Fatigue	8.00 min/day	Total cycle		0.889
Total allowance	62.00 min/day	14.80%		0.131572
		STD min/part		1.02057
		STD hours/100		1.70095
		STD parts/hour		58.79068

FIGURE 13.6
Final time study.

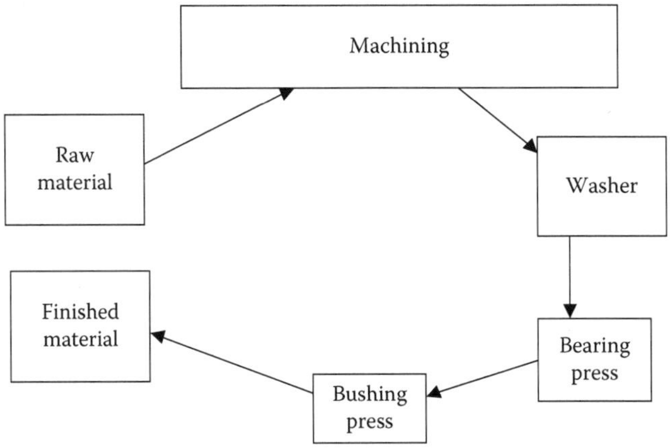

FIGURE 13.7
Final layout.

The production and scrap numbers were taken from two 20 consecutive production day periods. The initial numbers were taken from the last 20 days before the change, and the current numbers were taken starting 1 month after implementation. This gave the machine operators an opportunity to become familiar with the layout.

Conclusion and Summary

In summary, the production group wanted to develop a process to decrease product cost while achieving an increased quality level, decreased defects, increased production throughput, and improved production capacity, while still providing a design that would not limit the number of operators. After evaluating the process with the various JIT concepts along with the performance measures, a U-shaped cell was recommended.

The proposed layout was implemented and the results of the redesign are as follows:

- WIP decreased by 97%
- Production increased by 72%
- Scrap reduced by 43%
- Machine utilization increased by 50%
- Labor utilization increased by 25%
- Labor costs reduced by 33%
- Sigma level increased from 2.6 to 2.8
- Defects reduced by 43%

	Initial Production			Current Production		
Day	Daily Production	Shifts/Day	Scrap	Daily Production	Shifts/Day	Scrap
1	961	3	0	1291	2	33
2	571	2	0	1299	2	16
3	591	1	0	1134	2	0
4	630	2	78	850	2	0
5	640	2	9	910	2	22
6	951	3	21	1000	2	0
7	636	2	0	970	2	0
8	612	3	18	920	2	15
9	261	1	0	1035	2	0
10	865	3	0	431	1	0
11	261	1	28	1005	2	56
12	865	3	23	1353	3	0
13	848	3	0	970	2	55
14	570	2	9	1749	3	0
15	905	3	0	1782	3	25
16	355	1	21	852	2	0
17	1164	3	22	912	2	0
18	670	2	21	1030	2	0
19	570	2	0	910	2	0
20	591	3	0	875	2	0

	Initial	Current
Total Produced	13517	21278
Total Shifts	45	42
Total Scrap	250	222
Production per Shift	300.4	506.6
Scrap PPM	18159.4	10325.6

FIGURE 13.8
Production analysis.

This project resulted in reduced labor and scrap costs and allowed the process to better meet scheduled deliveries on time, while allowing a smaller finished goods inventory. Daily production numbers and single part cycle time served as a benchmark for monitoring progress toward the realization of the goal. Although the sigma level increase was not significant, the 43% reduction in defects, 97% reduction in WIP, and production increase of 72% made an impact on the project objective. The next step will be to target the specific defect types with additional Six Sigma projects or root cause analysis.

After seeing an immediate improvement in production numbers and a decrease in defects, the process will continue to use the implemented cellular layout. There is a definite savings associated with the implementation of this project in terms of Lean metrics. For example, capacity is met with only two-shift production as opposed to the original three-shift operation. JIT, kanban, and TOC implementation decreased product cost, improved process flow, decreased quality rejects, reduced labor costs through increased labor utilization, and minimized finished goods delivery while improving due date delivery.

14

Pump Teardown Review Process Improvement

Shrey Arora and Rodney Ewing

CONTENTS

Executive Summary

The purpose of this project is to determine the simplest options using Lean methods for producing a 50% time decrease in the pump teardown process for XYZ Company. The current process takes an average of 26 workdays to complete a single pump and provide the required information and feedback to the product development team. XYZ Company management was concerned that the discovery rate of new issues was extremely slow based on current lead time and the average backlog. Therefore, management chartered this project to reduce lead time in pump teardown from 26 to 13 workdays. XYZ Company's ultimate goal is to reduce the pump teardown rate to 7 workdays per pump, which becomes our working takt time.

The pump teardown process at XYZ Company documents the results of design validation testing for pumps completing rig, engine, and field tests. Test results are important to the product development team—the customer—in defining reliability and durability of pump designs. This project applies Lean principles to the teardown process and recommends changes that target improvements in workflow. However, the improvements can be applied to the teardown review processes of other components. The goal is providing faster feedback on the results of design validation testing to improve response time on performance and reliability issues.

The result of implementing Lean measures at XYZ Company was a 248% efficiency increase in the pump teardown process. At the beginning of the project, it took 26.56 workdays for a single pump to go through a six-step teardown process involving 24 separate activities; it now takes 7.64 days. XYZ Company's Lean efforts resulted in increased feedback on pump design validation tests, faster discovery of new issues and better understanding of pump reliability, and an overall greater contribution to pump development and manufacturing for XYZ Company.

Problem Statement

Our focus is on discovering the best cost-saving options resulting from analysis using Lean methods to produce a 50% time decrease in pump teardown for XYZ Company, which currently takes an average of 20–30 days to provide teardown feedback to the product development team on pump design validation testing.

Project Objectives

Today, teardown process flow takes an average of 26 workdays to complete a single pump and provide the required information and feedback to the product development team. Our team identified an average backlog of 10 pumps awaiting teardown as of January 15. XYZ Company management is concerned that the discovery rate of new issues is extremely slow based on current lead time and the average backlog and thus chartered this project to reduce lead time in pump teardown by 50%, from 26 to 13 workdays. XYZ Company management also requested proposed improvements in reducing the backlog of pumps awaiting teardown. The objective is for a single pump to complete the teardown process within 7 workdays.

This case study discerns the application of Lean principles to pump teardown using Lean tools such as value stream mapping and process flow improvements directed at lowering lead time by identifying and removing waste to more efficiently produce a pump. Our project focuses on pump teardown, which supports the design validation plan to define durability and reliability of the product. Pump teardown is a professional service in a multistage manufacturing industry, eventually producing a tangible product. The pump teardown procedure has service and manufacturing requirements. Thus, our project classification is both Lean manufacturing and Lean in a service industry. The resulting analysis presented in this case study provides greater tools in creating a shared understanding directed at improving XYZ Company's value stream. The benefit of this project is faster development cycles and better understanding of Lean methodologies. It is very difficult to apply a dollar value to this type of improvement. However, the project is important in providing quick feedback on pump design validation tests that are an integral part of developing new products for our customers. Thus, changes in efficiency throughout this case study describe project cost improvements. With that said, we start with a concise overview of the Lean methodology.

Lean Thinking

Lean thinking is a way to specify value (value), to line up value-creating steps or actions in the best sequence (value stream), to execute or perform these actions in an uninterrupted sequence (flow) at or when the customer requests them (pull), and to constantly search for improvements in effectiveness (perfection). James Womack and Daniel Jones describe the five principles of Lean thinking in their book *Lean Thinking* as "precisely specify *value* by specific product, identify the *value stream* for each product, make value *flow* without interruptions, let the customer *pull* value from the producer, and pursue *perfection*." Womack and Jones (2003) further state, "Lean thinking provides a way to do more and more with less and less—less human effort, less equipment, less time, and less space—while coming closer to providing customers with exactly what they want." Value and the customer are essential to Lean methods. The customer's view about specific products and services defines value. Identifying the value stream involves mapping out all linked actions from start to end. Value stream mapping creates a common vision—shared understanding—from the current condition to a future or ideal state. Making value flow continuously—uninterruptedly—implies that actions or processes are performed correctly the first time without deviation. Letting the customer pull value drives consumption where actions occur when needed by the next upstream customer. Finally, pursuit of perfection naturally occurs with a continual drive to eliminate waste using the aforementioned four principles (Carreira and Trudel, 2006).

Lean thinking is the solution to waste. It is a journey in understanding and completing incremental improvements from the ground up, involving employees and their ideas and supported by middle and upper management (Tapping, 2007). Waste increases production and service costs, lessens quality, and reduces capacity, which deprives companies of their competitive edge. Lean focuses on the identification and elimination of the seven types of waste described in Figure 14.1.

Womack and Jones (2003) define waste—also known as *muda* in Japanese—as follows:

Overproduction—producing more than needed or faster than needed

Correction/rework—repair of a product or service to fulfill customer requirements

Material movement—any material movement not part of a value-added operation

Processing—human or machine efforts that add no value (only cost) to the product or service

Inventory—any supply in excess of customer's just-in-time demand

Waiting—idle time produced when people wait on machines or machines wait on people or people wait on people

Motion—any movement of people or machines that does not contribute added value to the product or service

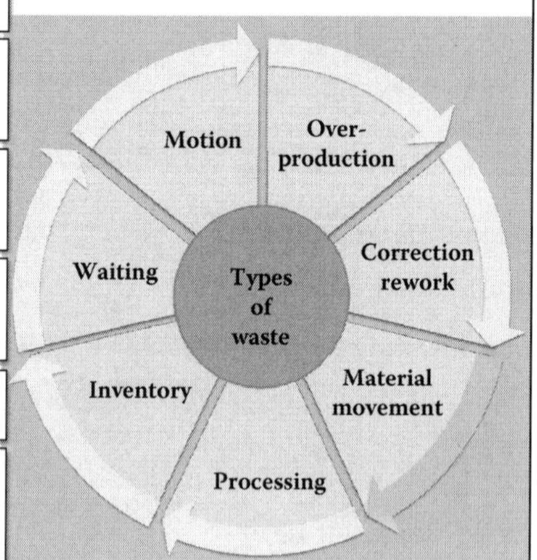

FIGURE 14.1
Seven types of waste.

> Any human activity which absorbs resources but creates no value: mistakes which require rectification, production of items no one wants so that inventories and remaindered goods pile up, processing steps which aren't actually needed, movement of employees and transport of goods from one place to another without any purpose, groups of people in a downstream activity standing around waiting because an upstream activity has not delivered on time, and goods and services which don't meet the needs of the customer.

Waste creates and provides no value; thus, if value is not created then it must be waste. The goal of Lean is identifying, analyzing, and eliminating waste throughout company operations and extending to suppliers all involved in the entire value stream.

Essential to Lean thinking in removing waste is harvesting vast numbers of small improvement ideas from employees doing the work in removing waste. A decisive moment—a moment of truth—occurs when employees have the knowledge and understanding of why change is needed, where respect is shown for experience on the floor by including them in the process. Harvesting great ideas—tapping a well of improvement ideas—requires management's chain of support in creating a contagious working environment where there is discussion and advancement of bottom-up ideas with management. Thus, Lean thinking is a customer-focused enterprise solution targeting waste elimination with the objectives of maximizing profitability, minimizing time, and minimizing cost without compromising value from the customer's perspective.

Our analysis presented in this case study follows the Six Sigma methodical approach to measuring products and processes against metrics and arriving at improvement activities through the formal Define–Measure–Analyze–Improve–Control (DMAIC) process.

Define

Define: Clearly define the problem or object.

The pump is an engine part. During development, endurance tests (rig, engine, and field) run as part of the design validation plan, which defines the durability/reliability of the product. Performance tests occur before engine start of test and after end of test (EOT). After the EOT performance test, the pump is disassembled and each component is inspected for performance and reliability concerns. The engineering team documents the issues identified during the teardown and initiates product improvement projects to resolve issues prior to engineering production launch.

Pump teardown involves 24 work tasks along nine locations—station A through station I—as shown in Figure 14.2.

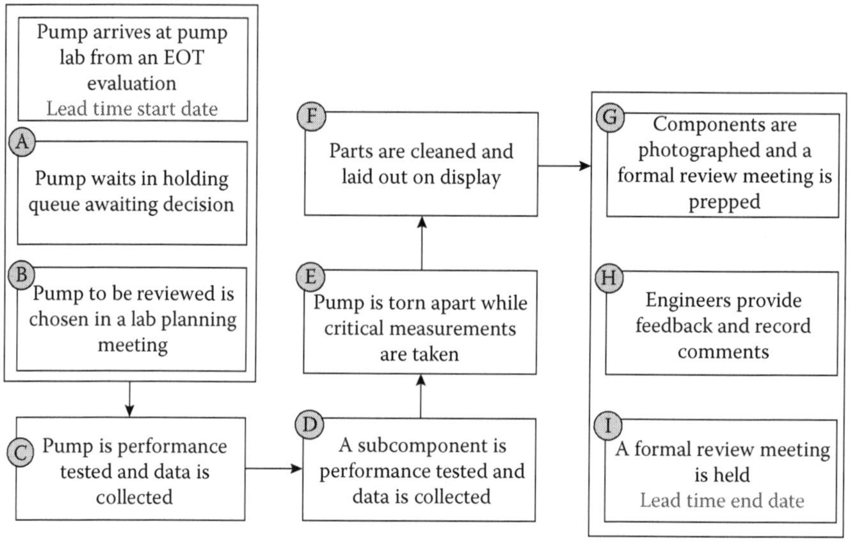

FIGURE 14.2
Pump teardown process flow diagram.

The work tasks explained in the next section of this case study include value-creating steps, time, and overall value-creating time. The overall process lead time starts with the arrival of the pump at the pump teardown lab from an end of test evaluation. Here, the pump waits at station A in a holding area or priority queue until planning priority decisions are made and then moves in process through the lab planning meeting at station B. Physical movement occurs between stations B and C where a pump test occurs. An operator moves the pump to station D for a subcomponent process and, similar to station C, test data is collected and documented. An operator moves the pump to station E located back in the teardown lab, tears the pump apart, and takes critical measurements. The pump is moved to station F for cleaning and display in preparation for future movement to the teardown room and microscopic photographs. When an operator is available, the pump components are photographed at station G in preparation for a formal pump teardown review meeting. Stations H and I host the formal review meeting where engineers provide feedback and documentation on the teardown results. The pump teardown lead time ends upon the completion of the formal review meeting. The timing of the entire pump teardown procedure is dependent on available resources at station G and priority queues at stations A, B, and G.

As previously stated, pump teardown takes an average of 26 workdays to complete the entire sequence of work tasks through the nine steps from station A to station I.

The goal is providing faster feedback on the results of design validation testing to improve response time on performance and reliability issues.

Our basic problem is to determine the best cost-saving options through an analysis using Lean methods in order to produce a 50% time decrease in pump teardown for XYZ Company. This teardown currently takes an average of 20–30 days to provide feedback to the product development team for pump design validation testing. The key objective is lowering the pump teardown lead time from 26.56 workdays a pump to 7 workdays a pump. Thus, our takt time—a measure of customer demand reflecting how often the customer requires a finished product—is 7 workdays per pump. An I-chart analysis was performed to understand the average time to complete the pump teardown process. An average time of 26.3 work days was found, shown in Figure 14.3. An initial baseline average number of workdays of 26.56 was determined, and will be used as the worst-case scenario.

There are several intangible benefits achieved due to increased communication between the pump teardown and product development teams, such as:

1. quick feedback on pump design validation tests—an integral part of developing new products;
2. discovery of new issues;
3. understanding pump reliability—the most important output of the process; and
4. promotion of faster issue discovery during the development cycle.

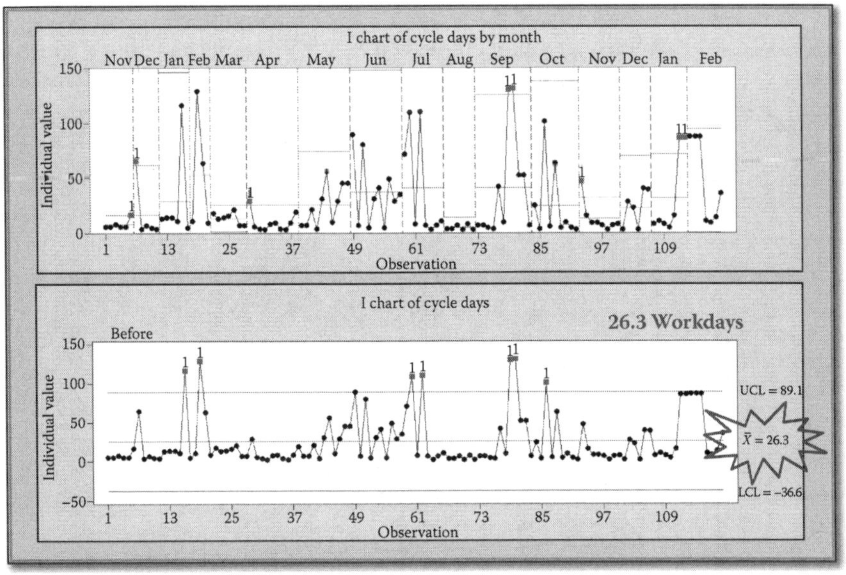

FIGURE 14.3
Baseline data—lead time for pumps reviewed.

Measure

Measure: Focus the improvement effort by gathering information on the current situation. The process is measured, data is gathered, and metrics are created.

We begin with creating a shared understanding of the current state. Our assumption is that descriptions of a particular situation within a snapshot of time are potentially changing as analysis occurs. We gain a better understanding of the situation and potentially reduce unintended consequences by looking at the problem from multiple perspectives. More importantly, multiple perspectives frame the situation toward producing a shared understanding. The most common frame of reference is each person's way of observing and interpreting expectations based on experiences or recent analogies relating to the situation. How each person understands the situation and interprets the problem is conceivably different. Reframing changes the recently created frame of reference. From a problem-solving perspective, reframing shifts the attention from trying to correctly solve the current problem to determining if we are solving the correct problem. Gaining a shared understanding between the various stakeholders on the problem and concurrence on the problem set is essential. Shared understandings on the current stance versus the objective stance are paramount to reduction in waste and the foundation of our methodology (Rein and Schon, 1991).

Value stream mapping creates a common vision—shared understanding—for everyone connected to the targeted value streams (for both the current and future states). It provides a visual map for ease of communication and allows waste to be seen by everyone so that improvements can be focused. Finally, it provides the foundation on which to base Lean initiatives from the customers' perspectives. According to Womack and Jones (2003), "Our initial objective in creating a value steam 'map' identifying every action required to design, order, and make a specific product is to sort these actions into three categories: (1) those which actually create value as perceived by the customer; (2) those which create no value but are currently required by the product development, order filling or production systems (type one *muda*) and so cannot be eliminated just yet; and (3) those actions which do not create value as perceived by the customer (type two *muda*) and so can be eliminated immediately."

Baseline Process

We described the pump teardown process in Figure 14.2; however, the plan view shown in Figure 14.4 provides clarity on the nine stations as we begin directly observing each activity toward building a current value stream map.

The layout also aids in describing the product family or product actions by location with the transition from one location to another by movement of the

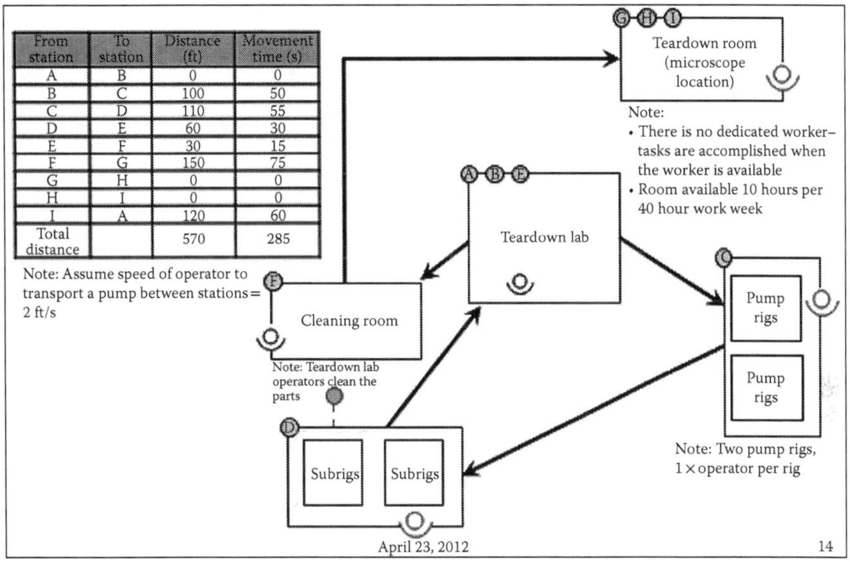

FIGURE 14.4
Plan view layout of current state.

pump. Each of the labs and teardown rooms are fixed locations and separate rooms within the building; thus adjusting the room arrangements to a more efficient layout such as a U-shaped room is not an immediate improvement.

Our walk-through of the pump teardown process proceeded along the entire length of the value stream from station A to station I. The total lead time is 12,749.75 minutes or 26.56 workdays, assuming 8 hour workdays. Our total value-creating time is 840 minutes or 1.75 workdays (1 day and 6 hours). Thus, approximately 93% of the work effort is waste—either type one or type two, as described by Womack and Jones.

Using the observed activities in a Pareto analysis (Figure 14.5), we begin to gain a picture of problem areas and activities where we should spend the majority of our effort to gain the greatest improvement. Roughly 88.5% of the non-value-added time is in the teardown review holding queue. The waste of waiting—idle time produced when people wait on machines or machines wait on people or people wait on people—restricts continuous flow.

Current Value Stream

The current state map was developed to gain consensus on the problem situation and provide a detailed analysis on physical actions and work locations required in pump teardown. The current state map was used to distinguish value-added steps, necessary non-value-added steps, and non-value-added steps. The current value stream map (shown in Figure 14.6) forms a visual representation of the pump teardown process.

Process set	Activity no.	Tasks	Value-added steps	Time (minutes)	Station cycle time (minutes)	Value-added time (minutes)	Value-added time (hours)
	Pump arrives into pump lab from an EOT evaluation						
A	1	Work order submitted by operator for pump		0.25	10225.00		
	2	Work order number is assigned		0.75			
	3	Pump waits in holding queue awaiting decision		10224.00			
B	4	Pump to be reviewed is chosen in a lab		120.00	120.00		
	5	Pump is moved by operator to station C for performance test		0.83			
C	6	Wait time before pump performance test		420.00	660.00		
	7	Pump is installed on rig		60.00			
	8	Pump performance test is conducted and data is collected	1	120.00		120.0	2.0
	9	Pump is removed from rig		60.00			
	10	Pump is moved by operators to station D for subcomponent testing		0.92			
D	11	Wait time before subcomponent performance test		60.00	180.00		
	12	Pump is installed on rig		30.00			
	13	Subcomponent performance test is conduced and data is collected	2	60.00		60.0	1.0
	14	Pump is removed from rig		30.00			
	15	Pump is moved by operator to station E for disassembly		0.50			
E	16	Pump is disassembled and critical measurements are taken	3	240.00	240.00	240.0	4.0
	17	Pump is moved by operator to station F for cleaning and display		0.25			
F	18	Parts are cleaned and laid out on display	4	60.00	60.00	60.0	1.0
	19	Pump is moved by operators to station G for photographs and formal review		1.25			
G	20	Components are photographed and a formal review meeting is prepped		240.00	240.00		
H	21	Engineers provide feedback and record	5	240.00	900.00	240.0	4.0
	22	Wait time (between pump teardown and formal review meeting)		660.0			
I	23	A formal review meeting is, conclusions are made, and component issues are documented	6	120.00	120.00	120.0	2.0
	24	Operators go back to their stations		1.00			
		Total time		12749.75			
		Value creating	6 steps			840.0	14.0

FIGURE 14.5
Pump teardown process current state value stream mapping.

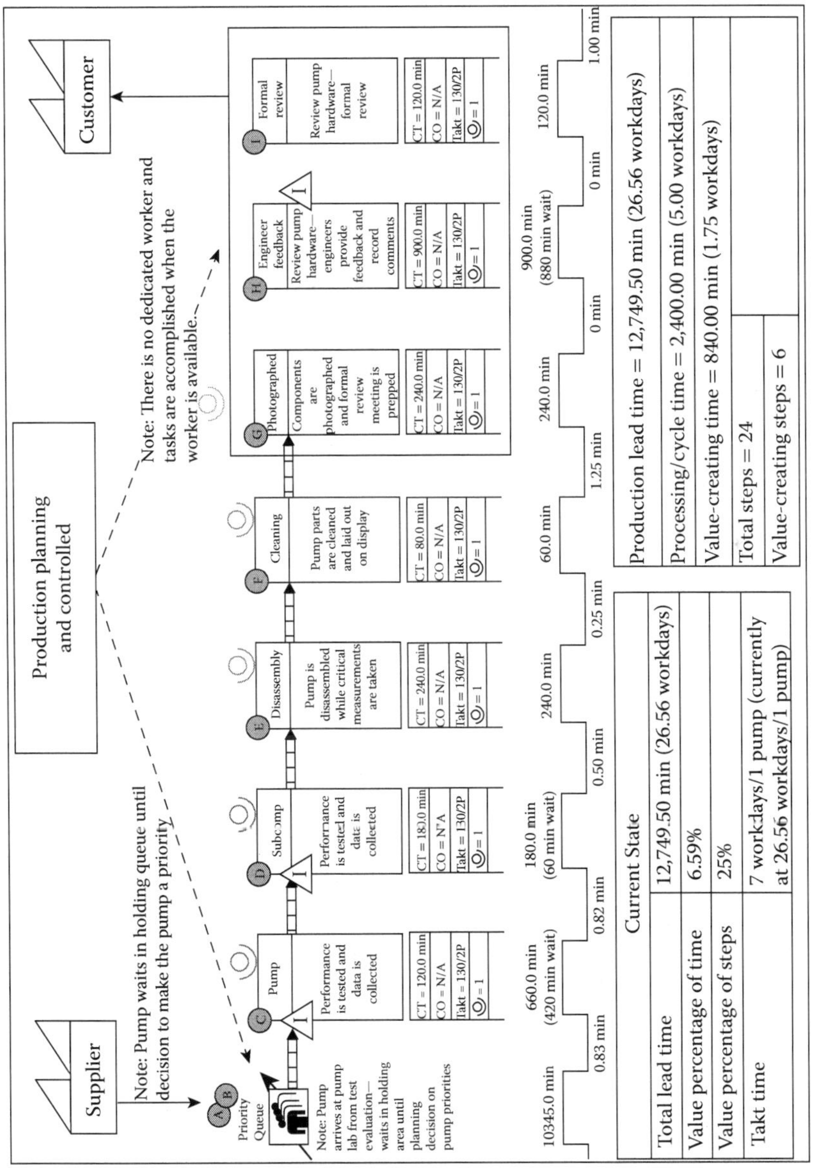

FIGURE 14.6
This is the current state value stream map.

We used value stream mapping icons to help standardize the articulation of flow (Rother and Shook, 2000). Mapping icons representing service-specific requirements, such as priority queues, were adapted from the article "Service Value Stream Management (SVSM)" (Bonaccorsi et al., 2011). These were used since the pump teardown procedure has both service and manufacturing requirements.

Analyze

Analyze: Identify root causes by analyzing the data. The output is a theory that has been tested and confirmed.

Teardown Constraints

Procedural or process restrictions were identified that included no dedicated operator in the teardown room at stations G, H, and I. Currently, the teardown leader reports to the lab operations supervisor and only dedicates 10 hours per 40 hour week to taking photographs and preparing for the teardown review. As a result, only two pumps are formally reviewed each week. Thus, waiting is the proximate cause in restricting continuous flow. The lack of a dedicated operator in photographing the components and preparing the teardown review materials is likely the bottleneck of the entire process.

Identified Gaps between Current Performance and Customer Expectations

The current teardown processing lead time between stations A and I must be improved to process more pumps per week. As previously identified, 88.5% of the non-value-added time is the teardown review holding queue. Waiting for decisions and waiting for resources is restricting continuous flow and is our primary target in eliminating waste between stations A and C and stations C and I:

- Waste in station A through station C:
  - Significant amount of time is spent waiting in priority queues, waiting on decisions based on downstream resource availability.
- Waste in station C through station I:
  - Time waiting for EOT performance test.
  - Time waiting for disassembling the pump and taking measurements.
  - Time waiting on photographing individual components and documenting results in the teardown database.

Sources of Variation

The work priority decision is a noise factor and involves two types of priorities: the priority of the teardown pumps in the queue, and the workload for the pump lab and performance rig. The business top 10 list will drive the priority of which pumps are chosen first. This priority decision has resulted in some pumps waiting in the priority queue for more than 100 days (Figure 14.3). The workload for the pump lab and performance rig is very high, with many competing project resources. Supporting the teardown process is not the only work the performance rigs and pump lab complete. There are a limited numbers of pumps that can be reviewed, and amounts greater than two pumps per week require overtime expenditures.

Why–Why Diagram

The why–why diagram asks the basic question of why something occurred. It is a repeated "Why?" question, thus breaking down the cause or solution into more explicit elements. The why–why diagram is a simple visualization of root causes and aids in creating a shared understanding. Figure 14.7 shows the failure mode of the lead time being too high and expended into proximate causes and ultimate causes.

Lean Tool: 5S

The implementation of the 5S Lean tool results in workplace organization and standardization by providing a structure and steps to organization, order, and cleanliness. 5S builds a foundation for delivering high-quality products and services in the right quantity at the right time to satisfy and exceed customer

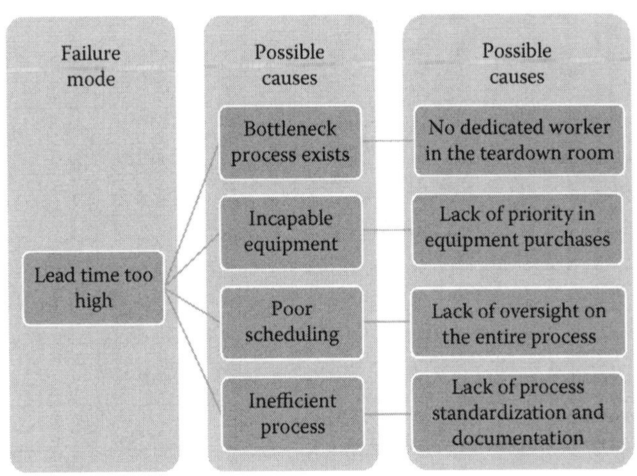

FIGURE 14.7
This is a Why–Why diagram.

requirements. Don Tapping describes 5S in his book *The New Lean Pocket Guide*, "To ensure work areas are systematically kept clean and organized, ensuring employee safety, and providing for the foundation on which to build a lean system" (Tapping, 2007). The 5S steps as defined by Tapping as follows:

- "Sorting" the necessary from the unnecessary (sort through and sort out)
- "Straighten" by setting things in order and setting limits (planning the best place to set items in order)
- "Shine" by cleaning and identifying items (clean the target area)
- "Standardize" by creating and setting the standard for cleanliness (standardized checklists)
- "Sustain" a discipline to maintain the first 4S over time (sustain through inspection) (Figure 14.8)

We begin with sorting the necessary from the unnecessary, separating needed from unneeded items. Our recommendations include using a "red tag" tool in identifying all the excess items in the lab and leaving only what is required—only the essentials. Efforts should focus on "what is needed, in what quantities it is needed, and where it should be located. Move all the unnecessary items to a different area. Items having no value and easy to dispose—throw away immediately. Items having some sale value—look for a buyer and sell them at the best available price. Items not sold and disposable is costly—work the least costly and safest way for disposal. Potential impacts include reduction in workflow problems, increase in product quality and

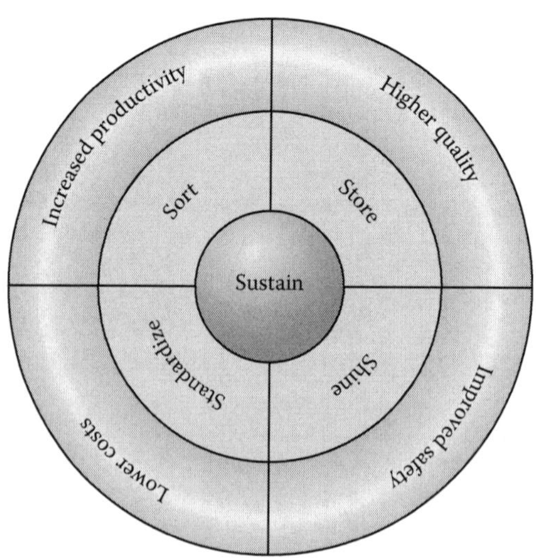

FIGURE 14.8
5S

productivity, as well as communication improvements. Management must manage the pump teardown inventory to keep pump idle time at less than 13 workdays. Today, the pump lab is cluttered due to the backlog of pumps ready for teardown.

The second S—straighten—reflects a very popular saying, "A place for everything and everything in its place." It emphasizes safety, efficiency, and effective storage and consequently improves the appearance of the workplace. After the sorting phase, whatever is left needs to be arranged so that there is ease of use and easy retrieval. Efforts should be made to arrange things based on the frequency of use. Items that are frequently used can be placed near the point of use, items that are sometimes needed can be placed further away, and items that are not used at all must be stored separately with clear identification. Racks and shelves are used to avoid stacking of items, and excessive visual symbols are used to make it easy to identify material at a glance. Potential impacts include the reduction of search time, orderliness is established, and there is no excessive inventory. Due to business needs, a first-in-first-out system is not used for the flow of pumps and hence a priority system is used.

The third S—shine—stresses cleanliness because it ensures a more comfortable and safe workplace, as well as better visibility, which reduces retrieval time and ensures higher quality work, product, or service. The performance rig and teardown lab already have processes to keep the engineering labs clean. Two hours have been set aside every Tuesday afternoon for lab cleanup. The process is documented, showing all the cleanliness activities and people responsible for them. Efforts are made to keep the workplace clean and functional. This is achieved through a combination of cleaning work and defect detection. Potential impacts include a more comfortable and safe workplace, higher quality goods, lower maintenance cost, and a good impression in safety audits.

The continuous employment of the first 3S will ensure a high standard in workplace organization and operations. The next S focuses on standardizing the best practices and properly maintaining the first 3S as a habit. The process is documented with ownership of each cleanup activity and is clearly displayed in the labs. A standard teardown review process would be documented in the engineering quality system, and teardown review metrics would be created, which will be made visible to the engineering teams in a weekly staff meeting. Potential impacts include better workplace standards, establishment of rules and standard procedures, and improvement in operation and workflow.

The fifth S is to sustain the habits—sustain—of maintaining the momentum of the previous 4S to ensure the sustainability of systems and make further improvements. Important steps include visually displaying the achievement of 5S, retraining of staff, and rewarding and recognizing the staff. The teardown leader must provide training to the engineering community on the teardown process, and management should recognize the team quarterly if the lead time is at targeted levels.

Improve

Improve: Test and implement solutions to address or eliminate root causes.

Sustaining Kaizen

Rother and Shook (2000) state that Kaizen is continuously improving by incremental steps. Essential to Lean thinking is harvesting a vast number of small improvement ideas from employees doing the work. Kaizen comes from the employees and is a process, not a result. Harvesting great ideas for incremental improvement—tapping a well of improvement ideas—requires management's support in creating a contagious working environment where there is discussion and advancement of bottom-up ideas. Improvement begins with ground-floor workers; however, continuous improvement starts with management and supervision. Hamilton (2012) states that Kaizen includes small changes for the better done by employees, supported by management, and not done to the employees.

Our recommendations to XYZ Company focused on six steps, targeting continued improvements and sustaining the gains already made. First, management commitment and ownership is essential for all improvements. Employees need to be encouraged to adopt this new way of working. Establish a 30-day follow-up session highlighting improvements and targeted areas. Second, create and maintain standards and make employees aware of their responsibilities and management expectations. Third, management should measure improvements and document data. We recommended that management log improvements on visible scorecards to amplify results and efforts for immediate visualization by the workers and management. Track key performance measurements including process cycle times, reduced work-in-process, and reduced bottlenecks. Fourth, maximize participation and improve ownership. Identify the process owners and assign responsibilities. Include ideas and improvement activities from team members. Fifth, manage should Kaizen the Kaizen by inviting ideas from workers and employees and reflecting on each stage and element in their Kaizen efforts. Sixth, inculcate the culture of Kaizen with monthly Kaizen activities and Lean workshops.

Identify Solutions in Eliminating Sources of Variation and Waste

Our recommendation to XYZ Company is to hire a dedicated employee on the pump team to manage the teardown review process. Currently, the teardown leader reports to the lab operations supervisor and can only dedicate 10 hours per 40 hour week. Thus, the capacity of formally reviewing pumps is two pumps each week, which creates a backlog given that

XYZ Company's demand for the entire process is 7 workdays per pump. Remember that the pump review process is only 5% of the total process effort (Figure 14.9) but causes a majority of the backup. Hiring a dedicated employee on the pump team will improve the available time to process pumps (i.e., move pumps out of priority queue). With a dedicated head-count, we could increase formal pump reviews from two pumps per week to four pumps per week.

The second recommendation is to implement an "informal" teardown review process with a decision matrix to determine which pumps need formal versus informal reviewing. For an informal review, there is no formal teardown meeting. The teardown leader works with the lead engineer to determine the extent of documentation in the teardown database including identification, documentation, and assignment of issue priority. The informal teardown review process is an enabler to pull pumps out of the holding queue and begin the teardown work process.

Our next recommendation is to reduce the 4-hour cycle time to photograph components and prepare for the formal teardown review, thus enabling more pumps to be documented per week. We suggested training new employees on properly using the microscope. We recommended that management review options for improving photographic equipment capability and component fixturing and implement progressive wear scales for pump interface

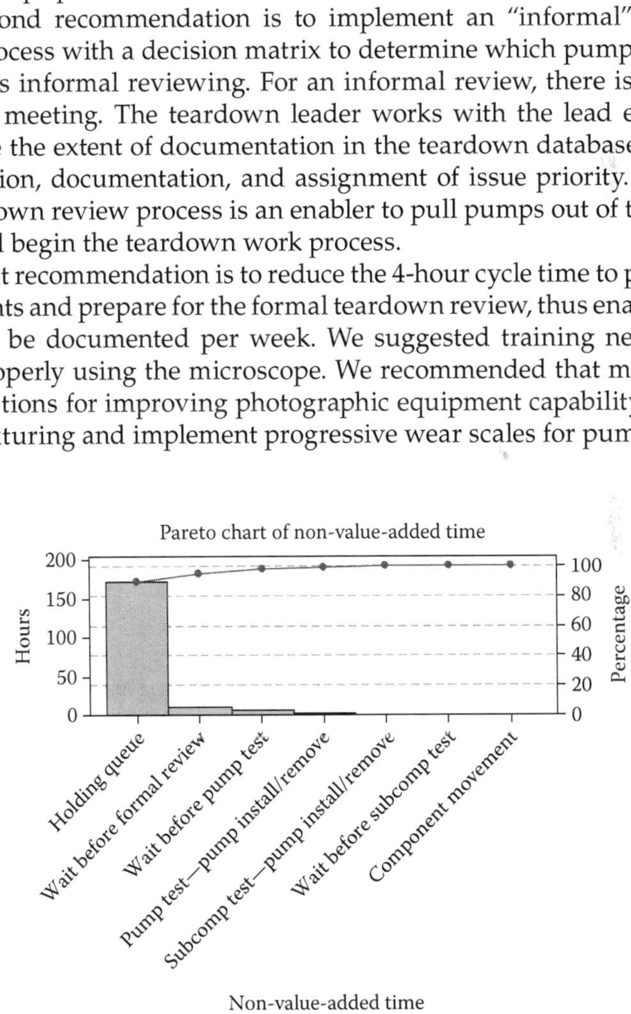

Hours	170.4	11.0	7.0	2.0	1.0	1.0	0.1
Percent	88.5	5.7	3.6	1.0	0.5	0.5	0.0
Cum %	88.5	94.2	97.9	98.9	99.4	100.0	100.0

FIGURE 14.9
Pareto analysis.

components to reduce the number of photographs. Also, we recommended the implementation of a teardown tracker spreadsheet/database for greater documentation. This included metrics to make the lead time and backlog visible, which includes presenting the metrics in weekly team staff meetings.

Solution Implementation Plan

It is our understanding that XYZ Company has hired a dedicated employee on the pump team to manage the teardown review process, thus increasing the formal review procedure from two pumps per week to four pumps per week. They have implemented an informal teardown review process with a decision matrix to determine which pumps need formal versus informal review. XYZ Company also reduced the 4-hour cycle time to photograph components and prepare for the formal teardown review, thus enabling more pumps to be documented per week. Finally, they developed and are currently using a teardown tracker spreadsheet and database for better documentation.

What-If Scenario

Improvement analysis and recommendations included a "what-if scenario," a small-scale simulated exercise to emulate the teardown process prior to actual Lean deployment as part of our improvement recommendations and future state plan. We used the lab logbook in determining the number of pumps received per week from January 2011 to the present time (Figure 14.10).

We then applied the decision matrix—a decision-making tool we identified as a potential solution—to determine whether the pump would be formally or informally reviewed. Next, we calculated the number of pumps each week in the priority queue based on the number of pumps received and the number of pumps processed. The teardown leader can review four pumps per week formally and four pumps per week informally. Finally, we calculated the lead time from January 2011 to the present.

Our simulated exercise revealed an average lead time improvement resulting in 7.64 workdays per pump, which meant XYZ Company was well on its way to achieving the customer requirement, the takt time of 7 days per pump. With the teardown review leader able to process up to four formal reviews and four informal reviews per week, the priority queue size is controllable and is zero in most instances (Figure 14.11).

This is a significant improvement over the preimprovement queue size. We also noticed the need for an extended value stream review, given the lack of continuous flow coming into the pump teardown process as in the variations in Figure 14.10. We then used the data from our simulated exercise to drive the future state 1 value stream map.

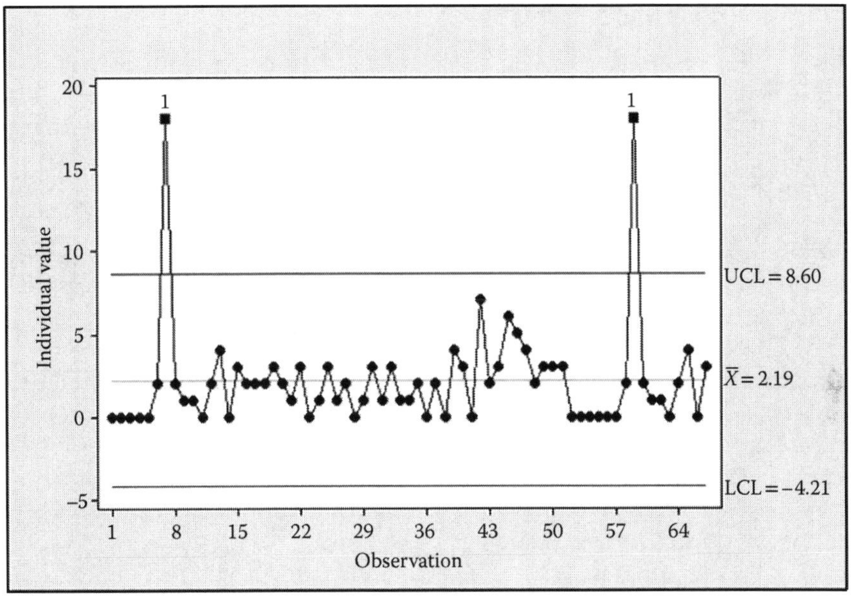

FIGURE 14.10
Number of incoming pumps.

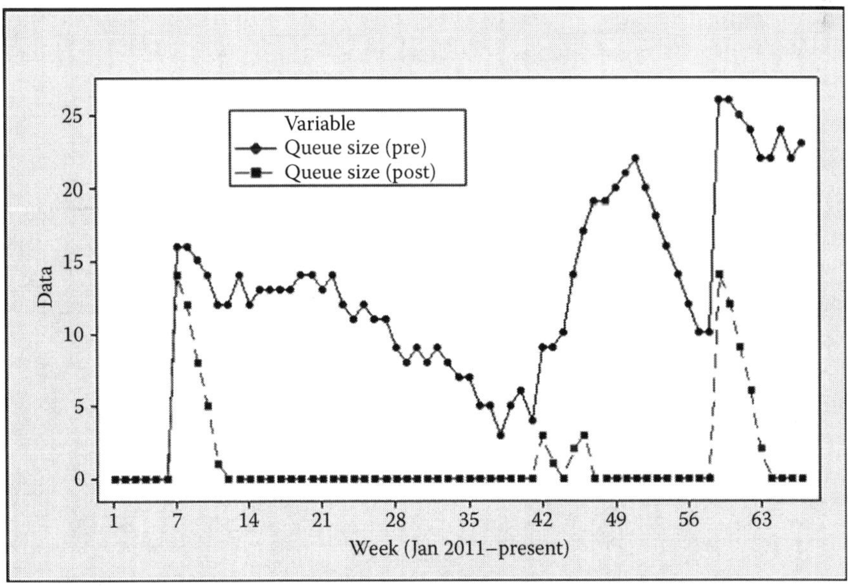

FIGURE 14.11
Queue size by week (pre- vs. post-improvement).

Future State 1 Value Stream Map

Future state maps do not signal the end of improvement; in fact, each future state map is a visualization of incremental improvements. The future state 1 map shows how value should flow by performing Lean changes from the current state value map.

Jones and Womack (2009) in their book *Seeing the Whole* note, "They [future state maps] are merely the first effort to introduce continuous flow with smooth pull while removing wasteful steps." We used data from our simulated exercise in populating future state activities including pump movement, priority queues, and activity time lengths. Figure 14.12 shows the list of actions by station, activity number, description of task, and time increments.

Improvements—reduction in activity time—produced a 248% increase in efficiency. Our value-added steps remained at six steps. However, the value percentage of time rose from 6.59% to 22.91%, thus producing the overall efficiency increase. The future state 1 pump teardown process is capable of managing incoming pumps—based on historical data—and keeping the number of pumps in the queue very low. Figure 14.13 represents the future state 1 value stream map with reduced activity time in station A in the priority queue and station H between pump teardown and formal review. There is an overall lead time change from 26.56 workdays (12,749.50 minutes) to 7.64 workdays (3,667.20 minutes).

Revising our Pareto analysis, Figure 14.14 reveals a more manageable pump teardown process and the likely leveling of flow given the somewhat even distribution of issues. The contributions of Lean efforts included reduced holding time at the priority queue due to an increase in the teardown leader's capacity and reduced wait time before pump testing.

Cost–Benefit Analysis

As previously stated, it is difficult to apply a dollar value to this type of improvement. However, the project is very important in providing quick feedback on pump design validation tests as an integral part in developing new products for XYZ Company's customers. The discovery of new issues and understanding pump reliability is the most important output of the process. Moreover, the project promotes faster discovery of issues during the development cycle.

Control

Control: Assess the proposed solutions and develop controls to put in place that will ensure the desired results and prevent future occurrences of the defects, problems, or unnecessary costs.

Process set	Activity no.	Tasks	Value-added steps	Time (minutes)	Station cycle time (minutes)	Value-added time (minutes)	Value-added time (hours)
	Pump arrives into pump lab from an EOT evaluation						
A	1	Work order submitted by operator for pump		0.25	1392.05		
	2	Work order number is assigned		0.75			
	3	Pump waits in holding queue awaiting decision		1391.05			
B	4	Pump to be reviewed is chosen in a lab		120.00	120.00		
	5	Pump is moved by operator to station C for performance test		0.83			
C	6	Wait time before pump performance test		420.00	660.00		
	7	Pump is installed on rig		60.00			
	8	Pump performance test is conducted and data is collected	1	120.00		120.00	2.0
	9	Pump is removed from rig		60.00			
	10	Pump is moved by operator to station D for subcomponent testing		0.92			
D	11	Wait time before subcomponent performance test		60.00	180.00		
	12	Pump is installed on rig		30.00			
	13	Subcomponent performance test is conducted and data is collected	2	60.00		60.0	1.0
	14	Pump is removed from rig		30.00			
	15	Pump is moved by operator to station E for disassembly		0.50			
E	16	Pump is disassembled and critical measurements are taken	3	240.00	240.00	240.0	4.0
	17	Pump is moved by operator to station F for cleaning and display		0.25			
F	18	Parts are cleaned and laid out on display	4	60.00	60.00	60.0	1.0
	19	Pump is moved by operator to station G for photographs and formal review		1.25			
G	20	Components are photographed and a formal review meeting is prepped		240.00	240.00		
H	21	Engineers provide feedback and record	5	240.00	650.40	240.0	4.0
	22	Wait time (between pump teardown and formal review meeting)		410.40			
I	23	A formal review meeting is held, conclusions are made, and component issues are documented	6	120.00	120.00	120.0	2.0
	24	Operators go back to their stations		1.00			
		Total time		3667.20			
		Value creating	6 Steps			840.0	14.0

FIGURE 14.12
Future state 1 process task and time.

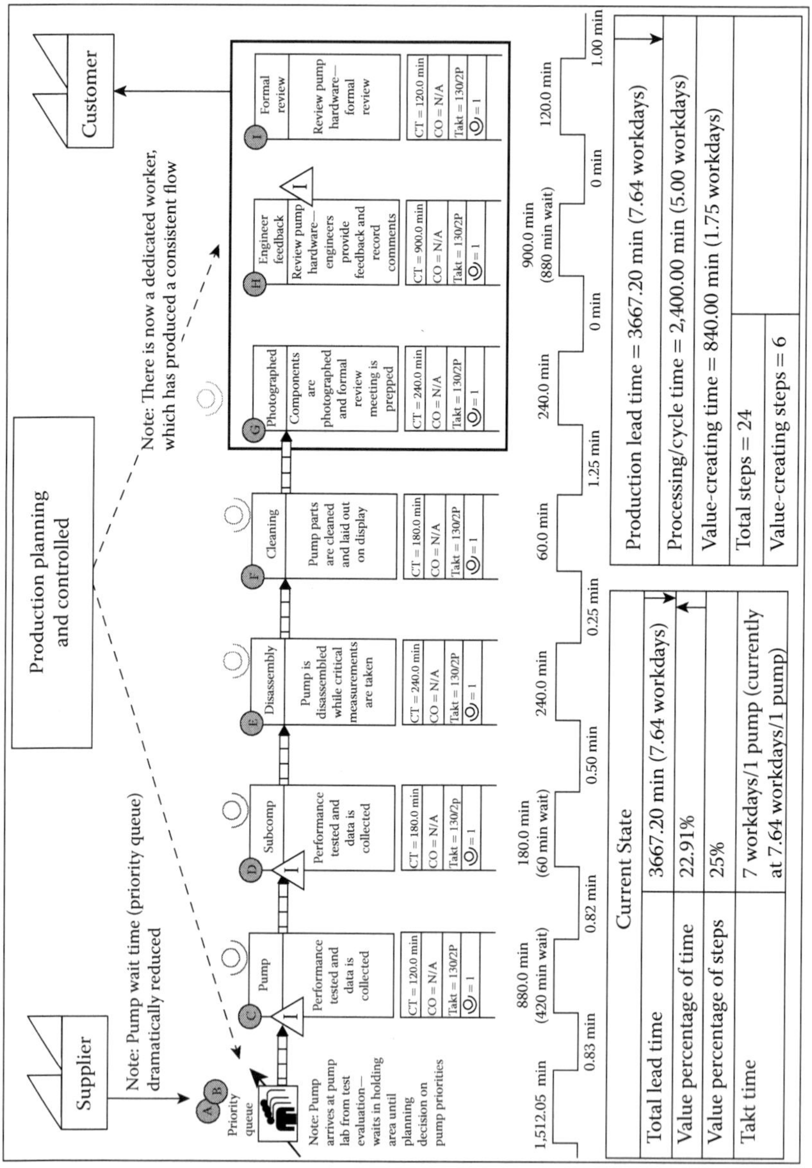

FIGURE 14.13
Pump teardown process future state 1 value map.

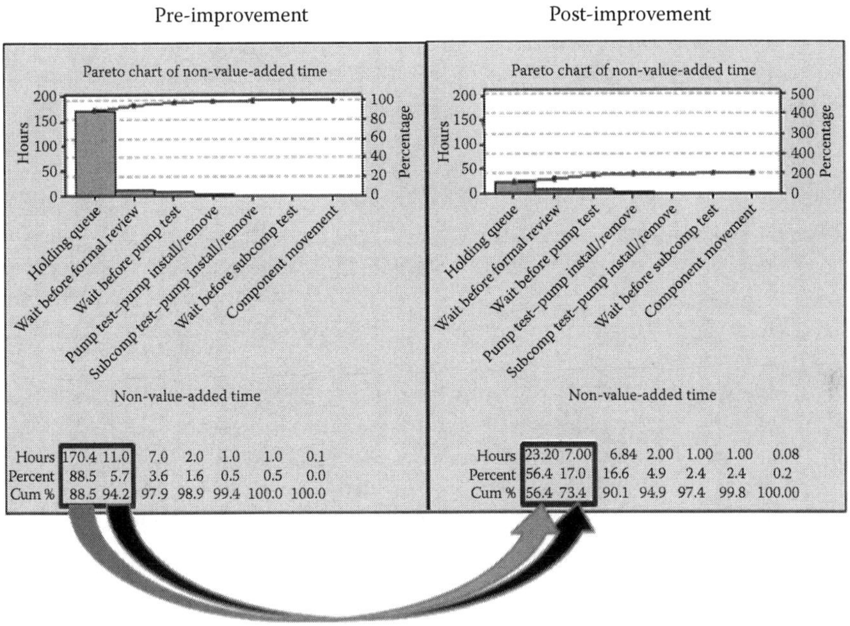

FIGURE 14.14
Pareto chart of non-value-added time (post-Lean).

Institutionalize the Improvements

As part of control mechanisms put in place to sustain the gains in waste reduction, a teardown tracker checklist macro prioritizes pumps for disposition of a teardown review. The process owner uses the checklist to select adjectives about a pump and the test it ran. The macro determines if a pump will be up for a "formal review" or an "engineering consensus." The tracker checklist maximizes the efficiency of the teardown process by choosing to review the whole pump, the pump head assembly, or the cam housing assembly. The checklist automatically calculates the cycle time (in business days), which is a control chart for complete visibility. Management fine-tunes the priority of test descriptors, if needed, thus allowing adjustments to formal or informal review criteria.

Control mechanisms also pertain to dashboards used in statistical processing control (SPC), documentation and standardization of work processes, and training. The tracker checklist automatically populates the pump teardown dashboard—used for statistical process control. The newly hired teardown leader monitors the lead time for completion for all pumps via SPC, including the number of pumps in the priority queue, and reports progress in the weekly team staff meetings. Documentation includes updated formal and informal reviews, and status and disposition of the project control plan. Finally, the teardown leader trains all XYZ pump engineers and technicians on teardown processes and Lean changes. The teardown leader also trains new hires as part of a standardized company orientation program.

Future State 2/Extended Value Stream

As XYZ Company achieves future state 1 their natural inclination is the identification of additional incremental changes in eliminating waste. The future state 2 and extended state maps then focus on looking for improvements between the company's suppliers and the pump teardown process in flattening flow variations.

For example, in 2011 there were two occurrences when 18 pumps were returned in a single week. Because the capacity is limited to four formal and four informal reviews, it took several weeks to recover. Why–why analysis determined that these 18 pump spikes occurred due to the customer holding all pumps until all testing was completed and then sending the pumps back in a single shipment. If the customer could send the pumps as the tests complete, this would reduce the occurrence of these spikes. Additional improvement could be implemented to recover more quickly when these spikes occur.

Internal to the pump teardown process is reducing the 4-hour cycle time to photograph components and prepare for the formal teardown review. This enables more pumps to be documented per week. Nevertheless, this likely requires improved photographic equipment capabilities, improved component fixtures to ease part placement for photographs, and implementing wear scales for the pump interface components to reduce the number of photographs (observations are classified according to a well-defined scale of 0 to 5).

Conclusion

The result of implementing Lean measures at XYZ Company was a 248% efficiency increase in the pump teardown process. At the beginning of the project, it took 26.56 workdays for a single pump to go through a nine-step teardown process involving 24 separate activities; now, it takes 7.64 days. Lean efforts are providing increased feedback on pump design validation tests, faster discovery of new issues and better understanding of pump reliability, and an overall greater contribution to pump development and manufacturing for XYZ Company. Our focus, as laid out in the charter and completed, was to determine the simplest and best cost-saving options using Lean methods to produce a 50% time decrease in the pump teardown process for XYZ Company. However, the work is not over as XYZ Company to continuously improve their processes and reduce waste.

References

Bonaccorsi, Andrea, Gionata Carmignani, and Francesco Zammori. Service Value Stream Management (SVSM): Developing Lean Thinking in the Service Industry, *Journal of Service Science and Management*, 4 (2011): 428–439.

Carreira, Bill, and Bill Trudel. *Lean Six Sigma That Works: A Powerful Action Plan for Dramatically Improving Quality, Increasing Speed, and Reducing Waste*. New York: AMACOM, 2006.

Hamilton, Bruce. Moments of Truth—Creating a Lean Chain of Support. *The e2 Continuous Improvement System*. Greater Boston Manufacturing Partnership (GBMP). Web accessed, 2012.

Jones, Daniel, and Jim Womack. *Seeing the Whole: Mapping the Extended Value Stream*, Version 1.1. Cambridge, MA: Lean Enterprise Institute, 2009.

Rein, Martin, and Donal Schon. Describe a frame as, "a perspective from which an amorphous, ill-defined problematic situation can be made sense of and acted upon." "Frame-reflective policy discourse," In *Social Sciences and Modern States: National Experiences and Theoretical Crossroads*, ed. Peter Wagner, Carol H. Weiss, Bjorn Whitlock, and Helmut Wollman, Cambridge, MA: Cambridge University Press (1991): p. 263.

Rother, Mike, and John Shook. *Training to See: A Value Stream Mapping Workshop Participant Guide*. Brookline, MA: The Lean Enterprise Institute, 2000.

Tapping, Don. *The New Lean Pocket Guide: Tools for the Elimination of Waste*. Chelsea, MI: MCS Media, 2007.

Womack, James P., and Daniel T. Jones. *Lean Thinking: Banish Waste and Create Wealth in Your Corporation, 2nd Edition*. New York: Free Press, 2003.

15

Improving Women's Healthcare Center Service Processes

Sandra L. Furterer

CONTENTS

Introduction

The original case study was published in *Design for Six Sigma in Product and Service Development: Applications and Case Studies*, CRC Press. It is adapted to focus on the Lean Six Sigma aspects applied for this case study (Cudney and Furterer, 2012).

Project Overview

A community women's healthcare center recently embarked on building and designing a new women's center to respond to the exponential growth of women's health-care needs in the market area by redefining, refocusing, reconfiguring, and expanding its women's health services line. The healthcare center had an existing smaller center, with fewer services, that was part of the women's center outpatient facility. It is recognized that a "one-stop shopping" model where women can receive a comprehensive range of services conveniently, comfortably, efficiently, and with speed and ease is desirable.

The goal of the new center was to provide women's services in a spa-like environment. The women's center will provide the comprehensive list of services shown in Figure 15.1. The services will be phased in as the facility is built and funds become available.

Project Goals

The intent of the women's center (WC) is to offer a diverse array of women's outpatient services in one location that provides convenient and easy access to multiple services including physician, diagnostic, ancillary, education, and outreach services. The vision of the WC is to build patient loyalty among women in the community and become the provider of choice for all their health-care needs. It is expected that the WC will not only increase market

Primary care	Gynecology	Diagnostic imaging
Urogynecology	Oncology	Breast surgery
General surgery	Endocrinology	Rheumatology
Cardiology	Osteoporosis treatment	Bone density screenings
Medical spa	Meditation chapel	Lifestyle center
Education center	Boutique	Heart-healthy café

FIGURE 15.1
Women's center services.

share for outpatient services but also increase inpatient volume and referrals back to the hospital.

Our new women's center business plan assumes potential process and staffing changes, such as the following:

- Having radiology results available while the patient is in the center
- Providing the ability to assess authorization of additional diagnostics
- Being able to meet with a physician if the patient requests

This project focuses on improving existing processes and adapting them for the new facility to meet the needs of the customers (women) who will receive services at the women's center. The processes should have workflows that optimize throughput, reduce wait times, and provide next business-day results to the patients who receive women's services.

Define Phase

In the Define phase, the team defines the problem to be solved and determines the scope of the project.

Project Charter

Overview

Many outpatient facilities are focusing on providing comprehensive services to women in a comfortable setting. In a qualitative study of women who had received a mammogram in the past 3 years, without a history of cancer, satisfaction was related to the entire visitation experience, not just the actual mammogram procedure. The authors found seven satisfaction themes from the focus groups: (1) appointment scheduling, (2) facility, (3) general exam, (4) embarrassment, (5) exam discomfort/pain, (6) treatment by the technologist, and (7) reporting results (Engelman, Cizik, and Ellerbeck, 2005). This supports the focus of designing a seamless experience for women in the women's center through applying the Lean Six Sigma methodology and tools.

Problem Statement

On average, the results for patients who visited the former women's center were transmitted to the primary care provider (PCP) within 57 hours. There was a large variability in how and when the test results were communicated to the patients.

Some of the questions that the team had to answer before the new center opened were the following:

- How do we measure the success of the new women's center?
- What processes will stay the same for the new center?
- What processes will be different in the new center?
- What is our current state and future state?
- Should we perform activities in parallel or series? Or both?
- What is our forecasted demand? Do we have different processes based on our volume?
- What is our goal for waiting time? Is it different for the different areas (i.e., registration, radiology reception, and gowned waiting room)?
- What should we prioritize patients by (i.e., appointment time, arrival time, or waiting time)?
- What is an acceptable wait time when having two tests done?
- Will patients have their results available after their tests? Will the results be available for all types of tests (diagnostic/screening)?
- Will patients have the ability to meet with a physician given the test result comes back requiring further testing?
- Will patients be able to get a diagnostic test?

Customers/Stakeholders

External: patients (women who receive services in the center), referring physicians, payers, and donors

Internal: imaging technologists, radiologists, administration, registration, centralized scheduling, information technology, physicians, and marketing and development

Goal of the Project

To design the processes and define the metrics that result in optimal flow for the new women's center phase 1.

Scope Statement

The scope includes the new women's center diagnostic services (mammography, ultrasound, stereotactic biopsy, and bone densitometry), registration, and appointment scheduling.

Projected Financial Benefits

Patient satisfaction, increased capacity due to efficient workflow, and resultant revenue are the potential financial benefits of this project.

Perform Stakeholder Analysis

Stakeholder analysis was performed to identify project stakeholders. The stakeholder analysis definition is shown in Figure 15.2.

Create Project Plan

A Gantt chart was used to keep the project development cycle on track and ensure that all key steps in the process were accomplished. The project started in March 2010 and finished in November 2010.

Stakeholders	Who they are	Potential impacts/concerns
Patients (women)	Receive services in the center	• Customer service • Quality • Efficiency
Referring physicians	Refer patients to the center and communicate results to the patients	• Quality of care
Payers	Pay for services, such as managed care, Medicare employers, self-pay	• Quality • Cost-effective care across the continuum
Donors	People who donate money to the center	• Quality of care • Meet patient requirements
Imaging technologists	Work in the new women's center and perform procedures	• Patient satisfaction • Improved/well-designed work environment • Associate satisfaction
Radiologists	Read and provide results of imaging procedures	• Reduced volume and revenue • Physician satisfaction
Administration	Manage the hospital and center	• Volume • Revenue • Patient satisfaction • Physician satisfaction • Productivity
Patient access and centralized scheduling	Registration department who registers patients and perform insurance authorizations, centralized scheduling who make the patient appointments	• Volume • Productivity • Timeliness of processes
Information technology	Provide phone and computer systems	• Meet requirements • On time • On budget
Physicians	Provide women's services	• Reduced volume and revenue • Physician satisfaction
Marketing and development	Perform business development, marketing, and fund raising	• New business • Funds available • Able to reach customers

FIGURE 15.2
Stakeholder analysis definition.

Measure Phase

The Measure phase of the project is designed to gain information on the voice of the customer (VOC) to understand the needs of customers and begin translating those customer requirements into the processes' technical elements, and to baseline the existing process. The main activities of this phase are to collect VOC and document and measure the existing process.

Collect Voice of the Customer

We first defined the market opportunities through a literature search and identified the following findings that support the importance of a women's center:

Market opportunities:

Specific areas in which women have high mortality and morbidity rates include the following:

- About 60% of hospital patients are women.
- About 50% of all women over the age of 50 will have osteoporosis.
- Hypertension is three times higher for women than men.
- Depression is more common in women than men, affecting over 7 million women in the United States.
- About 75% of Alzheimer's disease cases are women.
- About 90% of rheumatoid arthritis cases are women.
- About 90% of lupus patients are women.
- An estimated 10 million women in the United States suffer from urinary incontinence.
- Women have twice the incidence of multiple sclerosis than men.
- One in seven women aged 45 or older has cardiovascular disease, and 45% of all female deaths are cardiac disease related.
- Almost 80% of fibromyalgia cases are female.
- Over 80% of bariatric surgery patients are women.

The team then performed a literature search to identify customer requirements that impact satisfaction in a women's center. Engelman, Cizik, and Ellerbeck (2005) conducted focus groups to identify the factors and dimensions related to a patient's experience with a mammography procedure. They identified the number of coded text lines from the focus group transcripts for the factors and dimensions related to patient experience. A Pareto chart was developed, identifying the most important factors and the most important dimensions for each factor. The Pareto chart for the factors is shown in Figure 15.3.

Scheduling was the most important factor, with the most important dimensions being scheduling convenience, reasons to schedule a mammogram, and financial issues around paying for a mammogram. Results became the next most important factor, with the important dimensions being the manner in which the results are reported, the time it takes to receive the results, and the need to return for suspicious findings or additional procedures. The mammography technologist factor was the next most important element, with the important dimensions being the technologist's attitude toward the woman during the exam, the provision of the exam instruction, and the technologist's skill in performing the procedure. The facility was the next most important factor, with the most important dimension being the environment of the waiting room including cleanliness, having an educational video and magazines, noise level, decor, and beverage availability. The general exam was the next most important factor, with the issues surrounding the gown worn for the procedure being the most important, followed by instructions to prepare for the procedure and the cold or warm machine plate. Discomfort and pain was the next most important factor, with the most important dimension including uncomfortable pressure due to breast plate compression. The least discussed element of mammograms is the embarrassment factor, including the embarrassment of participating in an exam of a personal nature; embarrassment of discussing the mammogram with family, friends, or the patient's health-care provider; and self-consciousness due to being modest about personal health.

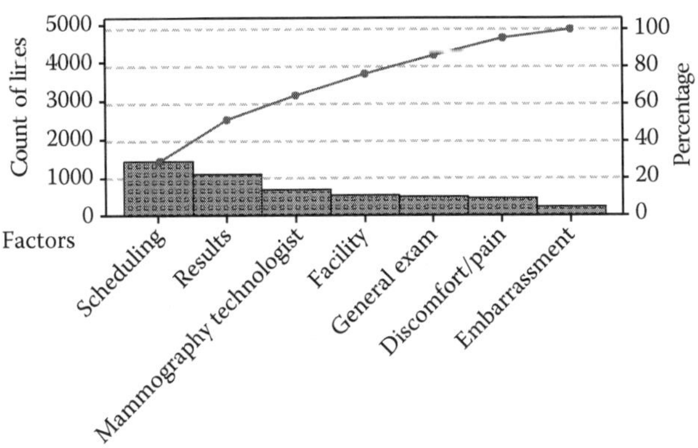

Count of lines	1415	1093	657	541	473	462	238
Percent	29.0	22.4	13.5	11.1	9.7	9.5	4.9
Cum %	29.0	51.4	64.9	76.0	85.7	95.1	100.0

FIGURE 15.3
Pareto chart of factors.

Identify Critical to Satisfaction Measures and Targets

Marketing held several focus groups to understand their critical requirements so that we could extract the critical to satisfaction (CTS) criteria from the focus group qualitative data. The CTS criteria are shown in Figure 15.4.

Create Operational Definitions and Metrics

The following operational definitions were developed for the results times:

Operational definition: Time from arrival at the imaging reception to exam.

Defining the measure: The focus is on the time it takes to present at the radiology reception desk until the procedure is initiated.

Environment
Comfortable
Ease of parking
Esthetic rooms geared psychologically to women
Creative access points to modern facilities
Guided through health issues in a nurturing, relaxing environment
Operational
Efficiently, with speed
Ease
Easy navigation throughout the system
Time saving, convenience
Combined appointments
Same-day results
Caring and competent professional staff
Customer-focused amenities
Advanced technology as an enabler of superior care
One visit, one stop
Guided through health issues in a nurturing, relaxing environment
Service
Seamless integration of service components
All ages and life phases
Comprehensive range of services
One visit, one stop
Guided through health issues in a nurturing, relaxing environment
Coordination of medical and health concerns and treatments, multidisciplinary team
Integrating between physicians, diagnostics, and ancillary
Functional medicine that combines traditional and integrative
Gender-specific medicine
Holistic care
Aligned with core values

FIGURE 15.4
Critical to satisfaction (CTS) criteria.

Purpose: We want to understand how long patients wait.

Clear way to measure the process: We will measure the time that a patient waits in the waiting area of the radiology department. This will be measured by calculating the elapsed time from when the patient presents at the Imaging reception desk to when he or she enters the procedure room. To baseline, we will use QueVision data from April 1 to June 24 and subtract CheckInTime from InProcessTime.

Operational definition: Time from procedure start to procedure completion.

Defining the measure: The focus is on the time it takes to complete procedures.

Purpose: We want to understand how long a procedure takes.

Clear way to measure the process: We will measure the time that a patient spends in a procedure room in the radiology department. This will be measured by calculating the elapsed time from when the patient enters the procedure room to when the procedure is complete. To baseline, we will use QueVision data from April 1 to June 24 and subtract InProcessTime from CompletedTime: Mammogram.

Operational definition: Time from procedure completion to when results are available.

Defining the Measure: The focus is on the time it takes read the image and transmit procedure results.

Purpose: We want to understand how long it takes for procedure results to be available.

Clear way to measure the process: We will measure the time that a radiologist takes to read the image and for the technologist to upload them into Meditech for transmission. This will be measured by calculating the elapsed time from when the procedure is complete to when it is available to the primary care provider (PCP).

Operational definition: Time from exam completion to review with patient.

Defining the measure: The focus is on the time it takes to communicate results to patients.

Current State Value Stream Map

We developed a current state value stream map using the data in the information system that tracked process times. The value stream map is shown in Figure 15.5. The main process steps are as follows:

1. A patient calls and schedules an appointment for a screening mammogram.
2. The center verifies the insurance and authorizes the screening.

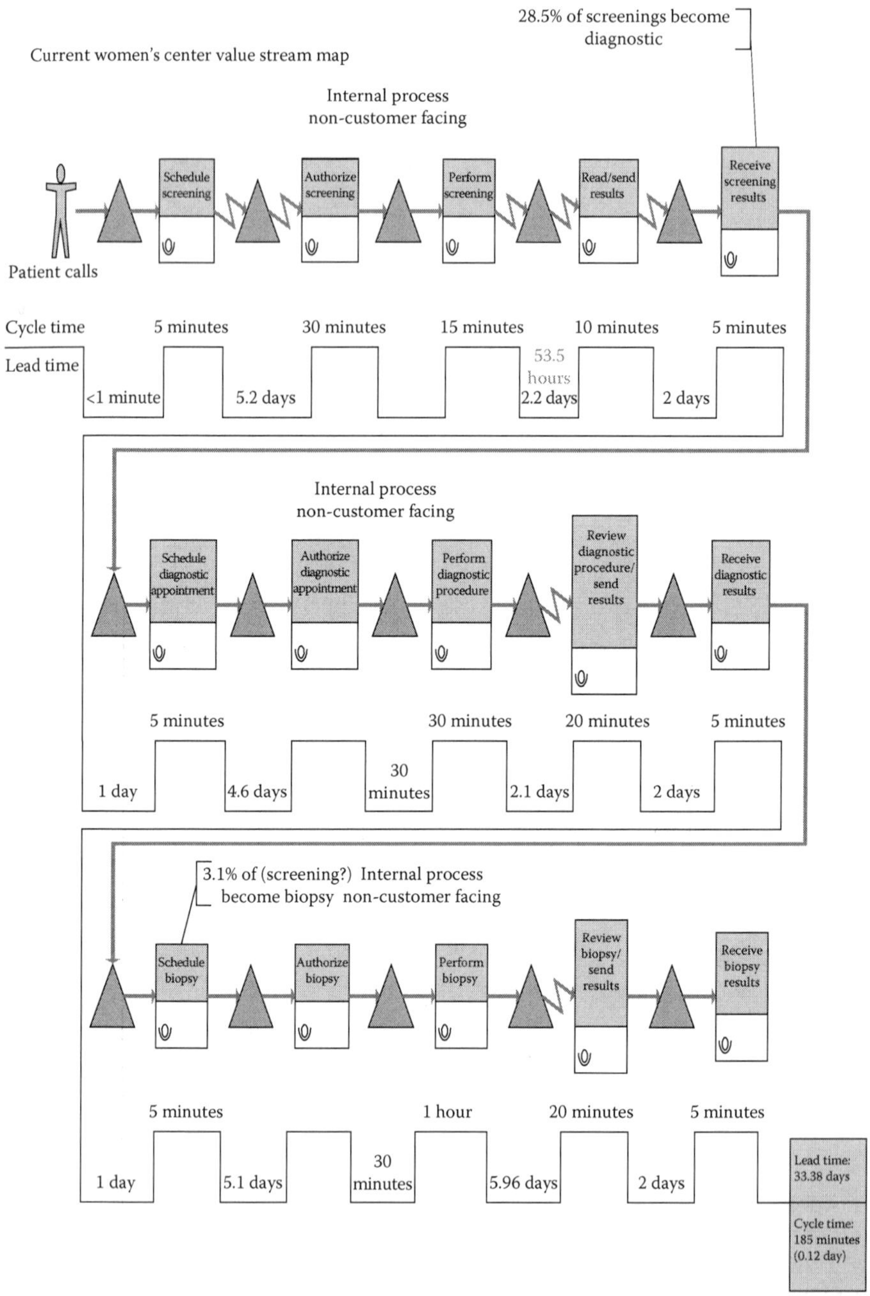

FIGURE 15.5
Current state value stream map.

3. The screening is performed at the women's center.
4. The results are read by the radiologist.
5. The results are prepared in a letter and sent to the patient.
6. Twenty eight and a half percent (28.5%) of the patients require a diagnostic appointment for further scans and views. The diagnostic appointment is made.
7. The center verifies the authorization for the diagnostic mammogram.
8. The diagnostic procedure is performed.
9. The results are read by the radiologist and sent to the patient.
10. About 3% of the patients require a biopsy. The patient calls to schedule the biopsy appointment.
11. The center verifies the authorization for the biopsy procedure.
12. The biopsy procedure is performed.
13. The results are provided to the patient.
14. The patient receives the results. Further surgical procedures would be scheduled at a surgical center.

The lead time for the entire end-to-end process is 33.38 days (from screening to biopsy [if necessary]). The value-added cycle time is 185 minutes, or 0.12 day. There is clearly a great deal of wait time and delays in the current process. Now, we move to the Analyze phase.

Analyze Phase

The Analyze phase for the women's center project is focused on understanding the factors that contribute to an efficient process and the potential root causes of inefficiencies so that they are reduced or eliminated with the potential to achieve a Six Sigma quality level. The main activities of this phase are as follows: (1) identify root causes and (2) identify potential risks and inefficiencies.

Perform Process Analysis

The process analysis defines the process inefficiencies for each proposed process, as shown in Figure 15.6.

Inefficiency
Process: Schedule service
Having to do a warm transfer from central scheduling to physician group
Process: Register patient
If patient is not preregistered, know late in the registration process that he or she is a VIP (not prior to arriving in Women's Center)
Should ensure all patient items are available and accurate prior to registration in Women's Center (labs, script, ID)
Ensure only needed patient information is provided from the patient when they register an existing patient (between imaging and doctor)
Authorize service
Multiple rework loops if authorization is not OK or correct
Having to obtain a changed order for a Medicare patient
Provide spiritual care
Resourcing for Women's Center needs to be identified
Connect to cancer center
Potential of not having preference card or not having it updated
Self-referral
Identify if accept self-referrals
Perform surgery
Rework getting labs, prep work, patient info, etc.
Multiple calls getting health history; informing of co-pay
Getting timely lab results
Having to bill from radiology to surgery and then have radiology reimbursed

FIGURE 15.6
Potential process inefficiencies.

The gaps identified between current and future state process maps are discussed in the following subsection.

Gaps between Current State and Future State Vision

- Lack of seamless integration between the hospital medical information system and the medical group information system.
- Triage for spiritual care.
- Referral physician preference cards.
- Concierge process.
- Patient navigator and patient navigator process.
- Patient navigator database and application to track patient interaction.
- Track utilization of spiritual care.
- Track connection to cancer center.

- Defined service levels for processes.
- Next-day results not meeting marketed expectations of same-day results.
- Seamless online appointment scheduling.
- Navigating through the women's center.
- Detailed future state process maps need to be completed.

The team developed and reviewed the process maps based on a business scenario analysis. A sample process map is shown in Figure 15.7. The following process maps were developed:

Schedule:

- Imaging service
- Medical doctor
- Cancel or reschedule
- Register patient
- Register imaging patient
- Register imaging walk-in
- Register physician patient
- Register physician patient walk-in
- Authorize patient
- Provide spiritual care
- Connect to cancer center
- Process self-referral
- Perform surgery
- Perform imaging service
- Perform physician service

Provide results:

- Screening mammogram
- Diagnostic mammogram
- Ultrasound results
- Biopsy results
- Bone density results
- Other results
- Patient navigator
- Request records/films
- Process VIP patients

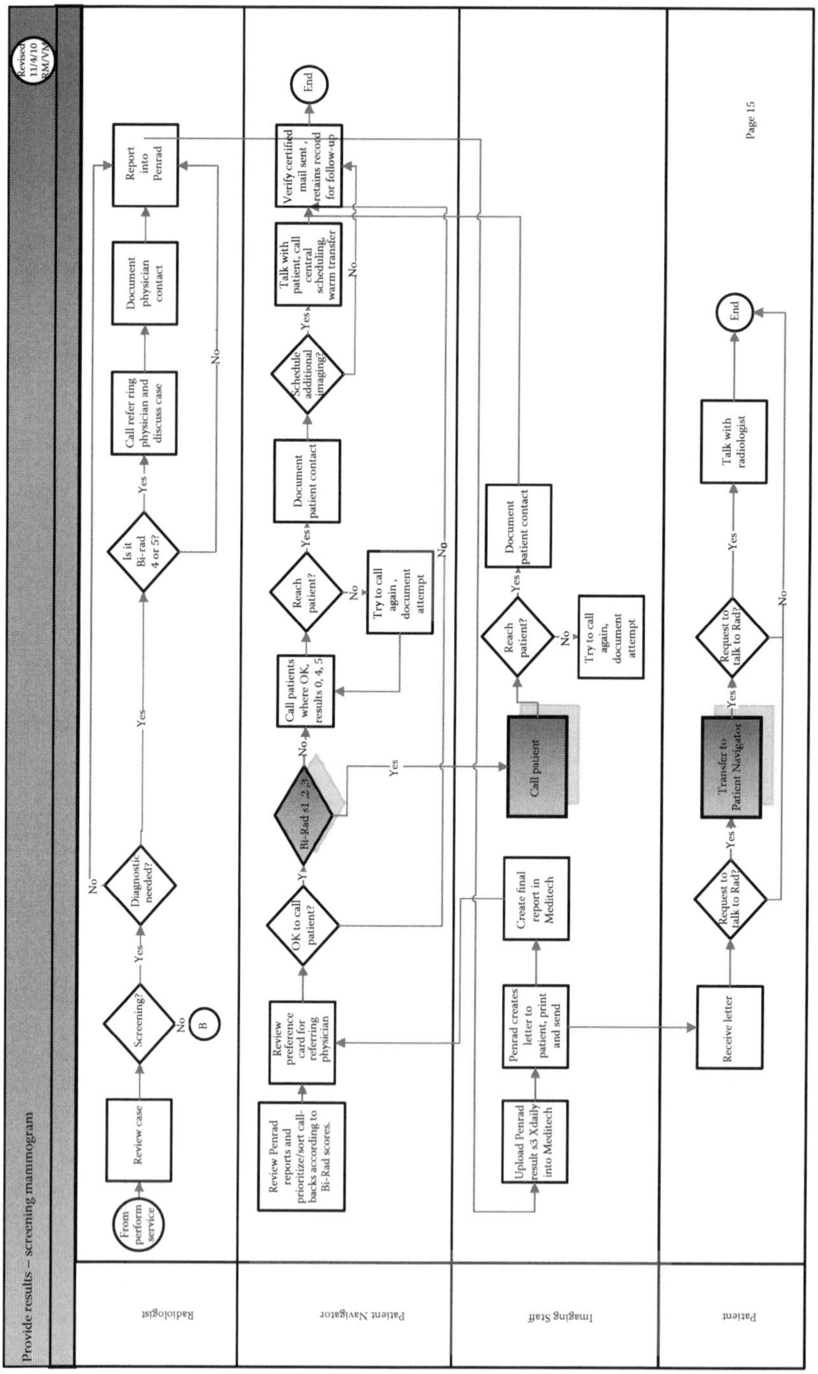

FIGURE 15.7
Sample process map: provide results—screening mammogram.

Perform Waste Analysis

The wastes shown in Figure 15.8 were identified for the future state processes.

We identified potential problems in the processes after they were piloted and created three why–why diagrams (Figures 15.9 through 15.11):

- Why are there excessive wait times in the Women's Center?
- Why does it take more than 1 business day to determine results for procedures?
- Why does it take a long time for the patients to receive the results?

Takt Time Analysis

A takt time throughput analysis was performed to assess whether the cycle times could achieve the needed takt time. Takt time is calculated based on the

Process	Types of waste
• Schedule	Overproduction: transfer between central scheduler and medical group Processing: duplicate information in medical information and physician system
• Register patient	Defect: VIP patient not identified prior to visit or during visit Defect: patient forgets prescription Processing and delay: need to call physician's office for prescription Delay: register walk-ins
• Authorize patient	Defect: wrong insurance Defect: wrong authorization Delay: verify medical necessity Processing: receive change order
• Provide spiritual care	Delay: availability of resources
• Connect to cancer center	Motion: patient going from WC to cancer center Delay: due to not having physician preferences to contact patient
• Process self referral	Delay: referral to physician
• Perform surgery	Overproduction: scheduling and patient information in surgery system
• Perform imaging service	Delay: wait times in lobby, procedure room
• Perform physician service	Delay: wait times in lobby, room
• Provide results	Delay: providing results Delay: physician providing results Delay: not having physician preferences for patient contact Delay: radiologist reading Delay: not reaching patient
• Patient navigator	Inventory: patient navigator capacity
• Request records/films	Delay: printing films
• Process VIP patients	Defect: VIP patient not identified as VIP Delay: difficulty fitting VIP in schedule

FIGURE 15.8
Waste analysis.

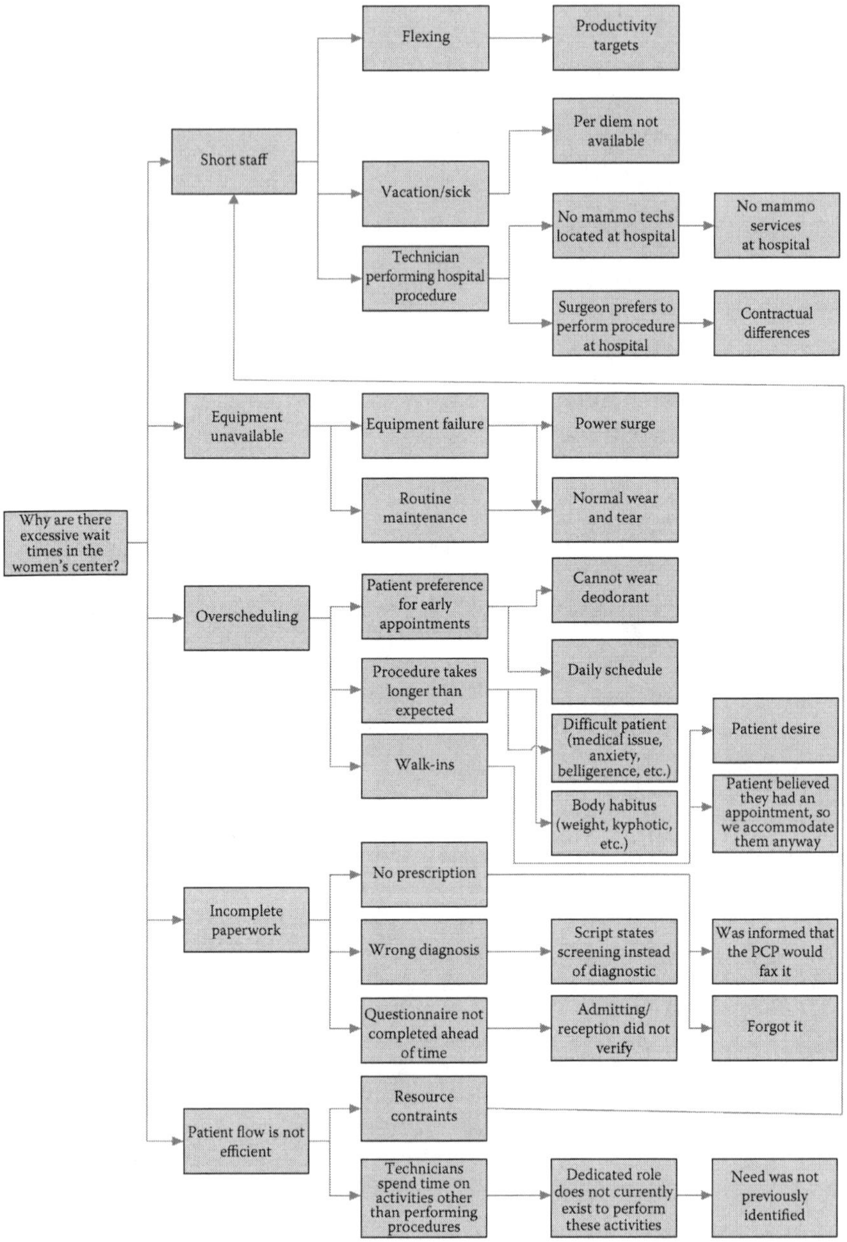

FIGURE 15.9
Why–why diagram: Why are there excessive wait times?

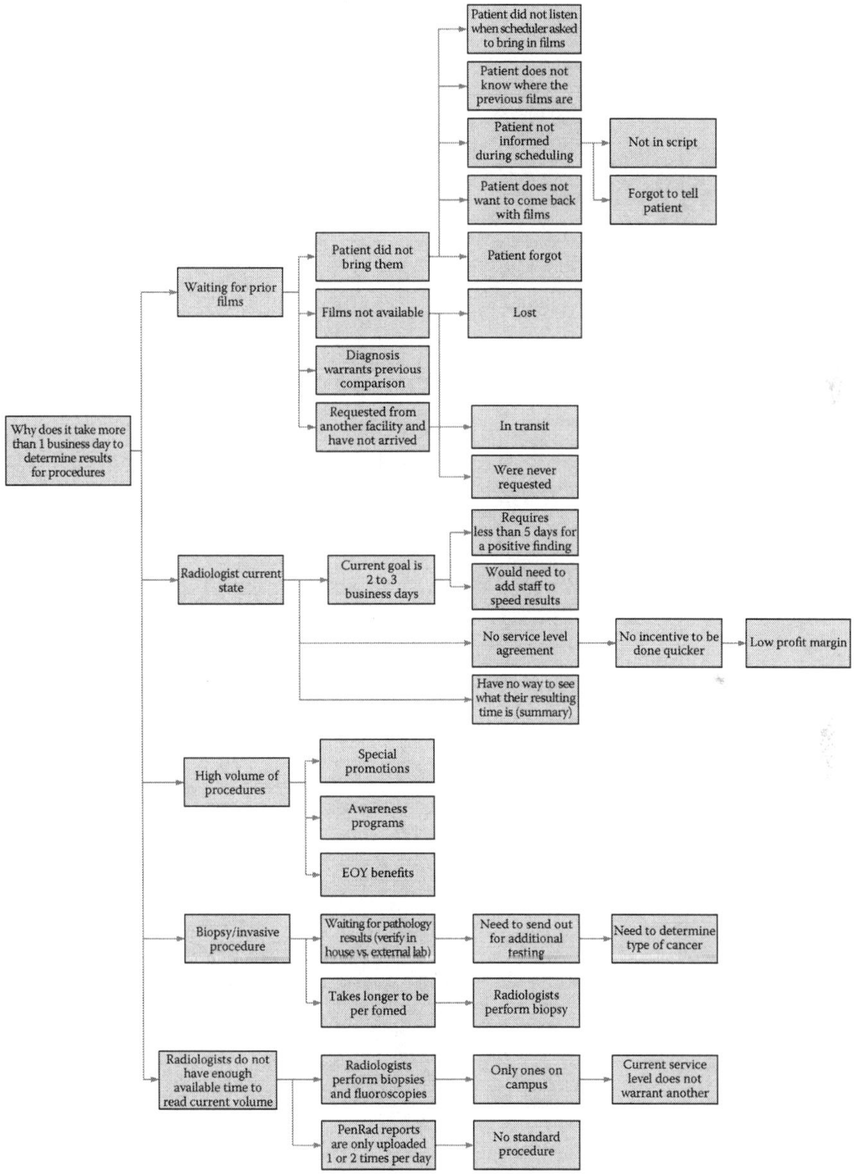

FIGURE 15.10
Why–why diagram: Why does it take more than 1 business day to determine results for procedures?

available time of scheduled resources divided by the number of scheduled procedures. Figure 15.12 shows there are times of the day when the scheduled procedures exceed staff capacity. This analysis helped to realign the staffing to meet the patients' demand. The continuous line shows the desired takt time, and the box plots show when the average cycle time exceeds the desired takt time.

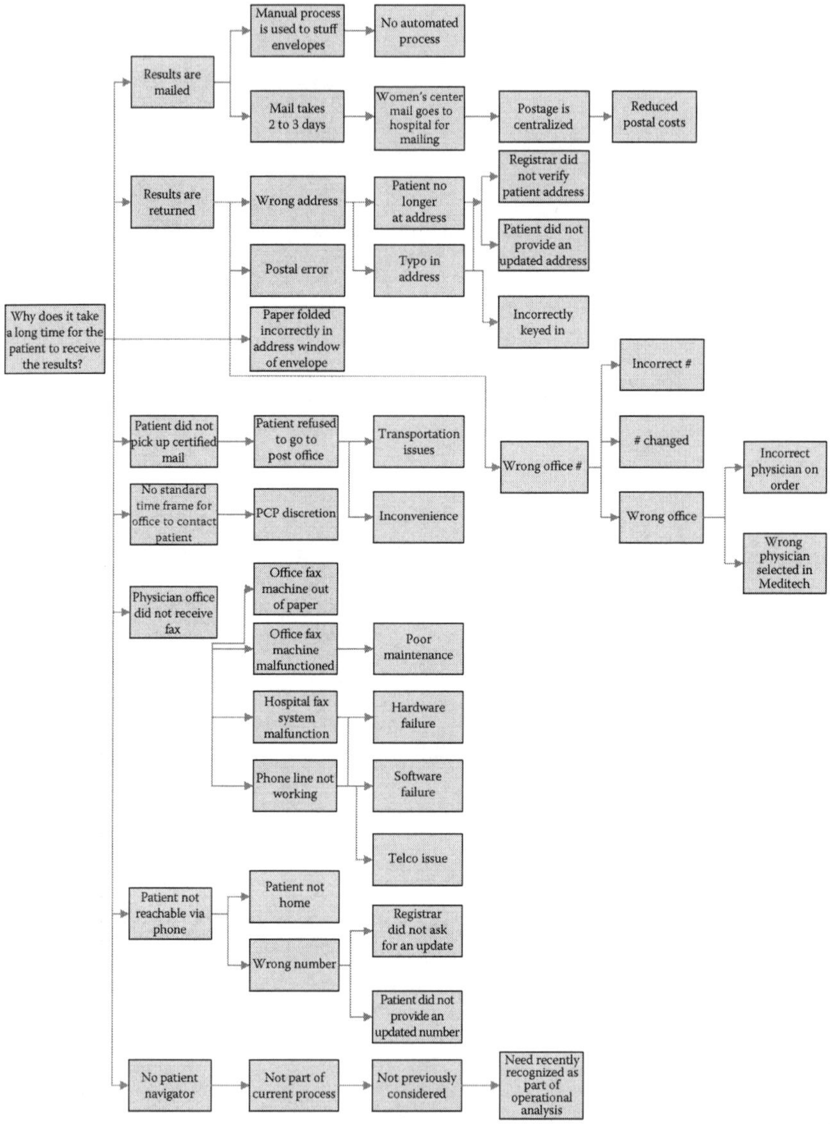

FIGURE 15.11
Why–why diagram: Why does it take a long time for the patient to receive the results?

Assess Performance, Failure Modes, and Risks

We assessed the potential failure modes in the processes and identified the potential risks:

1. Focusing on productivity as the key metric impacts our ability to meet the quality indicators of patient wait and throughput times.

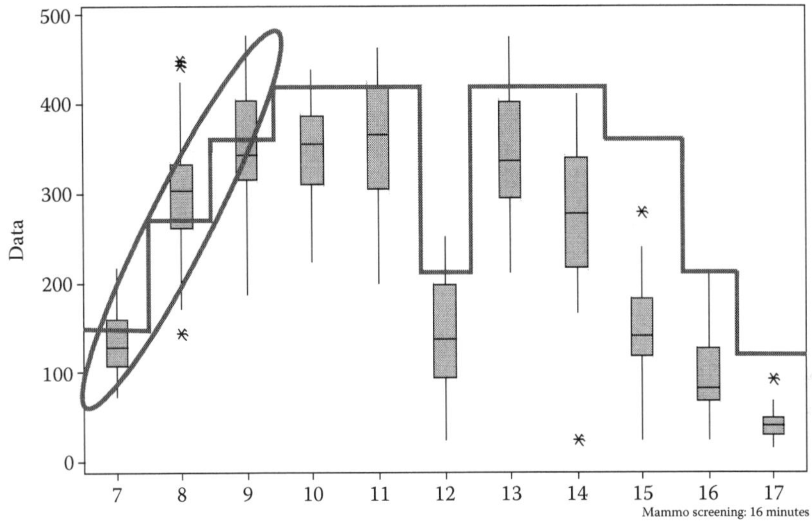

FIGURE 15.12
Takt time analysis: figure shows scheduled versus available time, all modalities. More time scheduled than what is available = longer wait; walk-ins not included; assumes no variability in test times; may be taking only machine time into consideration; no equal loading of resources.

2. The patient navigator continues to meet with referring physicians to describe the patient navigator services available and complete physician preference cards that identify when it is appropriate for the patient navigator to share the patients' results directly with the patients and the time frame to do so.
3. Identifying VIP members has been problematic with an information systems issue. This issue has been logged for the information technology help desk.
4. There are several outstanding action items that have not yet been resolved, and the executive team must embrace and resolve the more political issues related to growth, volume, and physician relationships.

Improve Phase

The purpose of the Improve phase is to identify improvements and pilot new processes and then assess whether they are capable of meeting the desired targets. The following activities are performed:

1. Identify improvement ideas.
2. Implement pilot processes.
3. Assess process capabilities.

Identify and Implement Pilot Processes

We implemented the processes when the women's center opened and compared the process times to the baseline in the previous imaging center. The proposed value stream map is shown in Figure 15.13. A detailed work plan was developed that included the activities required to open the facility and implement the newly designed processes. The key process elements implemented were as follows:

- Hire a patient navigator to guide the patient through the system and communicate the results of procedures.
- Hire an additional radiologist to provide results more quickly.
- New spa-like facility development.
- Incorporate additional services.
- Hire a women's services–focused physician.
- Train mammography technicians in customer service.
- Incorporate spiritual care resources into the center.
- Cross-train registration staff in imaging center and physician office scheduling systems.
- Hire a concierge to guide patients to their destinations.
- Have volunteers walk patients to their destinations.
- Incorporate a VIP service for donors and other important guests.
- Implement a process dashboard for metrics.
- Incorporate advanced imaging technology.

Assess Process Capabilities

The processes as designed were capable of meeting the proposed wait times, procedure times, throughput time, and results time. Further improvements were designed to streamline the processes, based on staffing, results turnaround, and throughput times. A patient navigator was

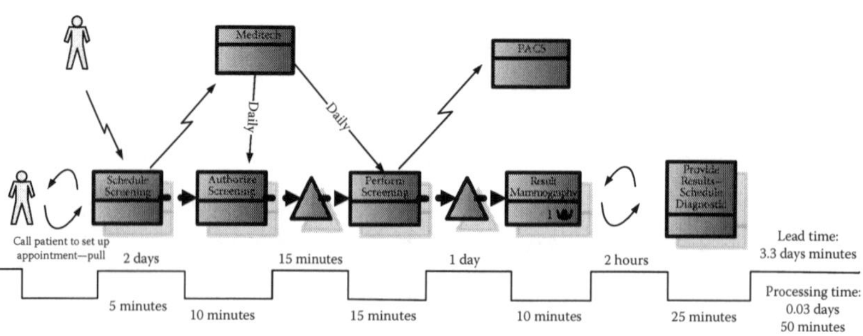

FIGURE 15.13
Proposed value stream map.

Metric	Prior imaging center (before)	Comprehensive women's center (after)	% Improvement
Imaging waiting area	Mean = 25 minutes; standard deviation = 30.6; count = 3219	21 minutes	16%
Procedure time	Mean = 13 minutes; standard deviation = 23.1; count = 3165	14 minutes	−7%
Total throughput time	Mean = 38 minutes; standard deviation = 36; count = 3165	35 minutes	8%
Time to determine results	Mean = 2.4 days (57.2 hours); standard deviation = 55.04 hours; count = 6693	1 day	58%

FIGURE 15.14
Improved process times.

hired in the first quarter of 2011. She has provided great value for our patients and referring physicians. She provides sociopsychological care to our patients by providing results in a more timely manner, guiding them through the system, and connecting them to spiritual care and to the cancer center when necessary. She has been a key differentiator in the process to provide the "delighters" of the patient being guided through the health system in a nurturing, relaxing environment; comprehensive care; holistic care; efficient processes; next-day results; and seamless integration of services.

The results for the newly designed processes are shown in Figure 15.14 and compared to the older imaging facility. The waiting time in the imaging area improved by 16%, and the procedure time increased by 1 minute (7%), on average. The total throughput time decreased by 3 minutes or improved by 8%, compared to the process times in the older imaging center prior to opening the new women's center facility. The time to determine test results, which was a delighter, improved by 58% from 2.4 days to 1 day.

Control Phase

The Control phase verifies that improvement was attained and ensures that control plans are implemented that will control the processes. The following activity was performed in the control phase: develop dashboards and control plan.

Women's center mammography throughput KPI dashboard				
	Target	**Jan**	**Feb**	**Mar**
Quality indicators				
Average mammogram procedure time (screening & diagnostic)	<=18 minutes			
Average mammogram wait time	<=25 minutes			
Average mammogram throughput time	<=45 minutes			
%Patient waiting > 45 minutes	5%			
%Mammo patients contacted by navigator within 5 days of date of service	100%			
Total patients				
	Green = meeting target			
	Red = not meeting target			

FIGURE 15.15
Dashboard control plan.

Develop Dashboards and Control Plan

A dashboard was developed that would be used to control the process and could be reviewed continuously during the day, to assess the wait times, procedure times, throughput times, patients waiting for procedures, volume, and result times. A more static dashboard was created that would track performance on a monthly basis, which is shown in Figure 15.15. The metrics were being met for average procedure time, average wait time, and average throughput time, but not for percentage of patients waiting more than 45 minutes and percentage of mammography patients being contacted by the navigator within 5 days of service. We continued to adjust staffing and improve the processes to better meet the targets. The new design has been stabilized, but it will be continually improved and monitored.

Conclusions

The Lean Six Sigma methodology was extremely successful in capturing VOC and translating customer requirements into process requirements and in improving a process that delights the customers of the women's center. The opening of the new women's center has surpassed the organization's and the customers' expectations. The facility provides comprehensive women's services in a nurturing and spa-like environment. The future looks bright for the next several phases of the center.

References

Cudney, E., Furterer, S., *Design for Six Sigma in Product and Service Development: Applications and Case Studies*, CRC Press, Boca Raton, FL, 2012.

Engelman, K., Cizik, A., Ellerbeck E., Women's Satisfaction with Their Mammography Experience: Results of a Qualitative Study, *Women & Health*, 42(4), 2005.

16

Application of Lean Tools in a Medical Device Company

Kelly M. Davis, Elizabeth A. Cudney, and Scott E. Grasman

CONTENTS

Executive Summary

By using Lean principles and tools, the product development cycle can be significantly shortened. Through an example, performed at a leading medical device company, this example of Lean tools demonstrates that by applying certain Lean principles and tools to the development of a medical device, shortened development times can be obtained. This case study uses value stream mapping, Kaizen events, and multiple Lean tools to create an artwork process for new medical devices going through the product development process (PDP). An example is given that illustrates how to take an artwork subprocess and reduce its overall cycle time to help reduce time to market.

Introduction

Overview of Lean Thinking

In an industry where 7500 new devices are marketed annually in the United States, reducing their average product life cycle time from an 18–24-month average seems to get little literary focus (ECRI, 2006). Perhaps this is because there are many unique requirements that drive this unusually long life cycle time: regulatory elements from the Food and Drug Administration (FDA), medical reimbursement, Group Purchasing Organization (GPO) policies, and strict global health-care regulations. Traditionally, long lead-time industries such as aerospace and automotive often have existing and mature Lean product development (LPD) programs. While striving to achieve decreased time to market, manufacturers will often look first at the assembly line. Although Lean manufacturing has had positive impacts on many industries (Bowonder et al., 2010; Haque, 2003; Kumar and Bauer, 2010; Panchak, 2003), the scope of time saving should be expanded and looked at more closely at the front end of the process used to develop a medical device.

From the perspective of both a medical device manufacturer and for the consumer, getting a technologically innovative or life-saving medical device to market quickly is imperative. Often referred to as the fuzzy front end, a slow and unsure beginning to the PDP can destroy a potential project from the very beginning of its life cycle. By incorporating development-specific Lean principles to the design and development process, it can become a process that is Lean, effective, and efficient.

Medical device firms are ever encouraging their management to release products faster. Yet the product development roadblock will often show itself immediately as it is believed to be the bottleneck in new product introduction (Martinez and Farris, 2009). According to Zak and Waddell (2011), "companies want their managers to be more entrepreneurial, but they expect that entrepreneurial spirit to thrive within the confines of reports, meetings, metrics, structures, and financial ratios that no entrepreneurs would ever waste a dime or a minute of their time implementing." Therefore, the medical device industry may find that implementing Lean into their development cycle may aid in their goal of reducing time to market because it is getting rid of many of those items in the process. Although approaches and tools for addressing this problem are continually taught, actual implementation plans and business examples do not readily exist.

Project Overview

Certain methods used in solving the problem presented in this case study stem from LPD, which has been in practice for over 20 years and was started by Toyota. In the world of product development, Lean represents a small fraction of the theories and tools available in the industry (Hoppmann et al., 2011). The tools being used today have morphed from the original Toyota ways and there are many different techniques, but there is no gold standard (Leon and Farris, 2011). Therefore, tools presented in this case study do not follow one strict method, but a large mix of Lean tools that include value steam mapping, Kaizen, 5S, visual management, hand-off diagrams, spider diagrams, standardization, and mistake proofing. Although this case study is specific to the medical device industry, the tools discussed are universal. They can be used as presented or modified to fit any industry, and they can be used in any business or development process to help obtain a quicker development cycle for varying types of products. No matter the industry, "an efficient productive development process enables a manufacturer to bring new products to the market more quickly than their competitors" (Nepal et al., 2011).

Selecting the Problem

The goal of LPD in the medical supplies sector is to shorten the development time to bring an initial concept to market. In an attempt to achieve this goal, many Lean concepts are being implemented: Obeya Rooms, Hoshin Planning, and Hansei. Lean expertise capabilities are being built as well. Specific to identifying and holding Medical Supply's Kaizen events, portions of the Toyota product development system and product development value stream mappings helped formulate the basis for the training and execution plans, although neither method was followed exclusively.

The LPD strategy and implementation plan followed these steps: (1) formulate and scope the Medical Supplies LPD strategy, (2) gain support from senior leadership, (3) teach a pilot LPD course, (4) value stream map the entire Medical Supplies PDP, (5) use the Medical Supplies PDP to help a cross-functional team determine a first tier of Kaizen events, and (6) execute the Kaizen events with the goal of reducing time to market. After obtaining senior leadership sponsors and support from upper management for the intended LPD implementation plan, a pilot LPD course was given an Operational Excellence (OpEx) team of Lean masters at Covidien. As part of this class, the first Kaizen event, artwork, was mapped and the Kaizen event was completed.

Define

Problem Statement/Scope

The following process improvement took place at Covidien, a global health-care products manufacturer and developer. The Medical Supplies segment accounts for approximately $1.7 billion of the company's total 2010 net sales of $10.4 billion. The key product lines in the segment include Nursing Care, SharpSafety, and Monitoring and Operating Room (OR) devices, whose products include enteral feeding, wound care, and operating room kits. Covidien has a mature OpEx program that began approximately 10 years ago. The main focus initially was on introducing the tools of Six Sigma and Lean in manufacturing, which was highly successful. As the program expanded, these tools were applied to business processes and product development. Recently, Covidien introduced the "Covidien Operating System" to holistically drive OpEx tools, mainly Lean, deep within the organization, with a strong focus on LPD. With LPD, Covidien's goal is to reduce time to market and time to peak sales. This is to be achieved by leaning out the processes supporting new product launches and by tackling problem processes with Kaizen events on products that will enter the PDP.

The artwork process is a complex and time-consuming effort that involves almost every function within the division. Artwork includes printing on the device itself, informational booklets, informational compact discs, and all packaging and shipping labels that are sent with the product. The process is guided by various regulatory, legal, and clinical requirements. The importance of mistake proofing is essential as a mistake could cause product recall or device misuse leading to injury or death. Further complications include multiple language translations per artwork piece, as most products are launched globally.

Problem Objectives and Goals

From 2004 to 2009, large projects in the PDP within the segment ranged in launch times from 14.5 to 39.1 months with an average launch time of 26.5 months. After interviews with multiple product functions, program managers, and senior management, the artwork process was picked as the process to be tackled first due to its reoccurrence on the critical path development timeline.

To ensure the Kaizen week for the artwork process was productive, a value stream map was created 5 weeks before the event. At the mapping session, current and future state maps were created. In the 5 weeks following the mapping session and leading up to the event, a large portion of the action items was completed to ensure the full implementation of the future state was complete by the end of the Kaizen week, namely, the

Step	Within the Division	Within the Artwork Process
1. Create a sense of urgency	Decrease launch times (current 14.5–39.1 months)	Rework is 100%, critical path reoccurrence
2. Form guiding coalition	Senior management involvement and dedicated resources	13 team members representing 9 functions
3. Identify key stakeholders	Senior management and customers	Individuals involved in any artwork process
4. Create a vision of the future	Value stream map the entire Medical Supplies product development process	Value stream map of the future state
5. Communicate the vision	In addition to identifying the artwork process, a cross-functional team identified an additional list of problem processes and presented and obtained approval from senior management	Training throughout the division to everyone with a function pertinent to the artwork process
6. Provide members with skills for the future	Pilot Lean product development course was given to an OpEx team of Lean masters	As part of the Lean masters class, the first Kaizen event, the artwork process, was chosen
7. Change systems and structures	Execute all planned Kaizens with the goal of reducing time to market	Change manual process to electronic and revise the communication system to train all persons involved
8. Monitor and manage progress	Monitoring of all of the pre-determined metrics for each Kaizen event	Monitor metrics discussed during the Kaizen event

FIGURE 16.1
Eight steps for managing organization change for the Medical Supplies segment and in the artwork process.

electronic routing and input systems. Before the mapping session, it was agreed that all changes that were to happen to the artwork process should be managed with respect to the ultimate goal for the supplies segment. The steps for managing organization changes (Kotter, 2007) were tabled with respect to both the Medical Supplies segment and the artwork process itself (Figure 16.1).

Measure

Baseline and Current State Value Stream Map

Before the Kaizen event, both a current value stream map and a future value steam map were created. Team members (Figure 16.2) were asked to come

Functions Involved in Mapping and Kaizen Events	
Kaizen leader (3)	Information technology (2)
Corporate OpEx	Intellectual property
Artwork (3)	Program manager (2)
Regulatory	Packaging engineer
R&D	Corporate marketing
Marketing	

FIGURE 16.2
Functions involved in the Kaizen.

Kaizen event:	Artwork				
Goal:	Improve time and accuracy in the long lead time artwork process				
Current State Minimum Lead Time	**Current State Processing Time**	**Takt Time**	**Rework**	**Deployment Plan**	**Lean Tools Used**
38 weeks	21 weeks	2.4 hours	100%	1. Only one input meeting 2. Approver elimination 3. New E-system for submissions and approvals 4. New marketing input checklist	5S Visual management Hand-off diagrams Waste walks Spider diagrams Standard work Mistake proofing

FIGURE 16.3
Current state map metrics.

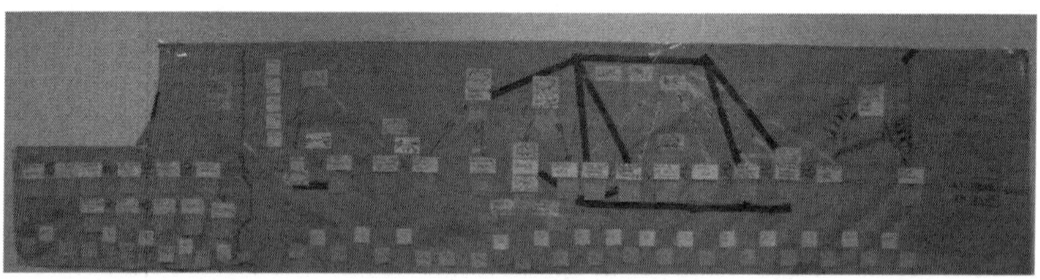

FIGURE 16.4
Final current state value stream map with 20 process steps.

to the event with notes on their ideas, thoughts, data, and suggestions for both the current and future states. The mapping event lasted 2 days, 1 day for the current state value stream map (Figure 16.3, current state metrics, and Figure 16.4, current state value stream map) and 1 day for the future state value stream map (Figure 16.5).

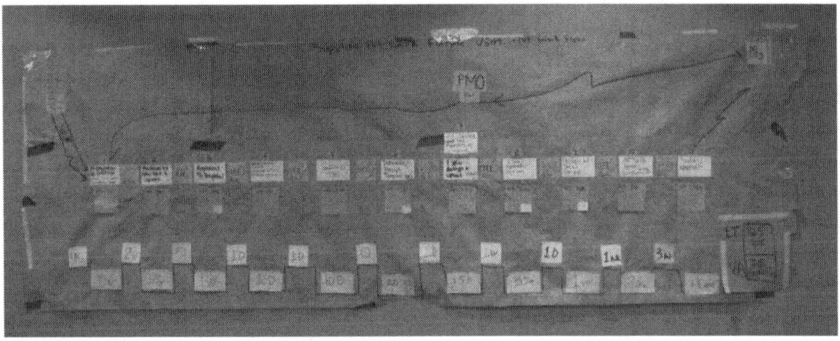

FIGURE 16.5
Final future state map with 11 processes.

Analyze

Five Principles of Lean

Before and during the course of laying out the future state map, the team took the five principles of Lean and constantly reiterated and evaluated them in terms of the artwork process. In accordance with the five principles and the artwork process, the five principles are as follows (Womack and Jones, 2003):

1. Specify value in the eyes of the customer: The value for the customer is the medical device being put to market quickly. The customer may pay for a high-tech device that is safe and effective, but they will not want to pay for and wait for the artwork that goes on that product.
2. Identify the value stream and remove the waste: Through value stream mapping of the future and current states, the waste was identified by the team in the current state and removed from the process in the future state.
3. Make remaining steps flow: In the current state, the hand-off diagrams were chaotic; removing the unnecessary steps allowed the hand-offs to be reduced by over half.
4. Let the customer pull just the value needed: Customers need the product as quickly as they can get it; therefore, everyone creating and initiating artwork must realize what must appear in the artwork through a more concise and fast process.
5. Pursue perfection: After the Kaizen itself, communication boards and continuing metric tracking allow senior management and all members involved to monitor progress of the PDP using a RACI (responsible, accountable, consulted, and informed) matrix.

Seven Tools of Lean Used in This Case Study

Seven tools, including a waste walk (Figure 16.6), were applied to help the team abide by the aforementioned principles and to help generate ideas for the future state map:

1. Hand-off diagram: Proposed a change from 48 to 28 hand-offs (Figure 16.7)
2. 5S: Proposed redesign of the artwork website and creation of a SharePoint site
3. E-system: Proposed electronic system to make all forms and sign-offs electronic
4. Mistake proofing: Train the trainer—helps train the entire Medical Supplies segment on the new changes
5. Spider diagrams: Aid in current versus future processes and changes (Figure 16.8)
6. Standard work: Include forms such as a marketing input checklist
7. Waste walk (Figure 16.6)

Team members performed a physical simulation of the artwork process to help come up with the current waste items.

Takt Time Calculations

For the artwork process, the Takt time is the rate at which artwork must be produced to satisfy average customer demand for new products and line

Waste Walk (Downtime Acronym)			
		Current Problem	**Future Solution**
D	Defects	100% rework	Create clear instructions for submitters—on artwork site
O	Overproduction	Designing working on all SKUs at one time	Create one code, check, then create the rest
W	Waiting	On all inputs from all different functions	Make all functions give their input in one meeting
N	Nonutilized talents	Busy on rework	Reduce rework percentage
T	Transportation	Physically going to meetings	One brand architecture meeting
I	Inventory	Not enough information to finish one piece	Make all inputs available before review by artwork
M	Motion	Many different meetings and sign-offs	Consolidate meetings
E	Extra processing	Phone calls, e-mails, meeting requests	Create electronic sign-offs and order for sign-off

FIGURE 16.6
Waste walk.

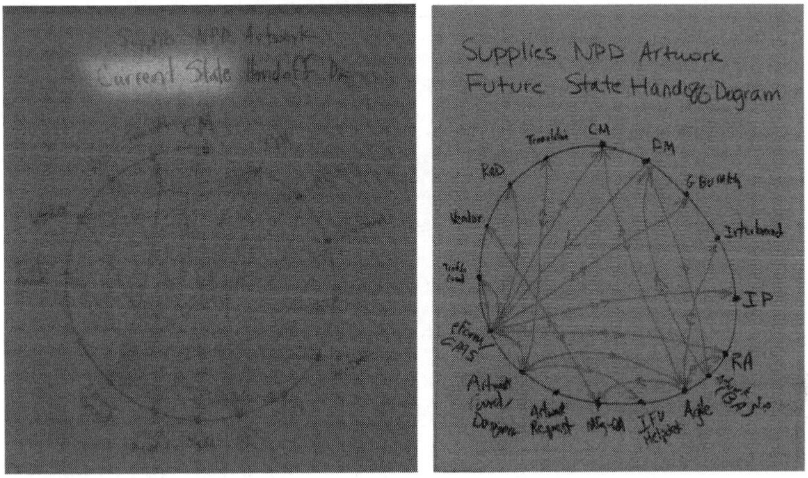

FIGURE 16.7
Hand-off diagram from current to proposed future state with a reduction of approximately 50%.

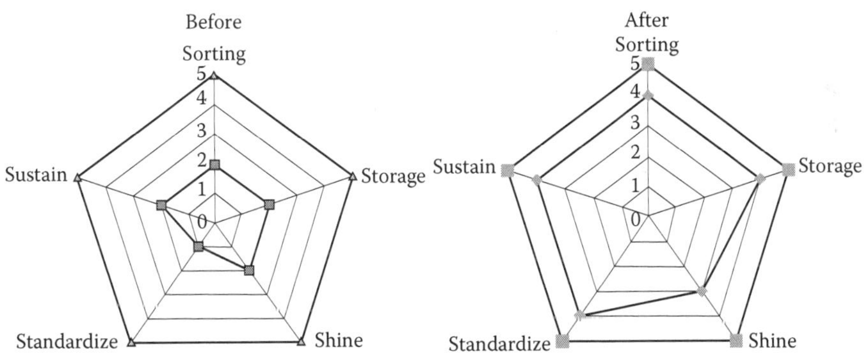

FIGURE 16.8
Spider diagrams.

extensions. In the current state, this was three pieces of artwork per day. However, a time calculation shows that one piece of artwork is needed every 2.4 hours. Therefore, the team must aim to better the future state to fulfill this requirement.

$$\text{Takt time} = \frac{\text{50 work weeks per year}}{\text{840 pieces of artwork per year}}$$

$$\text{Takt time} = \text{2.4 pieces of artwork needed per hour}$$

Average Lead Time	Maximum Lead Time	Minimum Lead Time
9.4 weeks	15.4 weeks	2.4 hours

FIGURE 16.9
Final metrics.

Improve

Implementation and Communication Plan

To ensure the new process (Figure 16.5) was effectively communicated across the business, a communication plan was created during the Kaizen and implemented before commencement of the new process. Training was conducted to ensure everyone who used the artwork process, and their managers, were trained on the following process.

Results

After 8 months of monitoring, the lead time was reduced to 9.4 weeks (Figure 16.9). The current state map was based on new products. A visual management board was put in place in the artwork department to track metrics. This board contributes to continued improvement and helps track statistics to ensure the process improvements that were implemented are working.

Conclusions and Results

The results of the Kaizen eliminated rework and reduced the cycle time to complete the deliverables. The artwork event reduced rework by having the process start earlier and ensuring that items needed later in the process were completed sooner. To this point, the long-term results of the artwork Kaizen indicate a significant decrease in cycle time (from 38 to 9.2 weeks). Although artwork is not always on the critical path of a project, it is recognized as an item that often appears on the critical path that is not necessarily critical to the product's success. When Lean tools are rolled out, sometimes the institutors use "new tools merely as substitutes, adding (instead of minimizing) interfaces, and changing tools, but not people's behavior" (Nepal et al., 2011).

It is important to keep in mind that this implementation plan only focused on the subprocesses of the Covidien Medical Supplies PDP. The strategy was intended to immediately address the low-hanging fruit, or the easiest and quickest problems to correct. Although very important in the development process, many of the Kaizen events focused upon were truly business processes. Future work should include scrutinizing the actual written PDP to try to remove any waste. In the future, it would be beneficial to apply all concepts of LPD to truly transform the culture to a Lean way of thinking.

References

Bowonder, B., Dambal, A., Kumar, S., and Shirodkar, A. (2010). Innovation Strategies for Creating Competitive Advantage. *Research and Technology Management*, May–June, 19–32.

ECRI (2006). Universal Medical Device Nomenclature System, http://www.ecri.org/Products_and_Services/Products/UMDSN/ (accessed August 2010).

Haque, B. (2003). Lean Engineering in the Aerospace Industry. *Proceedings Institution of Mechanical Engineers*, 217 (B), 1409–1420.

Hoppmann, J., Rebentisch, E., Domnrowski, U., and Zahn, T. (2011). A Framework for Organizing Lean Product Development. *Engineering Management Journal*, 23 (1), 3–15.

Kotter, J.P. (2007). Leading Change. Why Transformation Efforts Fail. *Harvard Business Review*, January, 92–107.

Kumar, S., and Bauer, K. (2010). Exploring the Use of Lean Thinking and Six Sigma in Public Housing Authorities. *Quality Management Journal*, 17 (1), 29–46.

Leon, H., and Farris, J. (2011). Lean Product Development Research: Current State and Future Directions. *Engineering Management Journal*, 23 (1), 29–49.

Martinez, C., and Farris, J.A. (2009). Lean Product Development: A Systematic Literature Review. *Proceedings of the IERC 2009*, Miami, FL, June.

Nepal, B., Yadav, O., and Solanki, R. (2011). Improving the NPD Process by Applying Lean Principles: A Case Study. *Engineering Management Journal*, 23 (1), 52–68.

Panchak, P., (2004). Pella Drives Lean Throughout the Enterprise; It's Not Just for the Shop Floor Anymore. *Industry Week*, June, http://www.industryweek.com/companies-amp-executives/best-practices-pella-drives-lean-throughout-enterprise.

Womack, J., and Jones, D. (2003). *Lean Thinking: Banish Waste and Create Wealth in Your Corporation*. Revised and Updated. New York: Free Press, a Division of Simon and Schuster.

Zak, A., and Waddell, B., (2011). *Simple Excellence: Organizing and Aligning the Management Team in a Lean Transformation*. New York: CRC Press.

17

Motor Grader Assembly Line Modification

Mujahid Abjul, Charlie Barclay, Nanday K. Dey, Amita Ghanekar, and Lynda Melgarejo

CONTENTS

Introduction

This chapter covers the work that was done to make the enclosure assembly line at Global Grader Company (GGC) a much more efficient Lean system. The team spent several months analyzing the current line configuration and then went through a process to optimize and improve the line efficiency, allowing for a reduction of waste, unneeded labor, idle time, and the space required to produce the enclosure. This case study will detail the process that the team went through in obtaining these results.

Project Overview

Global Grader Company (GGC) is a 50-year-old company located in the state of Arkansas that produces construction and mining equipment within the United States and international markets. With global sales increasing, GGC has identified that the current capacity for their enclosure assembly module of 5 per day in their motor grader production line is no longer able to procure enough modules to keep up with the new and forecasted capacity required in an 8-hour shift.

This project addresses improving the production efficiency of the GGC. The motivation for this project is to increase the operating efficiency of the enclosure assembly line at GGC, to maintain and improve their position as a global leader in the heavy construction equipment market.

Define

In this phase, a project charter was developed for the project team. In an effort to address the capacity restrictions of the current enclosure assembly line, GGC has tasked their manufacturing engineering (ME) team with the development of a preliminary larger assembly line that can sustain the production of 10 fully tested, enclosure modules per 8-hour day. The ME team has developed a new layout and assembly process breakdown to expand the current three work zones and one miscellaneous subtable to seven work zones and one miscellaneous subtable to accommodate all enclosure module sizes and attachments that are ordered per the customers' specifications.

With the facility's electrical and air installations scheduled to be ready in 60 business days (3 months), GGC has tasked an external contractor, the "Global Lean" Team, with reviewing the expansion plan and identifying improvements to the process efficiency for implementation of the new enclosure assembly line layout. The Global Lean Team must analyze the process and, through the use of Lean manufacturing tools, ensure that the new process breakdown provided by the ME team achieves an efficiency rated higher than the proposed 80%.

Problem Statement

The problem statement is to analyze the proposed fabrication line (aka 10EPP) utilizing Lean manufacturing tools for the enclosure assembly module and ensure the process breakdown is improved. The goal is to increase efficiency above the original 80% through the validation of layouts, process breakdowns, and line balancing activities for the implementation of a new enclosure assembly line.

The expansion plan should take into consideration the best locations for all torque tooling, lifting devices, standard work postings, special assembly fixtures, material presentation, and computer station installations to reduce assembly waste throughout the process.

Problem Objective, Goals, and Requirements

The final process breakdown analysis and layout proposal for this project were to be delivered in 40 business days. Project implementation was to include future value stream management (VSM) validation, Takt time versus cycle time analysis, and process breakdowns to achieve an efficiency level greater than 80%. The enclosure assembly module capacity for 10EPP was fully validated and ready to run full production in one 8-hour shift by the end of this project.

This process was initiated because GGC is experiencing a capacity growth that exceeds their daily production rates. To continue the forecasted production of 10EPP motor graders, it was imperative for GGC to ensure the enclosure assembly line is implemented to support the product flow since it had become a bottleneck to daily operations.

The current production constraint was driving the opening of a second shift to eliminate the enclosure module bottleneck that translates to an expense of $25,000 per week in variable costs for assemblers, electricity, supervisor salary, and other expenses. Finally, enclosure module production would ensure 10EPP motor graders were produced per day. The 10EPP motor graders have a margin of over 300% and a sales forecast that shows future growth of 100% (double today's volumes) in the next 2 years. This all leads to revenue of more than $10 million annually.

Project Boundaries

1. All tooling, fixtures, and equipment locations must be validated in the final layout proposal for installation.
2. Area layouts will include assembly work zones for material delivery—inbound/outbound locations.
3. All process parameters must meet safety requirements.
4. A final process breakdown with efficiency above 80% will be delivered along with a future VSM and a detailed schedule of project activities for timely project completion.

The material delivery, kits, and any other logistics and supply chain activities are outside the scope of this project and are to be coordinated by those respective departments.

A process flow diagram was created to gain an understanding of the product line layout. Figure 17.1 illustrates the process flow diagram, which describes the flow on the facility/machine layout.

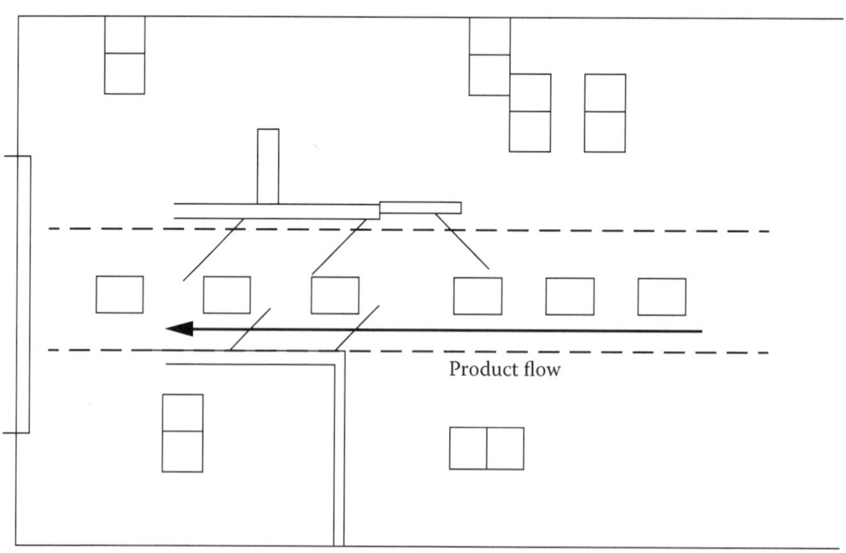

FIGURE 17.1
Process flow diagram.

Project Team

Amita Ghanekar has been selected as the lead for this project by the GGC Facility Manager. She has been appointed a team of four SMEs (subject matter experts) from different departments that will report directly to her for the duration of his project.

Their names and areas of expertise are listed as follows: Mujahid Abdul—Operations Management, Lynda Melgarejo—Manufacturing Engineering, Richard Barclay—Quality Control, and Nanda Kumar Dey—Facilities Engineering.

Project Plan

The new enclosure assembly line will be located in the same location as the current one. However, two more 20′ × 20′ bays will be added to increase the real estate for the new work zones. The ME team has already developed a preliminary base layout, performed time studies, and completed the initial work balancing content for each zone, as well as placed purchase orders for all added tooling and special fixtures needed to support the new line. The Global Lean Team will analyze those proposed breakdowns and will deliver a process breakdown validation and final layout design that has an efficiency above the initial 80% provided by the ME team.

Since the current enclosure assembly line is constrained, the Global Lean Team has proposed to have GGC running in two shifts to keep up with daily production needs and to start building enough enclosure module inventories to provide an opening of 5 business days to execute the new layout implementation. The timely scheduling and coordination of time and resources for implementation of these changes is vital for the continuous growth and success of the GGC business.

Lean Metrics Defined

The Global Lean Team has been contracted to improve the initial process breakdown proposed by GGC's Manufacturing team and improve the efficiency of the processes above 80%. The following Lean metrics will be used to assess the benefits from this project: (1) current versus future value stream maps; (2) cycle time versus takt time (41 minutes)—including identification of value-added and non-value-added activities; (3) process breakdown and line balancing; (4) current and future layout; and (5) project schedule.

Measure

The goal of the Measure phase was to learn about the current process and to fully understand both how it works and how well it works. This entails three key tasks: creating a detailed process map, gathering baseline data, and summarizing and analyzing the data. The process map was created first so that information gleaned from it can guide the data collection process.

Due to the nature of this process, a Six Sigma value stream map was created using direct input from the GCC representative who participated in the project. This is not the time to incorporate ideas for what can be done differently. The scope of the process under study was determined as part of the Define phase, so the process mapping effort during Measure should have those same delimiters.

The process map (Figure 17.2) shows the high-level current processes and the steps involved in the assembly of the enclosure. The process is divided into 13 broad steps; each step contains several substeps in it. In total, there are 95 process steps. The operations are performed in different zones,

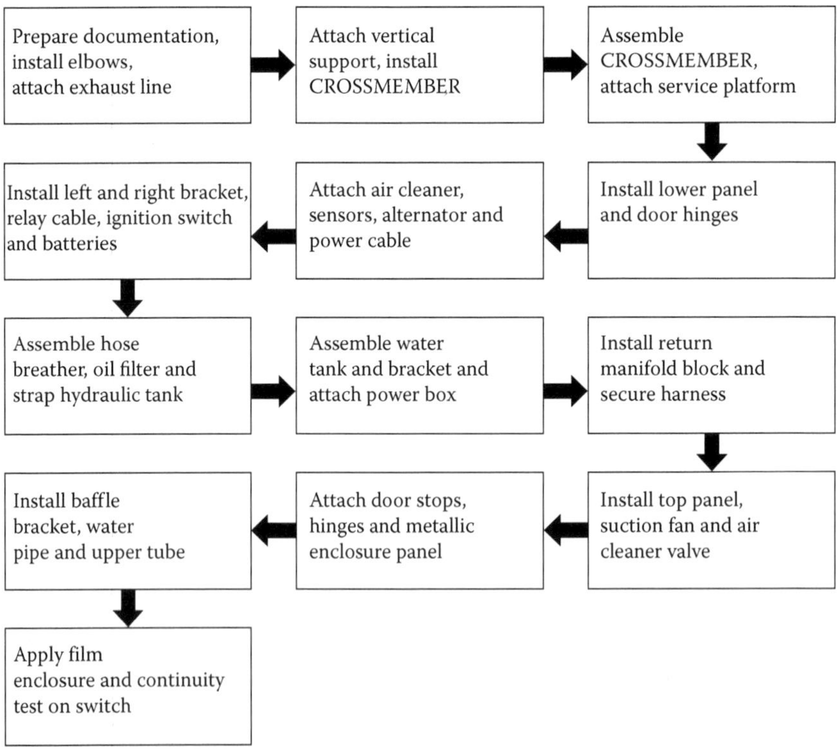

FIGURE 17.2
Process map.

totaling seven zones. The zones are further broken into suboperations. The process map in Figure 17.2 gives an understanding of the assembly process of the enclosure at the GGC.

Process Baseline

The baseline statistics are based on the current process. It was mentioned in the "Measure" section that the entire process is divided among seven zones. The entire process of the enclosure assembly consists of 629 steps. We identified the value-added steps and the non-value-added steps. There were steps that were necessary but still non-value-added.

Total Time—412.73 minutes
Total Value-Added Time—91.33 minutes
Total Non-Value-Added Time—21.27 minutes
Total Necessary Non-Value-Added Time—300.14 minutes

From the time study and analysis conducted on the baseline statistics, it was found that 22% of the time in the entire process is value-added, 5% of the time is non-value-added, and the remaining 73% of the time is considered non-value-added but necessary, as shown in Figure 17.3.

Total number of steps—629
Value-added steps—179
Non-value-added steps—39
Necessary non-value-added steps—454

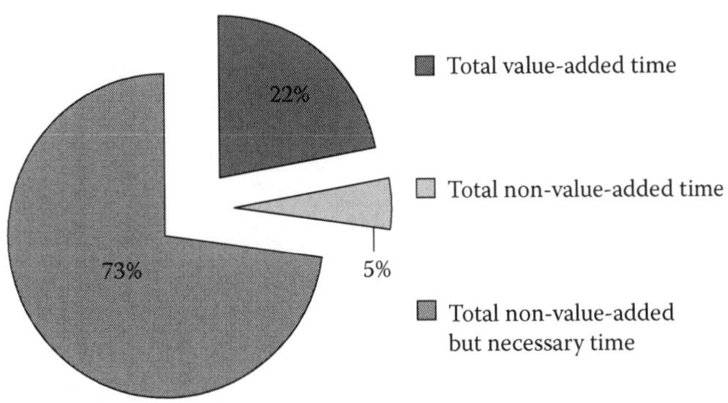

FIGURE 17.3
Total assembly time breakdown (min).

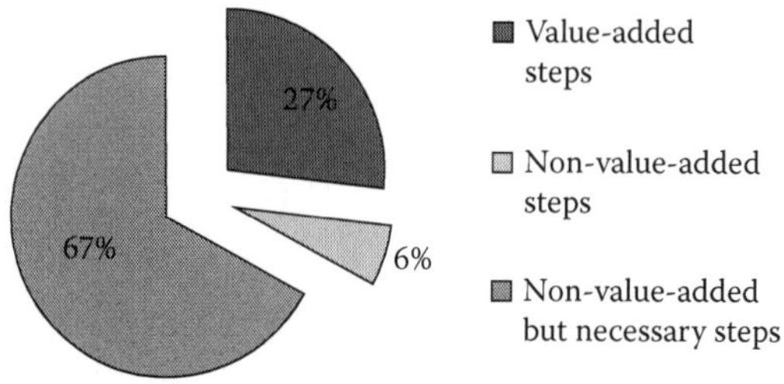

FIGURE 17.4
Breakdown of assembly steps.

Figure 17.4 shows a pie chart for the preceding data. From the total steps, 27% of the steps are value-added, 6% of the steps are non-value–added, and 67% of the steps are necessary non-value-added.

Current State VSM

Value stream mapping is a Lean manufacturing technique used to analyze and design the flow of materials and information required to bring a product or service to a consumer. It is a fundamental tool to identify waste, reduce process cycle times, and implement process improvement. The value stream map shown in Figure 17.5 helped in identifying the Takt time and cycle time and to make a comparative study of the available time for the entire process. It also aided in balancing the entire process in the later stages. The value stream map contains all seven zones along with the suboperation.

Analyze

The goal of the Define–Measure–Analyze–Improve–Control (DMAIC) Analyze phase is to identify potential root causes for the process problems being addressed and then confirm actual root causes with data. Having completed the Measure phase, our team had already established a clear problem statement that specifies what the problem is and under what circumstances it occurs. We had already gathered and analyzed data to establish the baseline performance of the process, relative to the critical to quality measures established based on customer input.

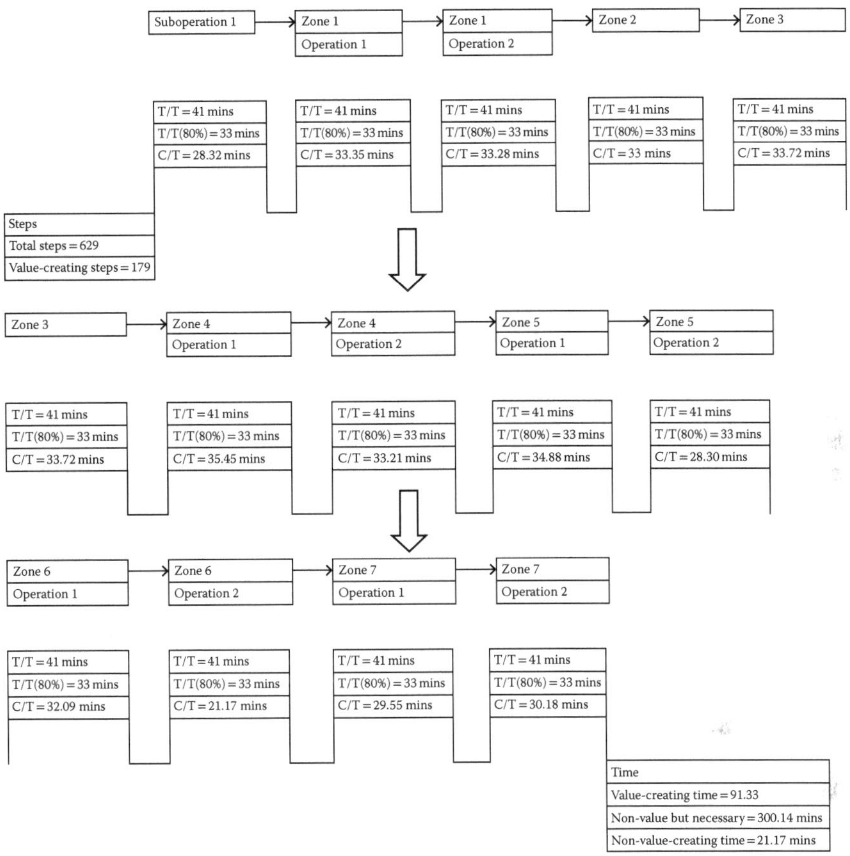

FIGURE 17.5
Value stream map.

Process Gaps and Sources of Variation

In many cases, clues to the factors affecting performance are already available based on the work that was performed in the Define and Measure phases. Perhaps the team demonstrated that the problem is isolated to one group, and they know that group is using older equipment. Or analysis of the process gaps may have revealed some fairly obvious sources of inefficiency and delays in the process.

However, this is not sufficient to confirm what is causing the problems for two reasons. One is that, as in all phases of the DMAIC, suspicions and hypotheses must be confirmed with data. Not only must the team confirm that these factors are present, they must also confirm that changes in these factors substantially impact the outcome. The other is that the goal of the Analyze phase is to determine root causes, which requires digging deeper than what is apparent on the surface.

As for our project, the process gaps were identifying the value-added and the non-value-added activities. However, there were some activities that were non-value-added but necessary for the process. We cannot ignore those activities as those are an important part in producing the motor graders.

As discussed in the "Problem Statement" section, we were trying to increase the available space for manufacturing the motor graders. The current space was three work zones and a single miscellaneous work table. To satisfy our goal of creating seven work zones and one miscellaneous table, we analyzed various problem areas. Analyzing the various sources of variation can be done as follows.

We assessed root causes related to the following categories of the 6M's: manpower, machine, measurement, mother nature, management, and material. Having performed a detailed study of these factors, we created a cause and effect diagram to give a clearer idea of the causes. This helped us to identify the sources of variation and helped in minimizing them by applying Lean tools.

Prioritize Improvement Opportunities

Looking at the process data collected, our team decided that line rebalancing will help achieve our objectives. We developed a breakdown of each single operation or process and calculated the total time taken. From the statistical analysis, we derived the total assembly time breakdown as shown in Figure 17.6.

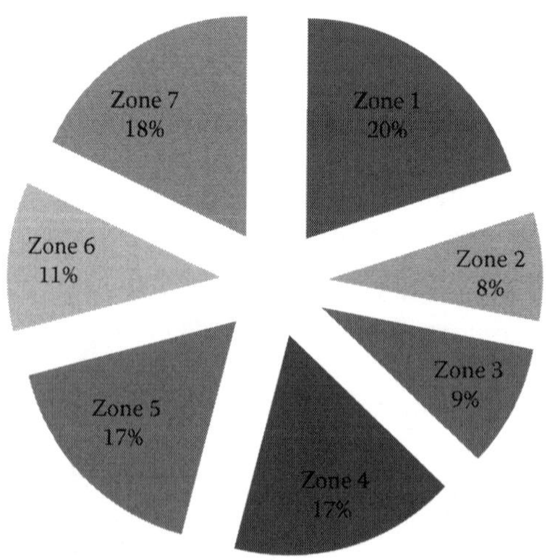

FIGURE 17.6
Total assembly time breakdown.

The pie chart in Figure 17.6 shows that the work is not fully balanced throughout the process. For example, Zone 1, Zone 4, and Zone 7 take more time than any other zones. This implies that we need to rebalance the work in different zones. This, in turn, will increase the efficiency of production operations and also reduce some dependency of operations on each other. For this purpose, we need to find the root cause of the time issues and split them wisely. Also, we can propose some changes in the current layout and implement it using Lean tools.

Cause and Effect

Once a list of potential root causes has been compiled, the next step is to organize them in a way that makes it easier to prioritize and assess them. Several tools can be used to accomplish this. The most popular is a fishbone diagram or Ishikawa diagram, which uses a display resembling the bones of a fish to categorize potential causes and illustrate the levels of causation. The main bones are used to reflect the 6M high-level categories. Once the diagram has been created, the project team reviews it to determine which seems to be the most likely potential root causes and to identify any that seem to have consequences in more than one area. These are good candidates for the validation process. Figure 17.7 depicts the cause and effects for our project.

In some cases, sufficient data is available from the Measure phase to conduct cause and effect analyses during the Analyze phase. Often, however, it is necessary to collect new data so that the relationship between the suspected root causes and the effect under study can be evaluated. As in the Measure phase, the methods used to analyze the data depend on the type of data collected.

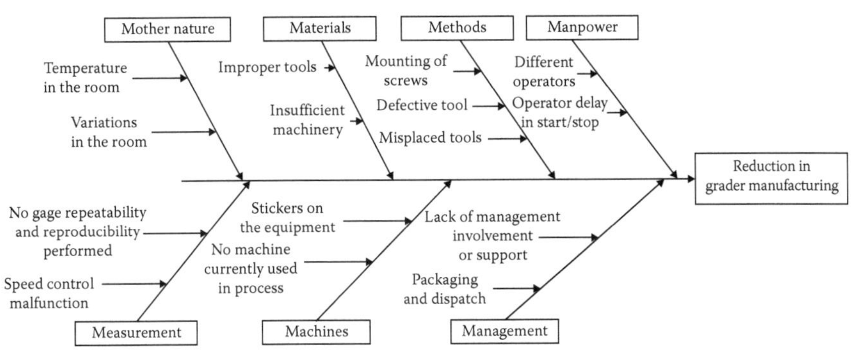

FIGURE 17.7
Cause and effect diagram.

Statistical and Graphical Analysis

Figure 17.8 represents the total time taken by different zones. The time is represented as value-added activities, non-value-added activities, necessary non-value-added activities, and total time. Thus, from this histogram it is clear that the necessary non-value-added activities take considerably more time than the value-added activities. Although we cannot avoid these activities, we can definitely try and reduce the total time consumed by applying Lean techniques.

For the graphical analysis, we calculated the time required for individual work zones. For example, consider Zone 2 as shown in Figure 17.9. As we can see in the pie chart, Zone 2 has 80% of the total time consumed for necessary non-value-added activities. On the contrary, only 5% of the total time is utilized for the value-added activities. Hence, by analyzing the process

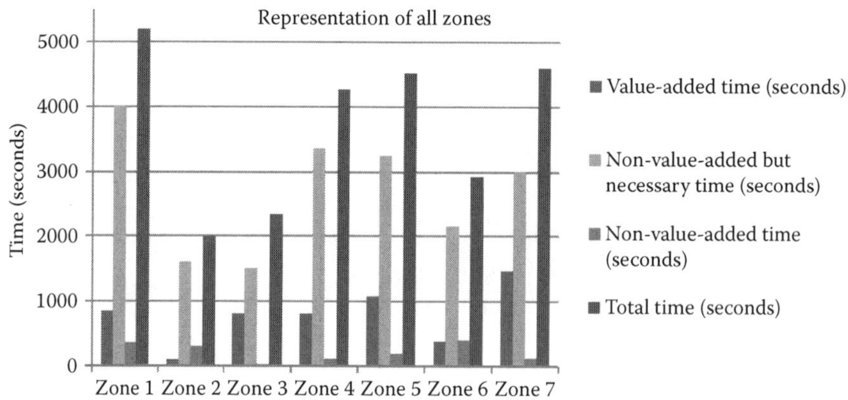

FIGURE 17.8
Zone classification.

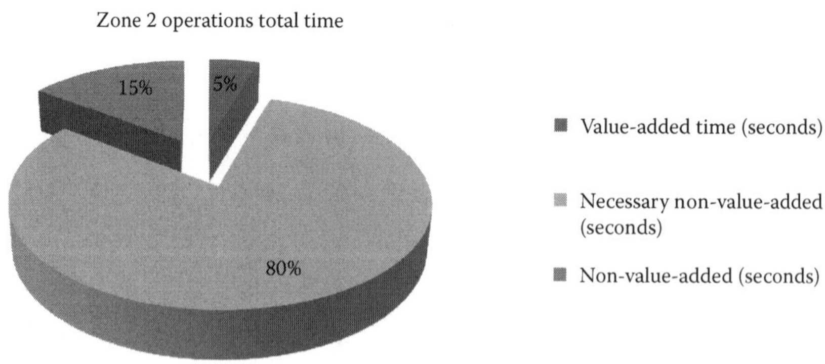

FIGURE 17.9
Pie chart for Zone 2.

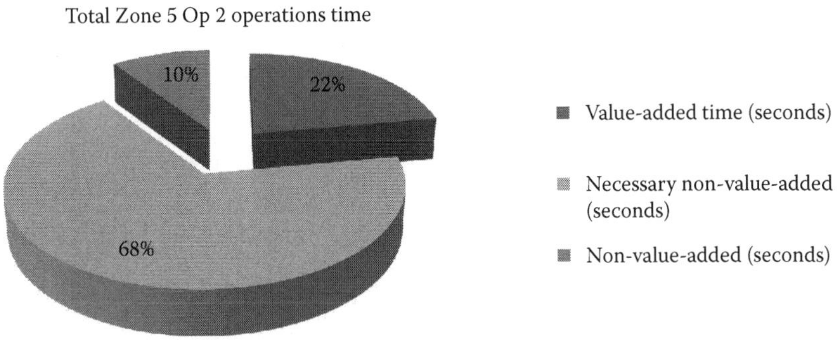

FIGURE 17.10
Pie chart for Zone 5 operation 2.

in detail, we conclude that we can reduce the time taken for necessary non-value-added activities as they cannot be avoided. Another example is Zone 5 operation 2 as shown in Figure 17.10. The total time taken by necessary non-value-added activities exceeds the time taken by value-added activities. Thus, doing graphical analysis helped us understand the time required for each process or operation in respective zones and where can we improve.

Improve

After careful examination of the data collected, it was determined that the amount of non-value-added and necessary non-value-added time was already minimal. There are some slight improvements that could be made in that regard, but our team made the decision to concentrate on the low hanging fruit, which we believe lies in the current assembly line unbalance. For this reason, we decided to move forward with the intention of eliminating waste by bringing the cycle times for all zones near the stated takt time.

To fully understand the problem, and to develop a plan for adjusting the zone loading, the first step was to study the current assembly line breakdown. Figure 17.11 is representative of the current assembly line setup. The areas listed to the right are the zones that are referred to in this case study. Each zone matches up with at least one operation on the left. The image only shows approximately one dozen operations, but there were, in total, nearly 100 operations. The takt time is shown on the right of the image, along with the current operating cycle time of 33 minutes, which is 80% of the takt time. In addition to the zones and operations, there was a detailed breakdown of each operation that listed all of the individual tasks that need to be completed to finish the operation and the time associated with each task. Figure 17.12 shows the breakdown of the "ASSEMBLE CROSSMEMBERS" operation, which is listed at the top of the operations list in Figure 17.11.

M2 Enclosure Sub						
	TIME (MINS)		1980 Target			
Operation	**Iter :1**	**Area**		**Cycle**	**Takt**	**80%**
ASSEMBLE CROSSMEMBERS	*379.50*	*Sub Op 1*	*1733.00*	28.88	41	33
ASSEMBLE CROSSMEMBERS	*379.50*	*Zone 1 Op1*	*2001.00*	33.35	41	33
ATTACH VERTICAL SUPPORTS	*120.00*	*Zone 1 Op2*	*1997.00*	33.28	41	33
ASSEMBLE RIGHT VERTICAL SKELETAL SUPPORT	*286.00*	*Zone 2*			41	33
ASSEMBLE RIGHT VERTICAL SKELETAL SUPPORT	*135.00*	*Zone 3*	*2023.00*	33.72	41	33
INSTALL FRONT CROSS MEMBER	*98.00*	*Zone 4 Op 1*	*2127.00*	35.45	41	33
INSTALL MIDDLE CROSS MEMBER	*133.00*	*Zone 4 Op 2*	*1992.80*	33.21	41	33
INSTALL MIDDLE CROSS MEMBER	*132.00*	*Zone 5 Op 1*	*2093.00*	34.88	41	33
ATTACH SERVICE CENTER PLATFORM	*583.00*	*Zone 5 Op 2*	*1698.00*	28.30	41	33
TORQUE SEQUENCE	*279.00*	*Zone 6 Op 1*	*1925.50*	32.09	41	33
MOVING STAND TO CART	*655.00*	*Zone 6 Op 2*	*1270.00*	21.17	41	33
FRONT - AIR CLEANER SUB ATTACH PLENUM	*169.00*	*Zone 7 Op 1*	*1773.00*	29.55	41	33
ATTACH SENSORS TO CLEANER GP	*73.00*	*Zone 7 Op 2*	*1811.00*	30.18	41	33
PREPARE CEM	*30.00*					
ATTACH EXHAUST LINE TO CEM	*217.00*					
PRECLEANER GROUP HOSE PLUG	*22.00*					
INSTALL FITTINGS TO CEM	*151.00*					
INSTALL AND SECURE AIR INTAKE TUBE	*355.00*					

FIGURE 17.11
Current assembly line setup overview.

Once all of the information was understood, a large amount of time was devoted to sorting and organizing the current workloads and in determining an appropriate redistribution of work. The goal of this project was to bring the current target cycle time of 33 minutes (80% of takt time) closer to the 41-minute takt time. This meant that the first thing to do was to determine what the appropriate number of zones would be to achieve the 41-minute takt time. This was done by simply dividing the total production time by the takt time. From this, it was determined that a total of 10 operators in 6 zones would be required, as opposed to the 12 operators in 7 zones previously needed. Once this information was obtained, then the line rebalancing could begin.

To facilitate the line rebalancing, a separate Excel sheet was created. This sheet was used to sort each operation by the zone that it was currently in and all of the associated tasks in the operation breakdown sheet. This allowed for easier identification of areas of imbalance and aided in the ability to quickly try different line setups. The original list that was created can be seen in Figure 17.13. Figure 17.14 shows the same operations after rebalancing. Note that several operations have been moved from Zone 1 operator 1 to sub operator 1. These images show just a snapshot of the operations analyzed. These images show approximately 30 of the nearly 700 operations that were analyzed in the completion of this project.

The process of rebalancing the lines consisted of a lot of time spent trying different configurations through many iterations before a satisfying result was found. The changes that were made to the assembly line work breakdown can be seen in Figure 17.15 compared to Figure 17.3. For instance, the "Attach hinges to frame" operation was moved from Zone 6 operator 2 to Zone 5 operator 2. Additionally, the "Secure fan wires"

Work center:		SAP OPERATION NUMBER:			5/010		
		OPERATION NUMBER:			UJ- 0000026300/UJ- 0000026301		
Operation	ASSEMBLE CROSSMEMBERS	Work area :			UJW00R24		
S.No	Elements	Tools used	Elemental time for activities	Value-added time (seconds)	Non-value-added but necessary (seconds)	Non-value-added (seconds)	Total time (seconds)
1	Pulling supports to build				53		53
2	Obtain bolts and parts				80		80
3	Install bolts					183	183
4	Remove bolts and paint					14	14
5	Install bolts				96		96
6	Get paint marker and paint bolts					55	55
7	Get tooling/painting					44	44
8	Torque bolts			95			95
9	Pulling parts to install				64		64
10	Install cross member to build stand				75		75

FIGURE 17.12
Operation breakdown for assembly step "ASSEMBLE CROSSMEMBERS."

	Zone-op	Elements	Value-added time (seconds)	Necessary non-value-added (seconds)	Non-value-added (seconds)	Total time (seconds)
Assemble crossmembers	*Zone 1 Op2*	Pulling supports to build		53		53
Assemble crossmembers	*Zone 1 Op2*	Obtain bolts and parts		80		80
Assemble crossmembers	*Zone 1 Op2*	Install bolts			183	183
Assemble crossmembers	*Zone 1 Op2*	Remove bolts and paint			14	14
Assemble crossmembers	*Zone 1 Op2*	Install bolts		96		96
Assemble crossmembers	*Zone 1 Op2*	Get paint marker and paint bolts			55	55
Assemble crossmembers	*Zone 1 Op2*	Get tooling/painting			44	44
Assemble crossmembers	*Zone 1 Op2*	Torque bolts	95			95
Assemble crossmembers	*Zone 1 Op2*	Pulling parts to install		64		64
Assemble crossmembers	*Zone 1 Op2*	Install cross member to build stand		75		75
Attach vertical supports	*Sub Op.1*	Obtain vertical supports and attach one side		95		95
Attach vertical supports	*Sub Op.1*	Walking		6		6
Attach vertical supports	*Sub Op.1*	Attach other vertical support		19		19

FIGURE 17.13
Step list created to aid in the line rebalancing.

	Zone-op	Elements	Value-added time(seconds)	Non-value added but necessary (seconds)	Non-value added(seconds)	Total Time (seconds)
Assemble cross members	*Zone 1 Op2*	Pulling supports to build		53		53
Assemble cross members	*Zone 1 Op2*	Obtain bolts and parts		80		80
Assemble cross members	*Zone 1 Op2*	Install bolts			183	183
Assemble cross members	*Zone 1 Op2*	Remove bolts and paint			14	14
Assemble cross members	*Zone 1 Op2*	Install bolts		96		96
Assemble cross members	*Zone 1 Op2*	Get paint marker and paint bolts			55	55
Assemble cross members	*Zone 1 Op2*	Get tooling/painting			44	44
Assemble cross members	*Zone 1 Op2*	Torque bolts	95			95
Assemble cross members	*Zone 1 Op2*	Pulling parts to install		64		64
Assemble cross members	*Zone 1 Op2*	Install cross member to build stand		75		75
Attach vertical supports	*Sub Op.1*	Obtain vertical supports and attach one side		95		95
Attach vertical supports	*Sub Op.1*	Walking		6		6
Attach vertical supports	*Sub Op.1*	Attach other vertical support		19		19
Assemble right vertical skeletal support	*Sub Op.1*	Walking gathering		0.59		1
Assemble right vertical skeletal support	*Sub Op.1*	Install		6		6
Assemble right vertical skeletal support	*Sub Op.1*	Walking/install bolts hand tight		279		279
Assemble right vertical skeletal support	*Sub Op.1*	Walking gathering		41		41
Assemble right vertical skeletal support	*Sub Op.1*	Install bolts	69			69
Assemble right vertical skeletal support	*Sub Op.1*	Walking		25		25
Install front cross member	*Zone 1 Op1*	Install front cross member		48		48
Install front cross member	*Zone 1 Op1*	Walking		10		10
Install front cross member	*Zone 1 Op1*	Install	40			40
Install middle cross member	*Sub Op.1*	Install		43		43
Install middle cross member	*Sub Op.1*	Paint		47		47
Install middle cross member	*Sub Op.1*	Walking/gathering		28		28
Install middle cross member	*Sub Op.1*	Preping	15			15
Install middle cross member	*Sub Op.1*	Torque				0

FIGURE 17.14
Future state step list resulting from the line rebalancing.

operation was moved from Zone 7 operator 1 to Zone 6 operator 2. The final results from the rebalancing of the assembly line can be seen graphically in Figure 17.16.

The top graph depicts the takt time breakdown for the assembly line in its current state, and the bottom graph depicts the takt time in the future state. You will notice on the top graph there are three separate series. The first series is the actual time required for each zone. The second series represents the target or takt time. And the third series represents the current operational target of 80% of the takt time or 33 minutes. The bottom graph clearly illustrates the gains that were made through the course of this project. The proposed cycle time for each zone is very close to the stated takt time of 41 minutes. As shown in the graph, Zone 7 Op 1 has a cycle time of 41.5 minutes. This could be further improved by eliminating some of the non-value-added time (get tools, parts, etc.) from the operator's responsibility and giving the responsibility to an assembly line service worker.

	Elements	Value-added time (seconds)	Necessary non-value-added (seconds)	Non-value-added (seconds)	Total Time (seconds)
Assemble crossmembers	Pulling supports to build		53		53
Assemble crossmembers	Obtain bolts and parts		80		80
Assemble crossmembers	Install bolts			183	183
Assemble crossmembers	Remove bolts and paint			14	14
Assemble crossmembers	Install bolts		96		96
Assemble crossmembers	Get paint marker and paint bolts			55	55
Assemble crossmembers	Get tooling/painting			44	44
Assemble crossmembers	Torque bolts	95			95
Assemble crossmembers	Pulling parts to install		64		64
Assemble crossmembers	Install cross member to build stand		75		75
Attach vertical supports	Obtain vertical supports and attach one side		95		95
Attach vertical supports	Walking		6		6
Attach vertical supports	Attach other vertical support		19		19
Assemble right vertical Skeletal support	Walking gathering		0.59		1
Assemble right vertical Skeletal support	Install		6		6
Assemble right vertical Skeletal support	Walking/ install bolts hand tight		279		279
Assemble right vertical Skeletal support	Walking gathering		41		41
Assemble right vertical Skeletal support	Install bolts	69			69
Assemble right vertical Skeletal support	Walking		25		25
Install front cross Member	Install front cross member		48		48
Install front cross Member	Walking		10		10
Install front cross Member	Install	40			40
Install middle cross Member	Install		43		43
Install middle cross Member	Paint		47		47

FIGURE 17.15
Changes made to the assembly line during rebalancing.

Solution Implementation Plan

The implementation of this plan will take place at the GGC's Little Rock Assembly facility. The plan will be handed off to the facilities engineering team to make the necessary changes to the layout and tooling locations to complete the changeover. The Global Lean Team will provide support and guidance whenever necessary.

Cost/Benefit Analysis

The cost of implementing this plan will be very minimal. All of the facility's changes will be completed by the facilities engineering team at GGC. The facility alterations will not affect the production schedule of GGC. Also, any labor

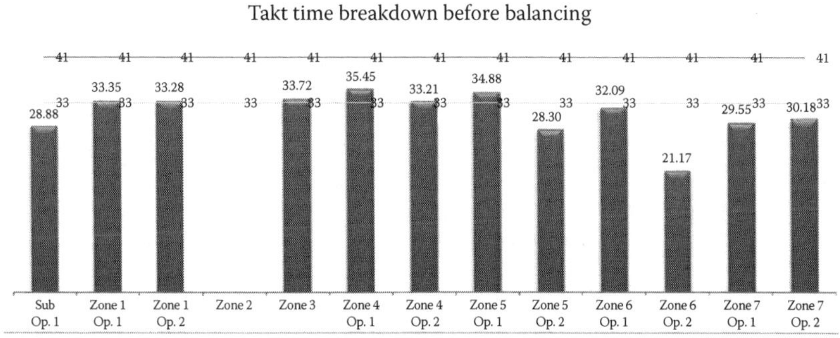

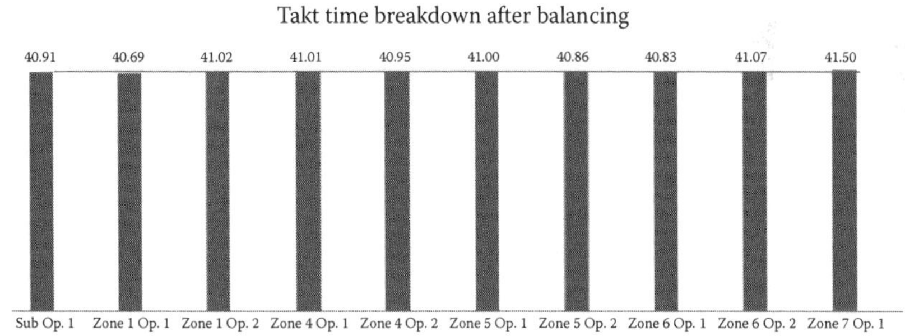

FIGURE 17.16
Takt time breakdown before and after balancing.

costs associated with these changes will be offset by the freeing up resources from the reduced labor requirements in the proposed layout. There will be three fewer operators required to complete the assembly in the proposed layout, and while we will recommend that the company relocate these employees to another department, they should be considered a cost savings of this improvement. Based on our estimations, the changes to the facilities should be completed in approximately 6 weeks, at a cost of approximately $80k. This cost will be offset with the savings incurred by the reduction of assembly personnel by three. At an average yearly salary of $45k, this savings will total to $135k/year, more than covering the implementation costs. On the contrary, these assembly personnel could also be used as extra resources elsewhere within the company.

Control

This is the fifth and final phase of the DMAIC process. It is crucial to ensure that the changes enforced in previous stages are in control; we cannot afford to lose the efforts and resources put in to bring the amendments. There cannot be enough emphasis placed on the importance of devoting the same high level of energy and commitment throughout this phase. Insufficient efforts in this phase may lead the process back to the former stage, losing all the success of precedent stages.

We can adopt significant key performance indicators (KPIs) to control the new environment. KPIs give the status of the process. KPIs could be as simple as periodic Takt time measurements. Our major objective in the project is to rebalance the manufacturing line of the GGC and maintain a constant takt time throughout the line. This has already been achieved during the Improve phase of the implementation. Even though there have been no major changes to the manufacturing method, there is a clear improvement before and after the changes as shown in Figure 17.16.

Dashboards

Figures 17.17 and 17.18 are examples of performance measurement where system activities have been stated with each activity classified into different categories. This study helped us to know the actual state of the system

Elements	Value-added time (seconds)	Necessary non-value-added (seconds)	Non-value-added (seconds)	Total time (seconds)
Assemble crossmembers	95	368	296	759
Attach service center platform	0	583	0	583
Moving stand to cart	0	655	0	655
Total Zone 1 Op 2	95	1606	296	1997

FIGURE 17.17
Details of an operation in the system.

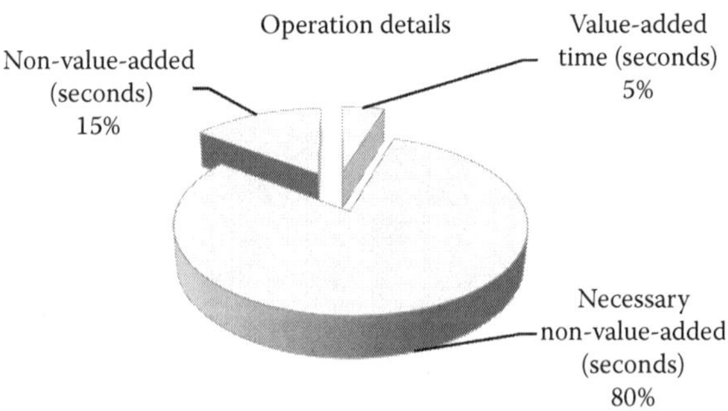

FIGURE 17.18
Pie chart showing the classification of operations.

FIGURE 17.19
Team doing a gemba walk in the GGC plant.

and the information we extracted is used to devise our action plan. After having the facts and performance reports in our hands, objectives and their respective benchmarks have been decided by dividing the system into further parts to execute a constant takt time plan. It was also decided to increase the production from 5 modules/day to 10 modules/day.

Before execution of any plans, all the information about the work culture and the environment of the system is extracted by multiple GEMBA walks. During this step, the team made sure to cover all areas of the plant including the assembly line. All the individual observations were gathered and documented. Figure 17.19 shows the team doing a gemba walk at the plant.

Cost estimation was completed to set up the new assembly line and resources were allocated through company management. The process is well-defined as explained in the previous stages, which completes the Hoshin planning.

Mistake Proofing

We have ensured that poka-yoke is used during production operations. Figure 17.20 shows a picture of a module frame. Looking at this frame closely one can see that holes are drilled using different types and sizes such that that no screw can be wrongly placed and also that holes are drilled in proper orientation to avoid any mistake from the operator. All the screw sizes and types are again mentioned in the standard work document for the convenience of the operator. The mistake proofing is successfully implemented (up to test) at the storage location where chances are very high that parts get interchanged or misplaced. As shown in Figure 17.21, a proper tray was ordered with the exact dimension of particular parts, which completely avoids misplacement of parts.

To act as a control plan, as per the mutual decision of the team, the assembly line will be further divided and a constant takt time will be maintained throughout. As shown in Figure 17.22, the operations are now divided into five work zones and one suboperation.

FIGURE 17.20
Frame of the module.

FIGURE 17.21
Example of poka-yoke at parts storage.

2460 Target

Area		Cycle	Takt	% Takt Deviation
Sub Op 1	2454.69	40.91	41	−0.215853659
Zone 1 Op 1	2441.18	40.69	41	−0.76504065
Zone 1 Op 2	2461.00	41.02	41	0.040650407
Zone 2				
Zone 3				
Zone 4 Op 1	2460.50	41.01	41	0.020325203
Zone 4 Op 2	2457.20	40.95	41	−0.113821138
Zone 5 Op 1	2460.00	41.00	41	0
Zone 5 Op 2	2451.50	40.86	41	−0.345528455
Zone 6 Op 1	2450.00	40.83	41	−0.406504065
Zone 6 Op 2	2464.00	41.07	41	0.162601626
Zone 7 Op 1	2490.00	41.50	41	1.219512195
Zone 7 Op 2				

FIGURE 17.22
Division of enclosure assembly line.

Standard Work

It is very important to ensure that standard work is being followed because it plays a vital role in the manufacturing plant. In the manufacturing companies, it is common to introduce new operators or move operators to new operations. Therefore, there should be a document with graphical pictures enabling properly trained operators to do any operation without confusion. Standard work is a simple guide for the operator, which is also very important when it comes to operator safety. Standard work was implemented at GGC and enabled smooth operation by the operators.

Results

The results of this project met the goals as outlined in the project charter. The team felt that the ability to balance the line in such an efficient manner was truly a job well done. The fact that the number of workers required to produce the enclosure assembly was reduced by three and the operating efficiency of the line was improved from 80% to nearly 100% speaks well to the abilities of the team. Again, these workers were transferred to other departments within the company.

GGC expects the facilities modifications to take about 6 weeks and cost about $80k. The removal of three employees from the assembly line will reduce the direct labor required to produce the graders by approximately $145k per year, freeing up those resources to be used elsewhere within the company.

18

Sunshine High School Discipline Process Improvement: A Lean Six Sigma Case Study

Marcela Bernardinez, Khalid Buradha, Kevin S. Cochie, Jose Saenz, and Sandra L. Furterer

CONTENTS

Case Overview

The Sunshine High School (SHS) is one of the largest high schools in the Orange County Public School system in Florida, with over 3400 students and over 340 faculty members. The student population is very diverse and made up of many nationalities and students from a variety of socioeconomic levels. The campus is divided into an East Campus that consists exclusively of freshman students and a West Campus that consists of sophomores through seniors. The leadership team consists of a principal, three assistant principals, and nine deans. The discipline program is charged with the responsibility of providing for a safe and effective learning environment. The discipline system is overseen by one assistant principal and three deans. This program is affected by many factors including student attendance, student code of conduct adherence, and classroom management and discipline. The discipline program is a system of subprocesses that work together to achieve an environment conducive to quality learning. The original case study was published in *Lean Six Sigma in Service: Applications and Case Studies*, CRC Press. It is adapted to focus on the Lean Six Sigma aspects applied for this case study (Furterer, 2009).

SHS's mission is to advance student achievement for all students with the education necessary to be responsible, successful citizens.

To ensure that all students succeed, they are committed to the following:

- Encourage students to develop pride in school and community.
- Recognize all students, faculty, staff, and community for their achievements.
- Create a culture of academic rigor and relevance.
- Use data to identify what is essential to know.
- Set high expectations that hold students and adults accountable for improvement.
- Create a curriculum framework that drives instruction.
- Provide students with real-world application of skills and knowledge.
- Create multiple pathways to rigor and relevance based on students' individual strengths.
- Provide sustained professional development focused on improving instruction.

- Obtain parental and community involvement.
- Establish and maintain safe and orderly schools.
- Offer effective leadership development for administrators, teachers, parents, and community.

One of the substitute teachers at SHS has noticed a lack of standardization in the discipline process across different classrooms. The administration is also concerned that the students who get referred to the office for discipline problems miss one to several class periods, while the paperwork is sent from the teacher to the office.

This Lean Six Sigma (LSS) project will look at the discipline system as a whole initially, then will focus in on key subprocesses in an effort to make recommendations for system-wide improvement, thus improving the overall academic environment. In this case study, the team applied both Lean and Six Sigma tools; however, mainly the Lean tools applied within the Six Sigma Define–Measure–Analyze–Improve–Control (DMAIC) framework will be presented here.

SHS process owners and administrators will assist in deployment of faculty, student, and parent surveys as well as interviews of key stakeholders within the administrative staff and leadership team. The SHS principal is new to LSS but is convinced of its value for improving the discipline process at SHS.

The LSS mentors will be available to meet with the Six Sigma team to provide background and process information on the discipline process, having worked with SHS in the past, as well as guide the participants with coaching on applying the LSS tools. A sample discipline referral form is included in Figure 18.1.

County Public Schools
Safety/Discipline Referral Form

Student Number: ________ Incident No.: ________ Student Name: ________
Sex: ____ Race: ____ Grade: ____ Date of Infraction: ________
Parent/Guardian Name: ________ Home Phone: ________ Work Phone: ________
Referred By: ________ Instructor/Staff #: ____ Bus Trip #: ____ Period ____
Time: ____ Location of Infraction: ________ Details of Offense: ________

Administrator #: ________ * Must be reported to Law Enforcement

Offense(s) pertaining to this referral

	1A Cheating		2A Destroy Prop/ Vand < $10		3A Battery*		4A Alcohol*
	1B Classroom Disruption		2B Disrespect		3B Breaking and Entering*		4B Arson*
	1C Disorderly Conduct		2C Fighting		3C Destroy Prop/ Vand ($10-$100)		4C Assault of Emp/Vol/Stdts*
	1D Disrespect for Others		2D Forgery		3D Disrespect		4D Battery of Emp/Vol/Stdts*

FIGURE 18.1
Safety/discipline referral form.

	1E Dress Code		2E Gambling		3E Extortion/ Threats		4E Bomb Threats/ Explosions*
	1F Failure to Report Detention		2F Insubordination/ Def		3F Fighting*		4F Drugs*
	1G False/Misleading Information		2G Intimidation/ Threats		3G Firecrackers/ works		4G False Fire Alarm*
	1H Insubordination		2H Misconduct on Sch Bus		3H Gross Insubordination/ Def		4H Firearms*
	1I Misconduct on School Bus		2I Repeat Misc/Less Serious		3I Illegal Organization		4I Incite/Lead/ Participate*
	1J Profane/Obs/ Abusive Lang		2J Stealing under $10		3J Possession of Contraband Material		4J Larceny/ Theft*
	1K Repeated Misconduct		2K Unauthorized Assembly		3K Repeated Misc/More Serious		4K Other Weapons*
	1L Tardiness		2L Bullying		3L Smoking/ Other Use Tobacco*		4L Repeat Misc/More Serious
	1M Unauth Abs School/Class		2M Other Serious Misconduct		3M Stealing over $10		4M Robbery*
	1N Bullying				3N Trespassing*		4N Sexual Battery*
	1O Other				3O Violation of Curfew		4O Sexual Harassment*
					3P Bullying		4P Sexual Offenses*
					3Q Other Serious Misconduct		4Q Violation Early Reentry
							4R Motor Vehicle Theft*
							4S Motor Vehicle Theft*
							4T Other

Action(s) taken for this referral

	A Parental Contact		F Return of Prop/ Pay/Restit.		L Referral to Intervention Program		R Suspend from School
	B Counseling and Direction		G Retention		M Confiscate Unauthor. Material		S Suspend 10 Days Exp/ Removal

FIGURE 18.1 ***(Continued)***
Safety/discipline referral form.

	C Verbal Reprimand		H Saturday School		N Special Program/School		X Probationary Plan (KG-05)
	D Special Work Assignment		I Behavior Contract/Plan		P In-school Suspension		Z Peer Mediation
	E Withdrawal of Privileges		K Alt Class Assignment		Q Suspension from Bus		

__________ Suspension Information __________________ Other Information__________
From: _____ # Days: _____ To: _____ Return: _____ Detention: _________
Sat Schl: _________ Other: _________ Ex Ed Student: Yes/No Sum Sch Susp: Yes/No Early
Re-entry: _____ From: _____ To: _____
Administrator's Comments: ______________________________

School Name: ______________ Time departed from office ____________
Administrator's Signature: ______________ Student's Signature: ______________

Parent's Signature: ______________

FIGURE 18.1 ***(Continued)***
Safety/discipline referral form.

Define Phase

Following is the Define phase of the LSS DMAIC methodology. The Define phase defines the problem to be solved and provides a scoping of the process to be improved. The first activity in the Define phase is to create the project charter, which defines the problem to be investigated. The project charter is shown in Figure 18.2 and a description follows.

Project Name

SHS Discipline Process Improvement

Project Overview

SHS is one of the largest high schools in the Orange County Public School system with over 3400 students and over 340 faculty members. The student population is very diverse and made up of many nationalities. The campus is divided into an East Campus that consists exclusively of freshman students and a West Campus that consists of sophomores through seniors. The leadership team consists of a Principal, three Assistant Principals, and nine Deans. The discipline program is charged with the responsibility of providing for a safe and effective learning environment. The discipline system is overseen by one assistant principal and three Discipline Deans. This program is affected by many factors including student attendance, student code

Project Name: Name of the Lean Six Sigma project

Project Overview: Background of the project.

Problem Statement: Business problem, describe what, when, impact, consequences.

Customer/Stakeholders: (Internal/External) Key groups impacted by the project.

What is important to these customers – CTS: Critical to Satisfaction, the key business drivers.

Goal of the Project: Describe the improvement goal of the project.

Scope Statement: The scope of the project, what is in the scope and what is out of scope.

Financial and Other Benefit(s): Estimated benefits to the business, tangible and intangible.

Potential Risks: Risks that could impact the success of the project. Can assess risk by probability of occurrence and potential impact to the project.

Milestones: DMAIC phase and estimated completion dates.

Project Resources: Champion, Black Belt Mentor, Process Owner, Team Members.

FIGURE 18.2
Project charter template.

of conduct adherence, and classroom management and discipline. The discipline program is a system of subprocesses that work together to achieve an environment conducive to quality learning.

Problem Statement

The discipline process at SHS is both inefficient and inconsistent. The students can wait from one to several class periods in the Administration office waiting for the paperwork to get processed by the referring teacher, or to be seen by the Discipline Dean. Additionally, the classroom discipline process varies as well as the discipline consequences given for various discipline infractions.

The SHS LSS team will work on improving the discipline program at SHS. The SHS discipline program consists of multiple subprocesses that affect one another. These subprocesses are complex in nature and are each affected by multiple factors that include student background, academic standing, and other variables that will be analyzed for correlation.

Ultimately, the customer of the discipline program is the parent or guardian of the student.

Improvements to one or more of the subprocesses will be made by this LSS team as well as control plans to assist in the implementation and control of the improvements. The project team has neither control over selected implementations nor ultimate control of the changes.

In the Define phase, the team began to understand the problem and process to be improved. They developed a detailed project description, or project charter, that describes the problem statement, project goals, and scope statement. A stakeholder analysis was also performed to identify the critical customers and stakeholders that are impacted by the process to be improved. The concerns and how the stakeholders are impacted are also defined in this phase. A detailed work plan was generated to provide guidance to the team for how they would successfully complete the activities of the DMAIC problem-solving methodology for this project.

Customers/Stakeholders

Faculty, students, parents, school administrators, school board, and security/police officers.

What Is Important to These Customers (Critical to Satisfaction)

Reduction in the number of discipline referrals, reduction in the number of classroom disruptions, consistency in application of the discipline consequences, and knowledge of the code of conduct.

Goal of the Project

Complete a comprehensive DMAIC analysis of this process/system using Six Sigma tools and methodology. The end state of the project will yield recommendations to the SHS administrative staff for improvements of the process.

Scope Statement

This project will analyze the discipline system of SHS. This analysis will map the subprocesses of the discipline system to include the following:

- Classroom Discipline Process: Defined as in-class discipline by a faculty member to include initiation of a Dean's office referral.
- The Discipline Action Process: Defined as the processes that occur once a student and/or referral arrives to the Dean's office through the discipline action with data input into the student database followed by feedback to the initiating faculty member.
- Attendance Contract Process: Defined as the process a student undergoes to receive and adhere to an attendance contract.

The project will map the subprocesses that are executed within the discipline system and the perceptions/satisfaction levels of these processes from the viewpoint of the system customers (administrators, faculty, students, and parents).

Upon completion of surveying the discipline system customers, the project team will focus on improving one or several of the discipline system's subprocesses with the intent of providing recommendations for improvement of the processes that will positively impact the overall discipline system and academic environment.

Principal Project Deliverables/Outputs

Define:

- Project charter
- Stakeholder analysis

Measure:

- Value stream map
- Process flow diagram (detailed process map)
- Critical to satisfaction (CTS): Key outputs of the process from the customers' view
- Key metrics: Key inputs of the process
- Pareto diagrams: Graphical depiction of target improvement areas

Analyze:

- Cause and effect diagrams
- Summary of data
- Summary of improvement areas (recommended)
- Waste analysis
- Process analysis

Improve:

- Recommended improvement plans
- Revised process flow and information flow diagrams (recommended)

Control:

- Recommended control plan

Potential Benefits

Potential benefits to improving the discipline process includes an enhanced academic environment to facilitate student learning, decreased discipline issues of repeat offenders, and decreased probability of potential liability issues regarding faculty usage of the disciplinary process.

Risk Management Matrix

The risk management matrix is shown in Figure 18.3.

Project Resources

Project Leader: Kevin Cochie

Division/Department: SHS Administration

Process Owner: Discipline Deans

Process Champion: Principal

Project Sponsor: Discipline Dean 1

CI Mentor/MBB: Sandra Furterer

Project Team Members: Marcela Bernardinez, Khalid Buradha, Kevin Cochie, and Jose Saenz

The estimated milestones are shown in Figure 18.4.

Potential Risks	Probability of Risk (H/M/L)	Impact of Risk (H/M/L)	Risk Mitigation Strategy
Data not available	H	H	Identify issues early to the principal; collect manual data
Resistance to change from students	L	M	Change strategy
Resistance from faculty	M	H	Change strategy
Time constraints	M	H	Good project planning

FIGURE 18.3
Risk management matrix.

Milestones	
Phase	**Estimated Completion Date**
Define	January 11
Measure	February 3
Analyze	March 2
Improve	April 6
Control	April 27

FIGURE 18.4
Milestones.

Critical Success Factors

- Partnership with SHS administration: The success of this project hinges on close partnership between the Six Sigma team and the SHS administrators and process owners.
- Complete process mapping: This process is detailed and complex. Successful data gathering and analysis is heavily dependent on the Six Sigma team becoming well versed with the procedures within the discipline process.
- Critical to satisfaction identification: Customer CTS variables must be identified from the standpoint of the customers identified in the stakeholder analysis. This may include more than one primary customer base.

Management Approach

Scope Management Approach

This project will be managed by the project leader, but responsibility of the success hinges upon a collective effort of all team members. Communication between the team members shall flow cross-functionally. Electronic mail will be a prime source of communication outside of class and group meetings; therefore, it is imperative that when communicating with other team members, the Master Black Belt, or a process owner from SHS, all other team members shall be copied on the communication.

Issues Management Approach

All issues will be documented through weekly team meetings by the team secretary. Issues for resolution at the team level shall be settled by the team collectively. Issues that rise above the team level will be settled by the Master Black Belt/Professor and/or the Project Champion.

Customer/Stakeholder Analysis

A critical part of the Define phase is to perform a stakeholder analysis, to understand the people impacted by the project. There are primary stakeholders, which are usually the main internal and external customers of the process being improved. The secondary stakeholders are affected by the project, but not directly manner. Figures 18.5 and 18.6 show the primary and secondary stakeholders for the discipline process and their major concerns. Note that + represents a positive impact or potential improvement, while – represents a potential negative impact to the project.

Figure 18.7 shows the commitment level of each major stakeholder group at the beginning of the project.

	Stakeholders	Who Are They?	Potential Impact or Concerns	+/–
PRIMARY	SHS Faculty	Customer: This includes all SHS permanent faculty and substitute teachers. They are customers of the discipline system. Their input into the system is referrals into the system with the expected output of a disciplinary action.	• Standardized processes • Reduction of errors and rework • Continuity of infraction enforcement • Resistance to enforcing codes	+ + + –
	SHS Students	Customer: This includes the over 3500 students that attend SHS. They are customers of the discipline system as they are the inputs to the system and the expected outcome is a fair and consistent reaction to infractions.	• Reduction of repeat offenses • Increase of academic performance • Resistance to imposition of strict policies	+ + + –
	SHS Parents/ Guardians	Customer: This includes all parents or guardians of the students of SHS. Their children are the inputs to the discipline system. The expected output is a safe environment conducive to a positive learning environment to their children.	• Increase of Knowledge of code of conduct • Reduction of communication gaps • Resistance to change current procedures	+ + –

FIGURE 18.5
Primary stakeholder analysis definition.

Team Ground Rules and Roles

The team brainstormed the following ground rules related to their attitudes and the processes or behaviors they would adhere to while working with each other.

Team Ground Rules

Attitudes

- Be as open as possible, but honor the right of privacy.
- Information discussed in the team will remain confidential. With regard to people's opinions, what is said here stays here.
- Everyone is responsible for the success of the meeting.
- Be a team player. Respect each other's ideas. Question and participate.
- Respect differences.

- Be supportive rather than judgmental.
- Practice self-respect and mutual respect.
- Criticize only ideas, not people.
- Be open to new concepts and to concepts presented in new ways. Keep an open mind. Appreciate other's points of view.
- Be willing to make mistakes or have a different opinion.
- Share your knowledge, experience, time, and talents.
- Relax. Be yourself. Be honest.

	Stakeholders	Who Are They?	Potential Impact or Concerns	+/−
SECONDARY	SHS Administration	Stakeholder: The assistant principals and deans are charged with a tremendous responsibility of educating young adults including quality academic programs and a safe learning environment free of classroom disruption. Oversight of the discipline system and its subprocesses.	• Decrease instances of classroom disruption • Resistance to change of discipline procedures that impact administrative focus areas	+ −
	SHS Security and Law Enforcement	Stakeholder: SHS security and law enforcement is responsible for the oversight of campus security. They require swift and consistent enforcement by the process owners of the discipline system in order to assist in maintaining good order and discipline on the school campus.	• Reduction of campus-related security issues • Resistance to change of discipline procedures that impact campus security	+ −
	County Public School System	Stakeholder: The school district is financially liable for the security and safety of all students within the entire school system. The public school system requires good order and discipline on all campuses and within all classrooms.	• Decrease instances of classroom and campus disruptions from discipline infractions • Potential public school system restrictions on recommended improvements (Bureaucracy)	+ −

FIGURE 18.6
Secondary stakeholder analysis definition.

Commitment Level	Administration	Faculty	Students	Parents	School system	Security/Courtesy Officers
Enthusiastic	★					
Helpful	★	★		★		★
Hesitant		★				
Indifferent			★		★	
Uncooperative			★			
Opposed						
Hostile						

FIGURE 18.7
Stakeholder resistance to change.

Processes

- Use time wisely, start on time, return from breaks, and end our meetings promptly.
- Publish agenda and outcomes.
- Ask for what we need from our facilitator and other group members.
- Attend all meetings and be on time.
- Absenteeism permitted if scheduled in advance with the leader.
- When members miss a meeting, we will share the responsibility for bringing them up to date.
- One hundred percent focus and attention while meeting.
- Stay focused on the task and the person of the moment.
- Communicate before, during, and after the meeting—to make sure that action items are properly documented, resolved, and assigned to a responsible individual and given a due date.
- Phones or pagers on "stun" (vibrate, instead of ring or beep) during the meetings.
- One person talks at a time.
- Participate enthusiastically.

- Do not interrupt someone talking.
- Keep up-to-date.

Project Plan and Responsibilities Matrix

The detailed project plan is shown in Figure 18.8, with tasks to be completed, due date, deliverables, and resources. It includes the person or people responsible for each activity.

ID	Task Name	Duration	Start	Finish
1	**SHS Six Sigma Project Start**	0 days	Mon 1/10/05	Mon 1/10/05
2	Define Phase	**18 days**	**Mon 1/10/05**	**Wed 2/2/05**
3	Define the Problem	2 days	Mon 1/10/05	Tue 1/11/05
4	Define the Scope of the Project	2 days	Wed 1/12/05	Thu 1/13/05
5	Define the Potential Benefits of the Project	2 days	Fri 1/14/05	Mon 1/17/05
6	Prepare Project Charter	2 days	Tue 1/18/05	Wed 1/19/05
7	Prepare Initial Process Flow Charts and Sar	2 days	Thu 1/20/05	Fri 1/21/05
8	Customer / Stakeholder Analysis	2 days	Mon 1/24/05	Tue 1/25/05
9	Make Work Plan	1 day	Wed 1/26/05	Wed 1/26/05
10	Responsibilities Matrix	1 day	Thu 1/27/05	Thu 1/27/05
11	Determine Next Steps	1 day	Thu 1/27/05	Thu 1/27/05
12	Prepare Team Member Participation Log	1 day	Thu 1/27/05	Thu 1/27/05
13	Prepare Define Phase Report and Presentat	4 days	Fri 1/28/05	Wed 2/2/05
14	Measure Phase	**20 days**	**Thu 2/3/05**	**Wed 3/2/05**
15	Meet with SHS Dean to finalize Process Flo	1 day	Thu 2/3/05	Thu 2/3/05
16	Determine Student, Faculty and Parent surv	1 day	Fri 2/4/05	Fri 2/4/05
17	Conduct Student, Faculty, and Parent Surve	1 day	Mon 2/7/05	Mon 2/7/05
18	Collect Survey Results	1 day	Tue 2/8/05	Tue 2/8/05
19	Analyze Survey Results	12 days	Wed 2/9/05	Thu 2/24/05
20	Prepare SIPOC	7 days	Wed 2/9/05	Thu 2/17/05
21	Draw Process Flow Charts	10 days	Wed 2/9/05	Tue 2/22/05
22	Make Pareto Charts using Survey Results	10 days	Wed 2/9/05	Tue 2/22/05
23	Identify Critical to Quality Characteristics	8 days	Wed 2/9/05	Fri 2/18/05
24	Key Metrics	2 days	Wed 2/9/05	Thu 2/10/05
25	Define items for Resolution	1 day	Wed 2/9/05	Wed 2/9/05
26	Process Capability Index	2 days	Wed 2/9/05	Thu 2/10/05
27	Determine Next Steps	1 day	Wed 2/9/05	Wed 2/9/05
28	Prepare Team Member Participation Log	1 day	Wed 2/9/05	Wed 2/9/05
29	Prepare Measure Phase Report and Presen	14 days	Fri 2/11/05	Wed 3/2/05
30	Analyze Phase	**25 days**	**Thu 3/3/05**	**Wed 4/6/05**
31	Determine Causes and Effects of the Proble	6 days	Thu 3/3/05	Thu 3/10/05
32	Define items for Resolution	1 day	Thu 3/3/05	Thu 3/3/05
33	Make Improvement Plans	3 days	Thu 3/3/05	Mon 3/7/05

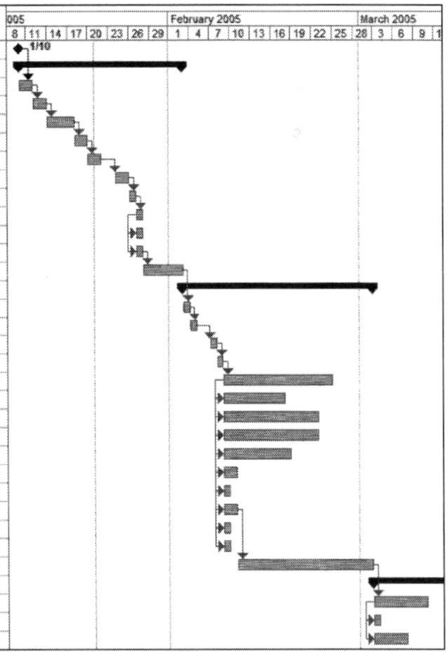

ID	Task Name	Duration	Start	Finish
34	Prepare Recommendations for Improvement	11 days	Thu 3/3/05	Thu 3/17/05
35	Cost Benefit Analysis	10 days	Thu 3/3/05	Wed 3/16/05
36	Prepare Team Member Participation Log	1 day	Thu 3/3/05	Thu 3/3/05
37	Prepare Measure Phase Report	21 days	Thu 3/3/05	Thu 3/31/05
38	Prepare Measure Phase Presentation	4 days	Fri 4/1/05	Wed 4/6/05
39	Improve Phase	**5 days**	**Thu 4/7/05**	Wed 4/13/05
40	Quantification of Improvement Plans	2 days	Thu 4/7/05	Fri 4/8/05
41	Define items for Resolution	2 days	Mon 4/11/05	Tue 4/12/05
42	Prepare Team Member Participation log	1 day	Wed 4/13/05	Wed 4/13/05
43	Control Phase	**5 days**	**Thu 4/14/05**	Wed 4/20/05
44	Proposed Control Mechanisms	2 days	Thu 4/14/05	Fri 4/15/05
45	Verification of Improvements / Results	2 days	Mon 4/18/05	Tue 4/19/05
46	Prepare Team Member Participation Log	1 day	Wed 4/20/05	Wed 4/20/05
47	Final Report	**5 days**	**Thu 4/21/05**	Wed 4/27/05
48	Team Assessment	1 day	Thu 4/21/05	Thu 4/21/05
49	Project Assessment	1 day	Thu 4/21/05	Thu 4/21/05
50	Document Lessons Learned	1 day	Fri 4/22/05	Fri 4/22/05
51	Prepare Team Member Participation Log	1 day	Mon 4/25/05	Mon 4/25/05
52	Prepare Final Project Report	1 day	Tue 4/26/05	Tue 4/26/05
53	Prepare Final Project Presentation	1 day	Wed 4/27/05	Wed 4/27/05
54	**SHS Six Sigma Project End**	0 days	Wed 4/27/05	Wed 4/27/05

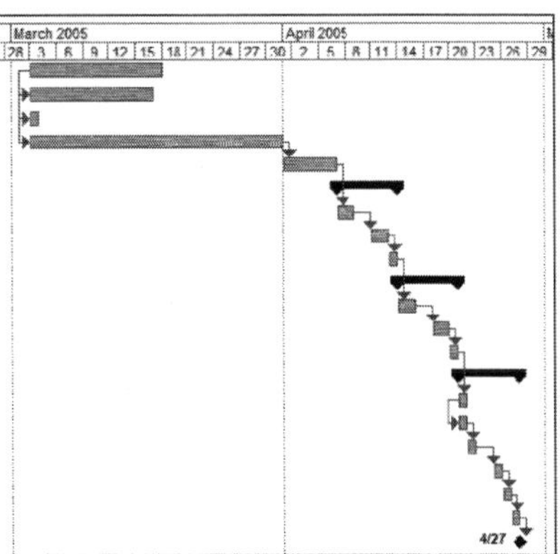

FIGURE 18.8
Project plan.

Team Member Bios

Each team member created a short biography describing their background and skills for the project.

Marcela Bernardinez

Marcela Bernardinez was born in 1980 in San Miguel de Tucuman, Argentina, but raised in Venezuela since her parents decided to move looking for new opportunities. After she finished high school in Venezuela, she decided to have a new experience, meet new people, find new opportunities, and discover a new world, so she came to the United States to study Industrial Engineering. She has been in the United States for 6 years, and it has been a challenge to arrive at where she is now. Marcela has a bachelor's degree from the University of Central Florida in Industrial Engineering and is pursuing her master's in Industrial Engineering at the same university. In addition, she is a member of the Institute of Industrial Engineers and Society of Hispanic Professional Engineers.

Khalid A. Buradha

Khalid Buradha is a Field Supervisor in the Inspection Department/Saudi Aramco Co. His contribution was focused on monitoring and managing the quality system of the final product (projects >$2 million). He holds a bachelor's degree in Electrical Engineering from Tulsa University in Oklahoma.

He was a maintenance engineer for both steam and natural gas plants. Part of his main duties was to put the faulty equipment back in service and to enhance the performance and the reliability of the plant's equipment. In addition to that, he developed a number of databases for the department.

Kevin S. Cochie

Captain Kevin S. Cochie is an active duty officer in the U.S. Army. His area of specialty is Special Operations Aviation. He pilots the MH-47E Chinook and MH-6 Littlebird and served combat time in both Afghanistan and Iraq. He holds a bachelor's degree in Design Engineering Technology from the University of Central Florida and a master's in Management from Troy State University.

Jose G. Saenz

Jose is native to Panama. He has a bachelor's degree in Industrial Engineering Management and is currently pursuing a master's degree in Quality Engineering at the University of Central Florida. He worked as a Cost Analyst for Towerbank, a local bank in Panama, where he contributed to the implementation of an Activity-Based Costing system. Jose is an active member of ASQ and the Treasurer of the ASQ Student Chapter of the Orlando Section.

Dr. Sandra L. Furterer

Dr. Furterer was the course instructor for the Total Quality Improvement course at the University of Central Florida as well as the Master Black Belt for the project. She has extensive consulting experience in process improvement, LSS, and information systems improvement. Dr. Furterer has a Bachelor and Master of Science in Industrial and Systems Engineering from Ohio State, an MBA from Xavier University, and a PhD in Industrial Engineering from the University of Central Florida.

Sunshine High School Discipline Process Improvement Lean Six Sigma Project

Measure Phase

Following is the Measure phase for the SHS Discipline Process Improvement project, including the key deliverables. The Measure phase of the DMAIC process is designed to gain information on the process performance and develop problem and/or process improvement statements. The objectives of this phase in relation to the SHS Discipline Improvement Project are as follows:

- Map the current process.
- Gather initial data and determine current process performance.
- Confirm key customer requirements of the process.
- Organize and stratify all data collected.

To define the current state of this process, face-to-face interviews and meetings with the Deans were conducted. Consequently, a process map of the entire discipline system was developed. This process map or workflow chart is a schematic picture of the process being studied. It also shows the major steps in the process. In the Analyze phase of the DMAIC cycle, each step in this process will be examined from a time, value, and cost perspective.

The discipline program is a system of subprocesses that work together to achieve an environment conducive to quality learning. The subprocesses are shown in the value stream map in Figure 18.9.

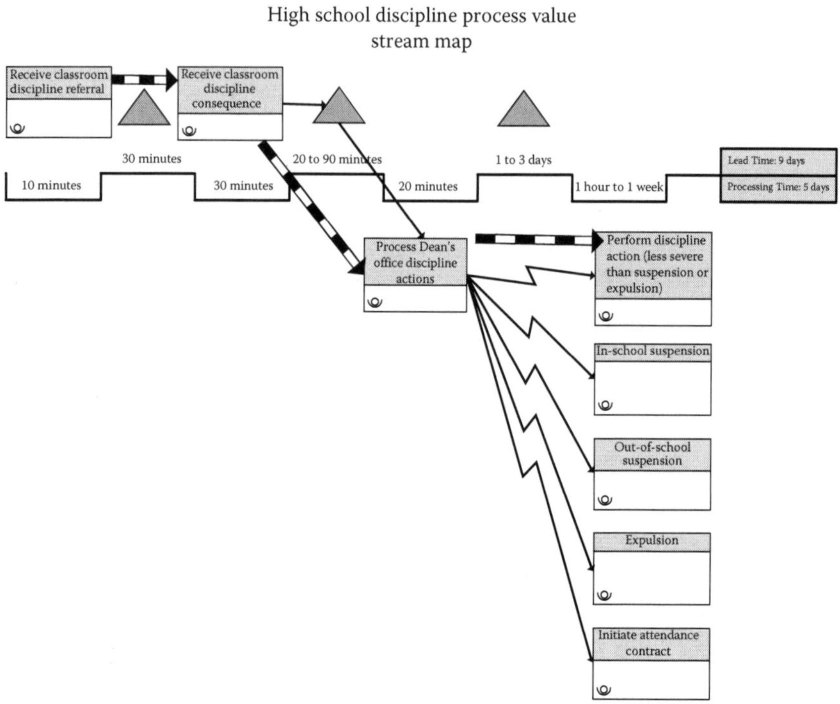

FIGURE 18.9
Value stream map.

The discipline system has been divided into the following subprocesses:

Receive classroom discipline referral:

This subprocess consists of the actions that occur in the classroom when a faculty member observes a student infraction of the code of conduct. This subprocess is initiated with a student infraction and ends with either classroom discipline imposed or the initiation of a referral to the Dean's office. Once a student misbehaves within the classroom, the teacher will decide based on the severity of the infraction whether they will give an in-class discipline consequence, typically an additional assignment, separate them from the class, or call the parent. If the teacher decides to call the parent, they will access the student's information card for the phone number; if they are able to contact the parent, they will talk to them and discuss the student's behavior. Sometimes the contact phone number is not correct, so the teacher will investigate to get the corrected phone number (e-mail the Discipline Dean, or ask the student for the correct phone number). The teacher will decide if they want to meet the parent and will meet as appropriate. The teacher will then give the student the discipline consequence. If the infraction is severe (level 2 or above), the faculty member will send the student to the Discipline Dean's office and complete a discipline referral form and then send this to the Discipline Dean.

Process Dean's office discipline actions:

This process begins from the student referral, includes the processing of that student and data entry of the completed referral, and ends with feedback to the faculty member who initiated the referral. There are severity levels of misconduct (refer to Safety/Discipline Referral Form, refer to Figure 18.1). If the student committed a misconduct of level 2, 3, or 4, or one that the teacher deemed necessary to send to the Discipline Dean they will receive a discipline referral. The teacher will complete the form, but may need to wait until the end of the class period. The student may wait at the Discipline Dean's office until the form is sent by the teacher, and the Discipline Dean is free. This wait can range from 20 minutes to 1½ hours. The Dean will pull the student's file and history of offenses. If the misconduct is level 2, 3, or 4, the Dean will contact the parent. If the misconduct is level 4 or higher, the Dean opens a police investigation. The Dean will complete the form and give the student the discipline consequence. The student should complete the discipline action. The Dean will follow up to ensure that the student completed the discipline action. If the student does not complete the action, then the Dean will assign another discipline action. If the misconduct is habitual, or the student refuses to complete the action, the Dean will institute a behavior contract. This is a contract with the student's the Dean, and the parent that provides for consequences and an agreement of improved behavior. The Dean will complete the referral form and give the form to the office assistant, who will enter the information into the student database. A copy of the form is put into the student's file and given to the faculty member.

Initiate attendance contract:

This process is initiated when a student is put on an attendance contract and includes the process to track the contract and subsequent discipline penalties if the contract is violated. When the student has more than 5 unexcused or more than 10 excused absences, the Attendance Dean will initiate a student attendance contract with the student. An initial meeting is held with the student, the Attendance Dean, and the parents/guardians. The student must get signatures from each teacher when they attend each class for the length of the contract (i.e., a month). The Attendance Dean will verify that the contract is filled out each month. If the student violates the attendance contract, the student will receive consequences and a discipline referral and consequence will be completed. Once the student completes the attendance contract, the contract is stored in the student's file.

Perform discipline action:

This subprocess includes having the student perform the discipline action. There are many types of discipline actions including calling and/or meeting with parents, lunch/after school detention, Saturday detention, in-school suspension (ISS) or out-of-school suspension (OSS), and expulsion. A critical subprocess is the determination of whether the student is put into ISS or OSS. It is important to determine if the students who spend time suspended out of school have lower academic performance than those with ISS. When the student receives a suspension, their teachers are e-mailed. The teacher is to provide the material that the student will miss during their suspension. The ISS teacher will verify that the teachers send the material; if they do not they will follow up with the teachers and Discipline Dean. The Discipline Dean will then follow up with the teachers to ensure that the students receive the missing work. Generally, the teachers provide the needed materials. The student will attend the ISS and the ISS teacher will log their attendance. The student who received the OSS will return to school. Their teachers will ensure that the work was made up.

Faculty and Student Focus Group to Determine Critical to Satisfaction Criteria

The LSS team held a focus group with a representative sample of faculty and another with a group of students. The purpose of the focus group was to identify what is important to the faculty and the students regarding the discipline process. There were four main CTS criteria derived from the focus groups:

1. Minimize classroom disruptions.
2. Minimize school discipline referrals.
3. Level of knowledge of the student code of conduct.
4. Consistency of offenses and actions.

Voice of the Customer Surveys

SHS process owners and administrators will assist in deployment of faculty and student surveys as well as interviews of key stakeholders within the administrative staff and leadership team.

Process Maps

The process maps for each subprocess of the discipline process are shown in Figures 18.10 through 18.12.

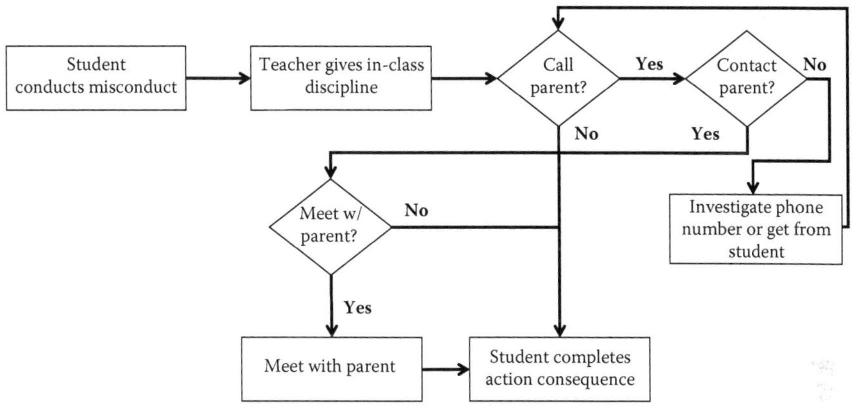

FIGURE 18.10
Classroom discipline process map.

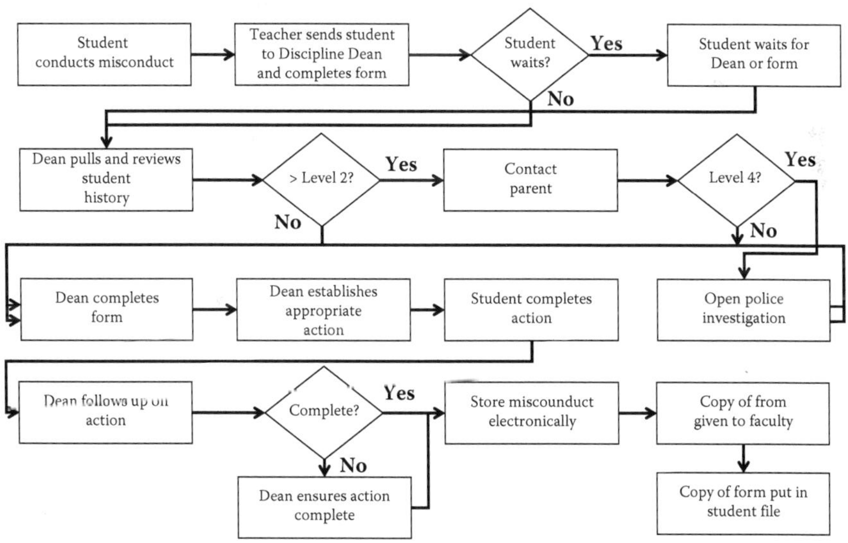

FIGURE 18.11
Dean's office discipline process map.

Operational Definitions

CTS: Minimize classroom disruptions

Defining the measure: Number of classroom disruptions per week

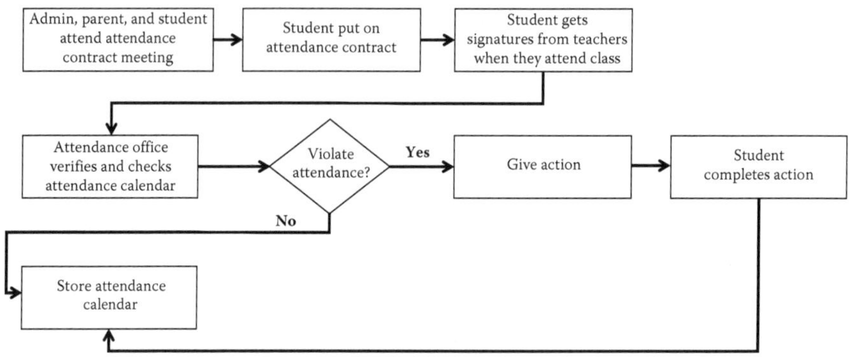

FIGURE 18.12
Attendance contract process map.

Purpose: To determine the number of discipline disruptions in the classroom

Clear way to measure the process: Determine a statistically valid sample of each type of classroom, by subject and class grade (9, 10, 11, 12) and level (i.e., Advanced Placement, International Baccalaureate, Honors, General, and Basic)

CTS: Minimize school discipline referrals

Defining the measure: Number of discipline referrals by week

Purpose: To determine the number of discipline referrals by period

Clear way to measure the process: Extracted from student database, average and standard deviation for time frame sampled

CTS: Level of knowledge of student code of conduct

Defining the measure: Number of discipline referrals by type

Purpose: To determine the root causes of discipline referrals and most frequently occurring types, to infer the level of knowledge of the code of conduct

Clear way to measure the process: Extracted from student database, average and standard deviation for discipline referrals by type

CTS: Consistency of offenses and actions

Defining the measure: Number of actions to offense types

Purpose: To determine the consistency of offenses to actions given

Clear way to measure the process: Extracted from the student database, listing actions to offenses in the sample; also, acquire the voice of the customer (VOC) from faculty and student surveys to understand perceived consistency of discipline process

Critical to Satisfaction (CTS)	Metric	Data Collection Mechanism	Analysis Mechanism	Sampling Plan	Sampling Instructions
Minimize classroom disruptions	Number of discipline referrals per week	Surveys, focus groups	Histograms, Pareto charts, basic statistics, hypothesis tests	Faculty (goal 50) Subset of students (goal 400)	Faculty: e-mail request sent by principal requesting participation. Students: select classes take survey in computer lab
Minimize school discipline referrals	Number of discipline referrals per week	School database	Basic statistics, hypothesis tests	All students	Extracted from database for year to date
Level of knowledge of code of conduct	Number of referrals by type	School database	Pareto charts, basic statistics, hypothesis tests	All students	Extracted from database for year to date
Discipline offense and action consistency	Number of consistent offenses to actions	School database	Pareto charts, basic statistics, hypothesis tests	All students	Extracted from database year to date

FIGURE 18.13
Data collection plan.

Data Collection Plan

Shown in Figure 18.13 is the collection plan for the project.

In the Measure phase of the DMAIC cycle, the LSS team collected the necessary data through the SHS student database to gain insight into the current discipline process performance and to identify areas of improvement.

In general, the data collection revealed the following facts:

- Most offenders are freshmen and represent 42% of the offender population.
- As repeat offenders (more than one referral, more than two referrals, etc.) are analyzed, the percentage of freshmen accounting for them increases.
- The grade point averages (GPAs) for students who have no discipline issues (2.99 GPA) are higher than the offenders (2.19 GPA).
- Students with discipline issues have more absent days (4 more on average) than students who have no discipline issues.
- Among all of the categories, repeat offenders who are part of the lower 30% Reading FCAT (Florida Comprehensive Assessment Test) have the lowest academic performance average of 1.53.

- Students who are given OSS statistically have lower GPAs (1.86 GPA) than students who receive punishment, but not OSS (2.26 GPA).
- Approximately 20% of the violation codes account for 80% of the referrals issued to students.

Voice of the Customer Faculty and Student Surveys

Survey Methodology

Surveys are powerful tools used to measure perceptions or importance of process characteristics. While perceptions are sometimes skewed to individual personalities, they are reality to those who work within the process. Process owners must be attentive to these perceptions because they influence the overall operation and efficiency of the system. Perceptions are key to workforce buy-in when collective efforts are required to maximize the efficiency of the process. This is particularly true for the discipline process because so many individuals influence the process. The students, faculty, and administration are composed of many different personalities, social backgrounds, and philosophies. For this system to operate efficiently and effectively, these individuals must buy-in to working collectively for a common goal of the entire organization. This includes consistency across classroom discipline, consistent backing of faculty discipline actions, and consistent enforcement of code of conduct violations.

The Six Sigma team developed two surveys to collect data from the faculty and students. The goal of these surveys was to capture the student and faculty perceptions of the discipline program at SHS.

The development of the survey was a very detailed process. The SHS Six Sigma project team first conducted a brainstorming session to decide what data needed to be collected from the surveys. The session was conducted by constructing two affinity diagrams. The purpose of the affinity diagram is to organize brainstormed data into categories. The first diagram was labeled "What data do we want to collect?" The second affinity diagram was then made by constructing questions that supported the "ideas" from the first affinity diagram. The end result of this was two draft surveys.

The draft surveys were then reviewed by the team's Master Black Belt. Once modifications were made to the format, order, and structure of the questions, the surveys were put in front of a group of administrators from SHS for review. The discipline and attendance deans reviewed the surveys and provided input for modification. This included adding several questions and modifying several answer responses for other questions.

The next step was to meet with a focus group of faculty members to ensure that the survey would collect appropriate key information pertaining to the discipline process. This session was very productive and was conducted by applying a formal brainstorming session. The affinity diagram that resulted yielded several areas they thought were important to classroom discipline. The major issue this group noted was that the faculty tended not to contact parents on discipline-related issues, a lack of parent involvement, and a lack of parent/teacher/counselor integration. This input was then incorporated into the surveys and final drafts were produced.

Once the final surveys were approved by the university's Internal Review Board and the SHS principal, the team established the logistics for administering the surveys.

It was determined that the entire faculty population of 173 teachers would be sampled as well as 500 students. The principal selected six teachers to solicit volunteers from their students to take the student survey. All six teachers were eager to participate with over 540 students respondeding to the survey. The team performed a statistical analysis on the sample to ensure that the demographics of the sample mirrored the student population demographic breakdown.

Both surveys were converted to web-based forms and posted on the web to streamline administration of the surveys and data collection time. The faculty was notified by multiple e-mails soliciting their participation. The students were sampled over a 2-day period by having their teachers bring them to a computer lab on the SHS campus.

Pareto Analysis

A Pareto analysis may be used to categorize and prioritize attribute data (qualitative). It helps prioritize improvement efforts and identify the *most frequent problems*. Moreover, it states that approximately 20% of root causes of problems account for approximately 80% of all problems encountered. Consequently, to significantly reduce problems in the organization, you can focus on a few key problem areas, that is, the 20%.

Most frequent offenses:

Based on the Safety/Discipline Referral Form, the team found that there are 15 possible level 1 offenses, 13 possible level 2 offenses, 17 possible level 3 offenses, and 20 possible level 4 offenses, which totals 65 possible offenses.

The team reviewed 2204 offenses from the database with the purpose of identifying what the most frequent offenses are. The dates of the offenses range from the beginning of the school year through February 23. A Pareto chart was developed, and is shown in Figure 18.14.

Data revealed that 14 out of 65 offense types (22%) account for 82% of all discipline issues in the school. The "other" category on the Pareto chart was large, but it was made up of a large number of discipline consequences that occurred very infrequently, so the team focused on the individual discipline issues with the largest percentages. Figure 18.15 shows the students' response when asked to rate the most effective discipline consequence. Attendance/

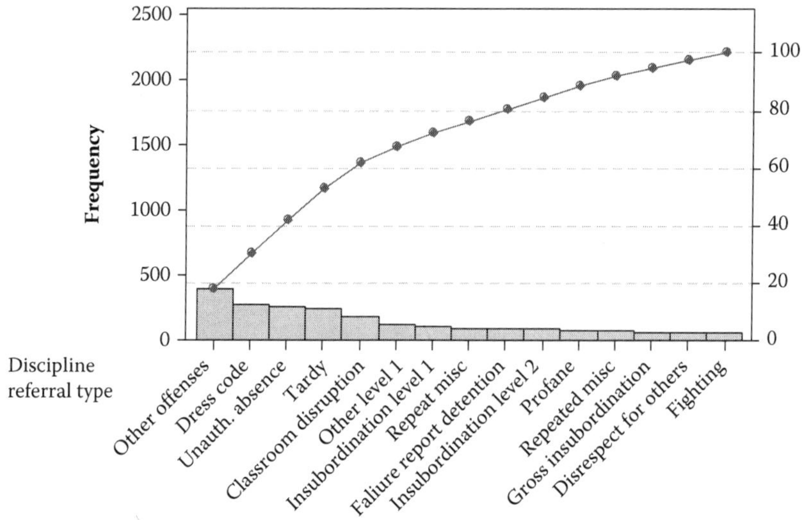

Frequency	395	274	255	247	188	120	111	97	91	91	83	73	64	58	57
Percent	18	12	12	11	9	5	5	4	4	4	4	3	3	3	3
Cum %	18	30	42	53	62	67	72	77	81	85	89	92	95	97	100

FIGURE 18.14
Pareto chart of offenses.

behavioral contract, lunch/after school detention, ISS, Saturday school, OSS, and special work assignments were the highest rated consequences.

Statistical Analysis

The average number of discipline referrals in the base for the 6 months of data for the entire student population was 1.536, with a standard deviation of 5.057. The histogram of data are shown in Figure 18.16.

Minitab was used to perform a test for normality. The p-value is less than 0.005, so the null hypothesis that data are normal must be rejected, so data data are not normal.

Analyze Phase

Following is the Analyze phase for the SHS Discipline Process Improvement project, including the key deliverables developed as part of the prior exercises.

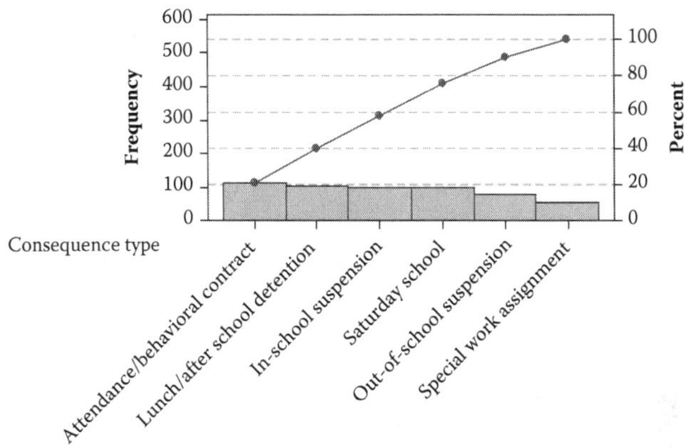

Frequency	112	102	100	98	76	52
Percent	20.7	18.9	18.5	18.1	14.1	9.6
Cum %	20.7	39.6	58.1	76.3	90.4	100.0

FIGURE 18.15
Pareto chart of most effective consequences.

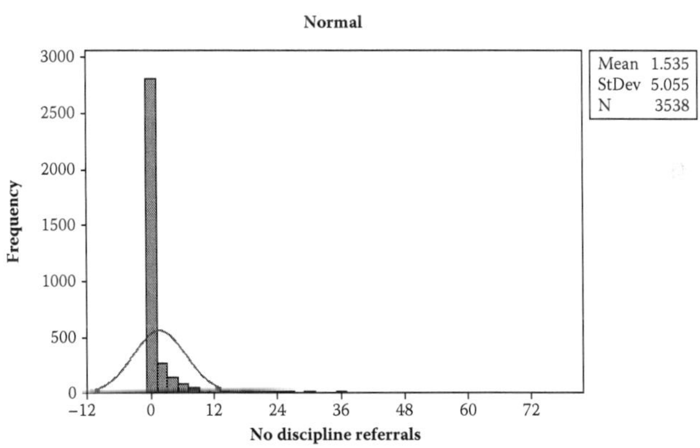

FIGURE 18.16
Histogram of number of discipline referrals.

The Analyze phase of the DMAIC process is designed to gain insight into the root causes of the problems as well as understand the process variables. The objectives of this phase in relation to the SHS Discipline Improvement Project are as follows:

- Understand the root causes.
- Understand the capability of the processes.
- Develop relationships between variables.
- Analyze the process for value-added and non-value-added activities.

- Identify and eliminate process waste.
- Understand the defects per million opportunities and the sigma levels.

Cause and Effect Diagrams

Root cause analysis is a very important activity in a LSS project. It is where the data is analyzed in detail and tools are used to determine the root causes of problems and inefficiencies. Too often, data are collected and project team members, champions, or knowledge workers jump to conclusions based on raw data. LSS and DMAIC prevent this from happening. Several tools are very useful in determining the root causes.

In this project, three areas of primary improvement were chosen to determine the root causes of inefficiencies. The team conducted brainstorming to determine root causes for the three areas. The areas of interest were based on data derived from the analysis of the school database and data collected from the faculty and student surveys. The three areas included the following:

1. Why is 42% of the offender population freshmen?
2. Why do 22% of the offense codes (14) account for 80% of the infractions committed?
3. Why do repeat offenders continue to commit code of conduct infractions?

The Six Sigma team first constructed fishbone diagrams (cause and effect diagrams) to get to the root causes of these three areas. As seen in Figure 18.17 the fishbone diagram for the effect labeled 42%, four branches were constructed as potential areas where causes exist. The branches were related to the students, training given to students, campus layout, administration, and faculty. From there, the Six Sigma team brainstormed potential causes to the effect. Several root causes were identified as the primary reasons freshmen account for so many of the offenders. First, by nature of the training, the teachers are not trained formally on the discipline program and the amount of training the freshmen receive is inadequate. This is also evident by data collected from both the faculty and student surveys. On average, the amount of training the freshmen receive per semester is less than 50 minutes. The freshmen are new to the high school and come from multiple different middle schools. The fact that over 40% of the offenders being freshmen could be a result of not understanding the code of conduct, the discipline policies of SHS, and lack of knowledge of appropriate behavior. Additionally, further analysis of this 42% population revealed that most of their infractions are level 1 and level 2 dress code and attendance violations. When looking for the root cause of the attendance violations, it was determined that several factors contribute to the effect. First, the students systemically malinger in

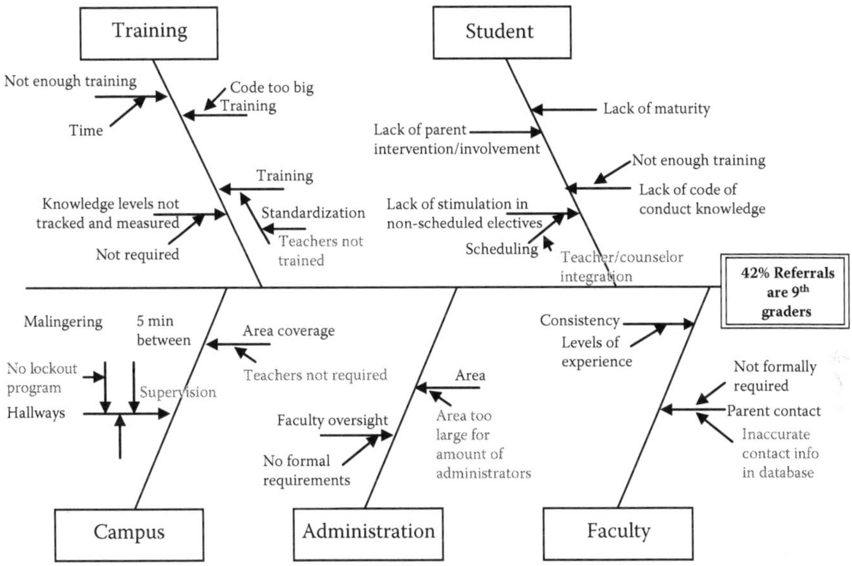

FIGURE 18.17
Cause and effect diagram: Why are most offenders freshmen?

the hallways between classes. There is little sense of urgency to move from one class to another. This coupled with the absence of faculty members in the hallways in between class and their nonenforcement of logging tardies into the system contribute to the problem of level 1 attendance infractions. A joint lockout program between administration and faculty is absent from the discipline/attendance system and potentially could significantly reduce the amount of attendance violators. Reducing attendance violations would have second- and third-order effects as the amount of constructive classroom time would increase while the amount of students receiving attendance contracts decrease.

When looking at the next area of potential improvement in Figure 18.18, common infractions, many of the same root causes contribute to 22% of the offense codes making up 80% of the number of offenses that are committed. The root causes for common offenses among the entire school lie with the training of students in their behavioral practices. Research proves that adolescents can be taught appropriate social behavior (Metzler et al., 2001). Of the 14 offense codes that account for the preponderance of offenses, most are level 1 and level 2 offenses. Attendance violations and dress code account for 40% of these offenses. When looking at the fishbone diagram for this effect, the root cause is drawn to the basic fact that there is no formal training program for faculty or students on the code of conduct. The administration of SHS must give consideration to the possibility that the county code of conduct guide is not enough guidance for this size of school with the demographics involved. The county code of conduct broadly covers unacceptable behavior

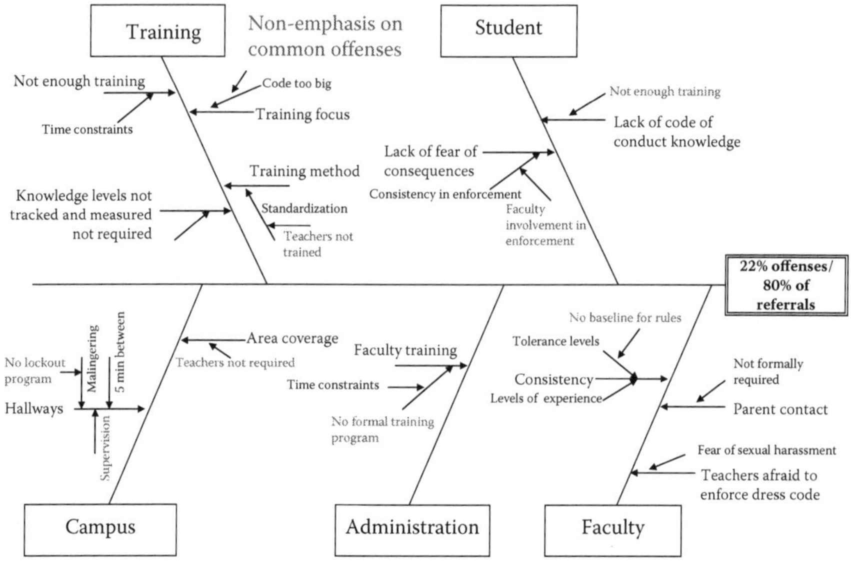

FIGURE 18.18
Cause and effect diagram: Why do 14 offense codes account for 80% of referrals?

for the entire county school system. SHS does not possess a SHS-specific guide for behavior within the high school. Absent with this document is a specific training program for faculty or students that emphasizes instruction on the common types of infractions that account for most offenses that are committed.

The last area of potential improvement the LSS Team focused on was the area of repeat offenders, shown in Figure 18.19. Repeat offenders are defined as those students who commit two or more violations of the student code of conduct. A cause and effect diagram for this area was constructed to determine the source of causes that contribute to the repeated noncompliance of the rules.

Of the repeat offenders, the number of freshmen violators increased from 42% to over 60% when looking at students with five of more violations. Again, this root cause returned us to the freshmen class and their lack of adherence to the code of conduct partially due to inadequate training. Additionally, the Six Sigma team determined that many students commit repeat offenses because of the lack of positive reinforcement in their home environments as well as no special programs developed for high-risk students and repeat offenders. Aside from developing specialized programs to reform the repeat offenders, the team also looked at the sociological aspect of the student. Lack of parental involvement is a significant factor in the reinforcement of appropriate behavior in any student. Students in today's society lack much of the parental involvement that students once had. Thirty percent of SHS students who responded to the student survey said their parents have no

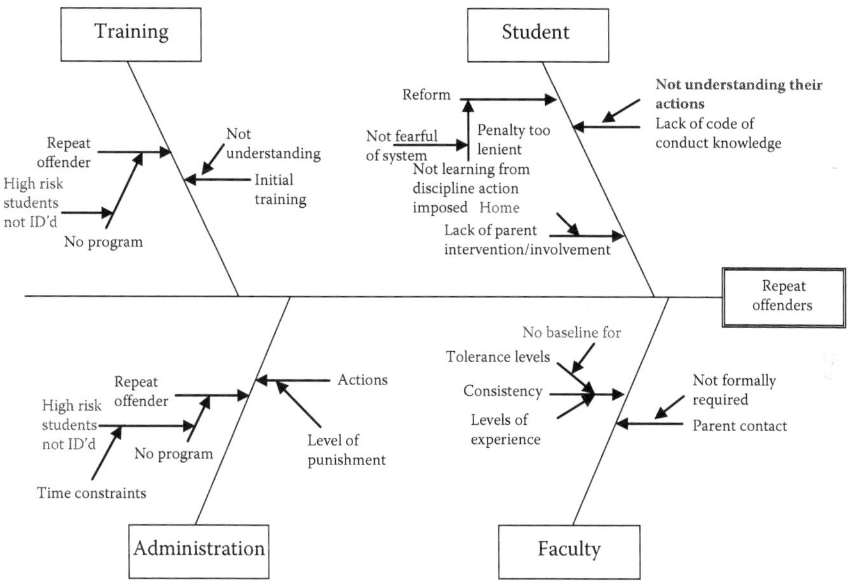

FIGURE 18.19
Cause and effect diagram: Why do students continue to commit offenses?

opinion of them getting into trouble at school nor do they have an opinion on the type or amount of punishment imposed. These students are disciplined the same way other students are disciplined, which is a reaction to negative behavior. This brings the question of whether or not a positive behavior support system would be appropriate for these students as well as all students at SHS. This cause and effect diagram depicts that students do not reform because they do not understand their actions/behavior are inappropriate. Additionally, no positive behavioral support system is in place that *teaches* the students appropriate behavior. Such a system would be proactive in lieu of a reactive negative support system.

Cause and Effect Matrix

A cause and effect matrix was used to understand if the same root causes contribute to multiple effects. It establishes the relationship Y = F(X), where Y equals the output variables, and X represents the input/process variables or root causes. The cause and effect matrix for the discipline process is shown in Figure 18.20. The total score can be used to understand where process improvement recommendations should be focused in the Improve phase. The consistency of the process, lack of training on the student code of conduct and the discipline process, and student lack of maturity are the top three causes that should be focused on in the Improve phase to identify improvement areas that can eliminate these root causes first.

	Effects				
	42% of offenders are freshmen	**22% of offense codes = 80% of infractions**	**Why repeat offenders?**	**Total**	**Relative weighting**
Causes/ Importance	10	3	5		
No training	9	3	9	144	2
Lack of parental involvement	3		9	75	5
Faculty experience	1	1	3	28	8
Lack of resources	9	3		99	4
Lack of consistency	9	9	9	162	1
No ID of high-risk students			9	45	6
Students not understanding appropriate behavior			9	45	6
Student lack of maturity	9		9	135	3

FIGURE 18.20
Cause and effect matrix.

Why–Why Diagram

An additional tool that was utilized to help determine root causes was the "5 Whys." The 5 Whys is a tool that causes a team to continue to ask the question "Why?" three to five times to drive the team deeper than a first-order cause to a deeper root cause. Again, in this project, we chose the three major areas of improvement discovered during the database analysis to derive root causes. The 5 Whys is used in conjunction with the cause and effect diagrams to determine the root causes.

The first 5 Whys analysis shown in Figure 18.21 correlates to the first cause and effect diagram. Why do 9th graders account for the bulk of the offenders in the high school? When looking at the Pareto diagram of the offenses they commit, they are mostly level 1 and level 2 offenses that largely consist of dress code violations, attendance violations, and disrespect. Why is this? They are new to the high school and to the discipline policies and procedures at this high school. Why? Because their level of knowledge is lower than the upperclassmen who have institutional knowledge from years prior, yet these freshmen get the same amount of training as the upperclassmen receive. Why do they receive the same amount of training? The root answer to this is because there is no special program in place for the freshmen to educate them on the appropriate behavior at this school.

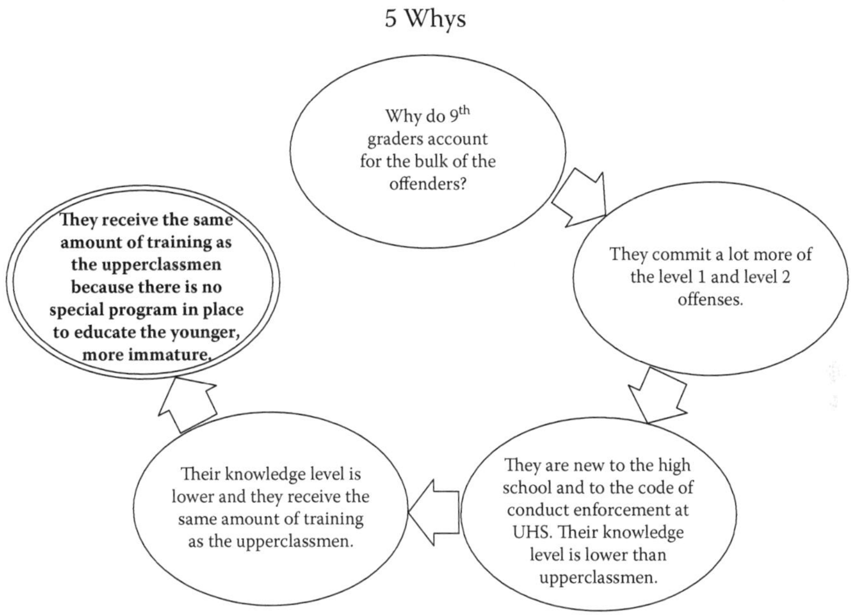

FIGURE 18.21
Why–why diagram: Why do 9th graders account for the bulk of the offenders?

The second 5 Whys analysis in Figure 18.22 correlates to the second cause and effect diagram and looks at why 14 of the 64 offense codes account for 80% of the offenses that are committed. The Six Sigma team only went through four iterations of this model to derive the root cause. Among other reasons noted in the cause and effect diagram, the team determined that the reason why these common offenses are so frequently committed is because these are the offense codes that have the highest opportunity for occurrence. For example, at every moment of the day, a student can violate the dress code or have the opportunity to skip class while other offense codes are not as opportunistic. However, these offense codes are given the same amount of focus when training the students on the code of conduct. This is because there is no special policy in place that first trains the faculty on the code of conduct or the SHS policy and then there is no program in place that trains the students in a consistent manner. The absence of such a policy neglects the potential of focusing on the common infractions when giving instructions on proper behavior. For example, if the school had a specific written policy on discipline at SHS, this policy might give the faculty a framework for instruction that emphasizes the common offenses. This framework is ideally written under the umbrella of the county's code of conduct.

Lastly, a 5 Whys analysis was conducted on why repeat offenders do not reform, which is shown in Figure 18.23. This analysis took the team beyond

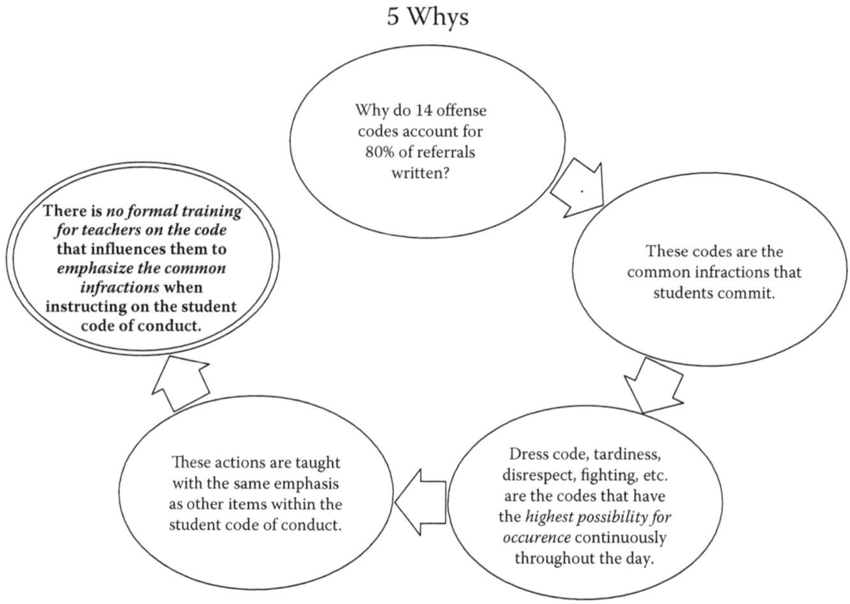

FIGURE 18.22
Why–why diagram: Why do 14 offense codes account for 80% of referrals written?

five iterations of why to arrive at a basic root cause. Many factors influence why or why not a student is reformed after an act of misbehavior. SHS has multiple students that are repeat offenders as defined by students who commit two or more offenses. Factors that range from not understanding their actions to inadequate training contribute to repeat offenses by the same student. This analysis took the team to the root cause that no program is in place at SHS that either identifies high-risk behavioral students before they become repeat offenders or once a student becomes a repeat offender, no monitoring program is in place to continuously monitor these students to track their behavior and help them reform. A program, through design, could identify students by race, gender, grade point average (GPA), and middle school referral history to identify potential behavioral problems before them committing multiple offenses during the school year. Not in place, but possible, a high school could complete an analysis of the incoming freshmen class to identify "high-risk" students for behavioral problems. By identifying this high-risk group, faculty members could be notified in an effort to focus behavioral training and policy to this subgroup.

Process Analysis

A process analysis was performed to identify the non-value-added activities in the classroom discipline process, the Dean's discipline process, and

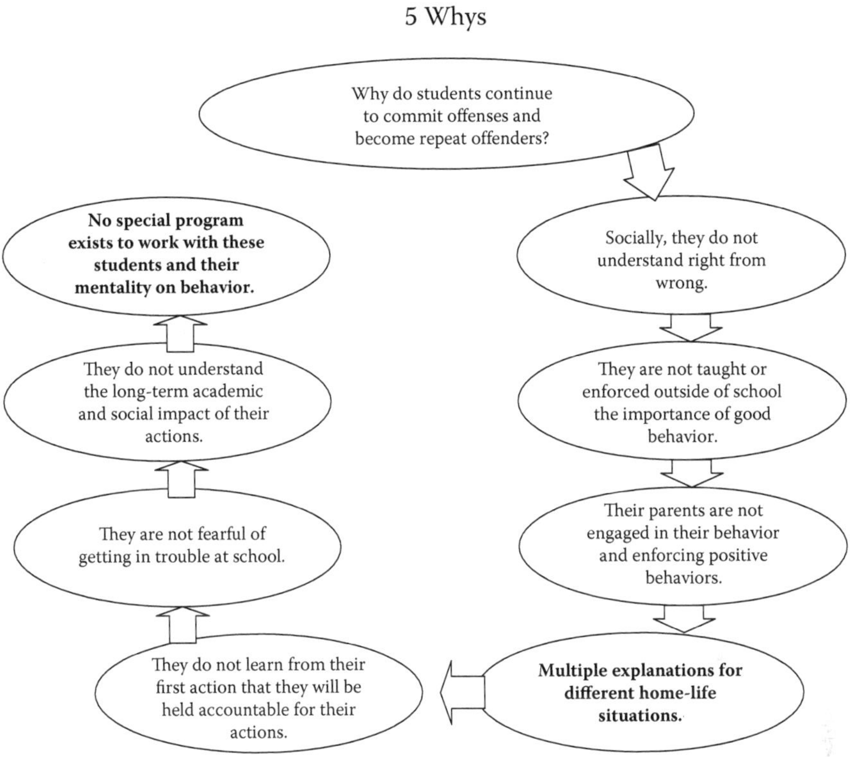

FIGURE 18.23
Why do students continue to commit offenses?

the attendance/behavioral contract process. This analysis can be used to focus improvement activities in the Improve phase. Although one could argue that the entire discipline process is non-value-added and that students should just behave, we can still differentiate between value-added and non-value-added activities. A value-added activity could be that a student receiving and serving their discipline consequence can provide value if he or she corrects the behavior. Another value-added activity might be meeting with parents to resolve the discipline issue. Figure 18.24 shows the process analysis.

Histogram, Graphical, and Data Analysis

Data Analysis

The majority of the offenders are freshmen, who represent 42% of the offender population. To comprehend these offenders' behaviors, further analysis was performed on this sample utilizing the school database inputs.

Process Step	Non-Value-Added	Value-Added
Classroom Discipline Process		
Student performs misconduct	X	
Teacher gives in-class discipline consequence		X
Call parent	X	
Contact parent		X
Look for phone number	X	
Meet with parent		X
Student completes consequence		X
Dean's Discipline Process		
Student performs misconduct	X	
Teacher sends student to Discipline Dean		X
Teacher fills out form	X	
Student waits for Dean	X	
Dean pulls up student info		X
Contact parent (level 2 or greater)		X
Open police investigation (level 4)		X
Dean completes discipline form	X	
Dean assigns consequence		X
Student completes consequence		X
Dean checks that student completes consequence	X	
Store discipline form electronically		X
Give copy to faculty	X	
Put copy in student folder	X	
Attendance/Behavioral Contract Process		
Attend meeting		X
Student put on contract	X	
Student gets signatures	X	
Attendance officer verifies	X	
Student violates attendance	X	
Give action		X
Student completes action		X
Store attendance calendar	X	

FIGURE 18.24
Process analysis for discipline process.

Unique characteristics of this sample are noted as follows:

- The average number of absent days is 9.
- Fourteen percent of the offenders received OSS and the average time is 3 days.
- When looking to the number of offenses committed by an offender in this sample, the average is two offenses.

- Attendance issues account for 22% of the offenses.
- Dress code violations account for 21% of the offenses.

Figure 18.25 shows the number of offenses by offense type.

Freshmen Offenders (Last Year)

The following analysis was performed on data from August 1 through February 23 to compare with the sample pulled from the same date range for this study. A list of offenders from the database was generated on this period; however, the graduated seniors who committed offenses during this time frame were not found in the school database.

A total of 918 offenses plus Senior offenses occurred during that time frame last year. The data revealed that most offenses were committed by Freshmen students, who represented 44% of the offender population. Although the number of offenses is reduced by 156 comparing to the current data, the freshmen offenders' percentage did not change significantly. Even if you added senior offenders of approximately 100 offenders, the percentage of freshmen offenders would only drop to 42%, which is the same proportion of offenders for this study.

In summary, it is a positive observation that the total number of referrals has decreased, but the high percentage of freshmen offenders did not decrease.

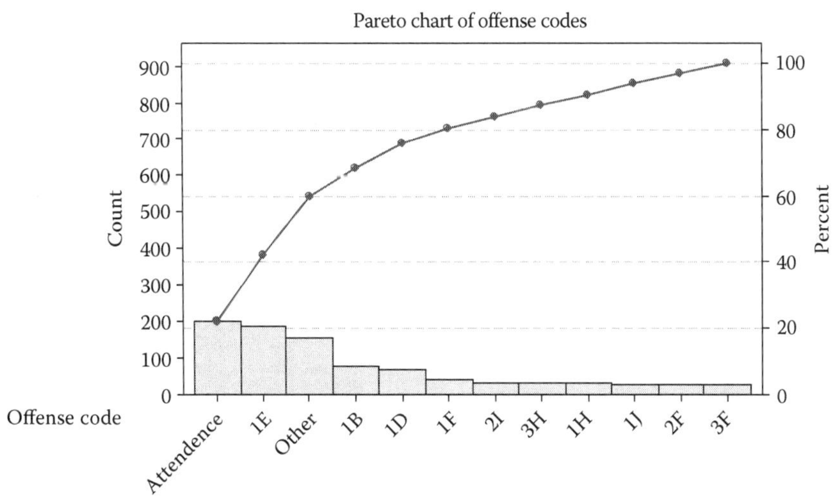

Offense code	Attendence	1E	Other	1B	1D	1F	2I	3H	1H	1J	2F	3F
Count	199	186	157	76	70	40	33	31	30	29	29	27
Percent	21.9	20.5	17. 3	8.4	7.7	4.4	3.6	3.4	3.3	3.2	3.2	3.0
Cum %	21.9	42.4	59.8	68.1	75.9	80.3	83.9	87.3	90.6	93.8	97.0	100.0

FIGURE 18.25
Pareto chart of the number of offenses by offense type.

Waste Analysis

To identify possible areas for improvement, the SHS Six Sigma Team used a systematic approach to identifying and eliminating waste through continuous improvement.

The LSS team identified the following types of wastes in relation to the SHS disciplinary process. Efforts to reduce these wastes will Lean the discipline process. The waste analysis is shown in Figure 18.26.

Description	Type of Waste	Explanation
Unnecessary Referrals	Overproduction	Unnecessary referrals are those offenses that could have been handled in class by the faculty member such as sending a student to the Dean's office right away instead of imposing lunch detention on the student or calling the student's parent.
Data Entering Errors	Defects	Since the information from the student referrals is uploaded by a person, errors may be present. There might be errors in the input of data into the database causing misleading information.
Repeat Offenders	Correction	Students are constantly being corrected for repeat violations of the code of conduct. This is a result of not enough instruction on proper behavior. Repeat offenses occur on common types of level 1 and 2 infractions.
Filling 1 Referral Forms	Inventory	This type of waste might be present when hard copies such as student calendars are stored in files. This is considered to be an inventory waste since this information would be pulled from the school database.
Student Signing Off Calendars	Motion	This process requires a great deal of motion in order to sign off the attendance calendar. By enforcing the classroom management program system, students will be logged during class periods, reducing the number of students walking in the hallways.
Referral Routing Procedure	Processing	By creating an electronic referral system, the Dean's secretary can be eliminated from the processing procedure all together. Instead of the Dean writing the offense action on the referral form and giving it to the Dean's secretary for input, he/she could simply input the offense action into the electronic referral system themselves.
Student Signing Off Calendars	Processing	When students are placed on attendance/behavioral contracts, they are required to obtain signatures from their teachers and parents. This task is time consuming for the students, teachers, and parents.

FIGURE 18.26
Waste analysis.

Referral Routing Procedure	Transportation	The referral system is antiquated considering the school and school system has a fairly modern and extensive IT network.
Data Input in the Database	Waiting	When referral is filled out by the Dean it is then handed to his/her secretary for data input. The referral has to wait until the secretary is not busy working on another task to upload it in the database.
Classroom Disruption	Waiting and People	When the class time has been interrupted by misconduct, students have to wait for the professor who is writing a referral to continue the lecture.
Students Malingering in the Hallways	Waiting	When students fail to get to class on time and faculty members do not strictly enforce the tardy system, teachers and students both end up waiting for the entire class to get seated and ready to conduct class.

FIGURE 18.26 (***Continued***)
Waste analysis.

Basic Statistics

From the student population, we identified 743 students who have one or more offenses in their record.

- Twenty-one percent of the students have one or more discipline referrals.
- Forty-two percent of the offenders are freshmen (see Figure 18.27).
- Forty-nine percent of the offenders are of Hispanic race (see Figure 18.27).
- Sixteen percent of the students are repeat offenders with more than one discipline referral; 9th graders make up the highest percentage of repeat offenders (see Figure 18.28).

The mean and standard deviation for the following variables are as follows:

- GPA: mean = 2.82, standard deviation = .97.
- Unexcused absences: mean = 5.42, standard deviation = 7.84.
- Excused absences: mean = 1.48, standard deviation = 2.58.
- Number of discipline referrals across all students: mean = 1.53, standard deviation = 5.06.
- Number of discipline referrals across students with discipline referrals: mean = 7.37, standard deviation = 8.93.

Grade	% Students by Grade		Race	% Students by Race	% Students in Student Population
9	43		Black	13	8
10	25		Hispanic	49	39
11	20		Caucasian	34	45
12	12		Other	4	8

FIGURE 18.27
Percentage of students with discipline referrals by grade and race.

Grade	% Students by Grade		Race	% Students by Race	% Students in Student Population
9	38		Black	12	8
10	28		Hispanic	50	39
11	21		Caucasian	34	45
12	13		Other	3	8

FIGURE 18.28
Percentage of students with discipline referrals that are repeat offenders (>1 referrals) by grade and race.

Confidence Intervals

The team calculated confidence intervals about the mean for the following variables:

- GPA: (2.79, 2.85).
- Unexcused absences: (5.16, 5.68).
- Excused absences: (1.39, 1.56).
- Number of discipline referrals across all students: (1.37, 1.70).
- Number of discipline referrals across students with discipline referrals: (7.07, 7.66).

Hypothesis Testing

We analyzed the following hypotheses using hypothesis tests:

- Is the GPA different for students with discipline issues and those without?
- Is the GPA different for students suspended versus not suspended?
- Is the average number of discipline referrals different by gender?

We found that there is a significant difference in GPA for students with discipline issues (2.17) and those without (2.99). The GPA for students with and without discipline issues is shown in Figure 18.29, with the boxplot of GPA in Figure 18.30.

We found that GPA is different for students suspended (1.90) versus not suspended (2.86), as shown in Figures 18.31 and 18.32.

	GPA
No Discipline Referrals	2.99
Discipline Referrals	2.17

FIGURE 18.29
GPA for students with discipline issues and those without.

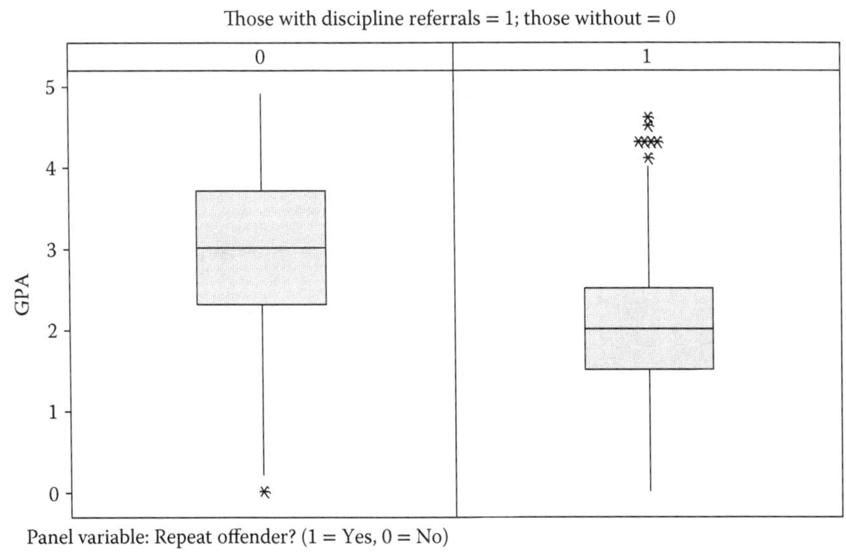

FIGURE 18.30
Boxplot of GPA for students with discipline issues and those without.

	GPA
Not Suspended	2.86
Suspended	1.90

FIGURE 18.31
Average GPA for students suspended and not suspended.

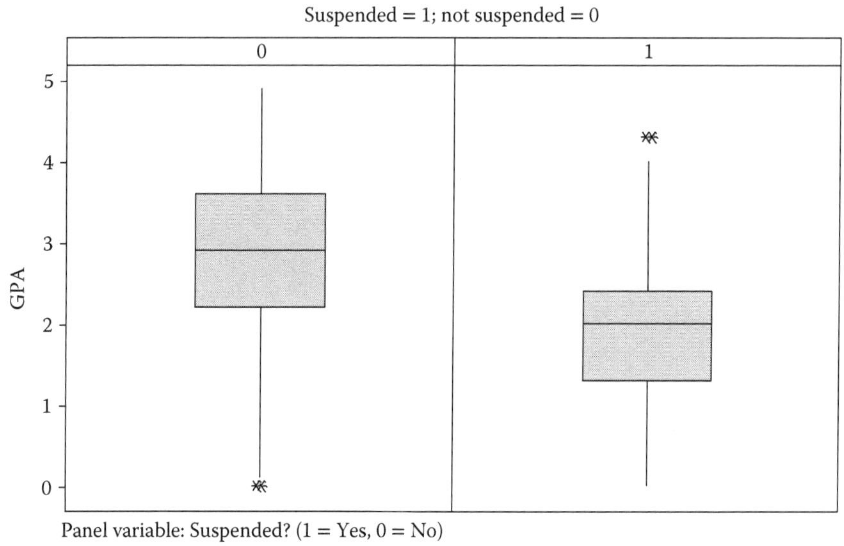

FIGURE 18.32
Boxplot of average GPA for students suspended and not suspended.

Analysis of Variance

We analyzed the following hypotheses using analysis of variance:

- GPA is the same by race.
- Average number of discipline referrals is the same by grade.
- Average number of discipline referrals is different by race.

We found that the GPA is significantly different by race. See Figure 18.33 for the average GPAs by race and Figure 18.34 for the boxplot of GPA versus race.

The average number of discipline referrals is different by grade. See Figure 18.35 for the average number of discipline referrals by grade and Figure 18.36 for the boxplot of discipline referrals by grade.

The average number of discipline referrals is different by race. See Figure 18.37 for the average number of discipline referrals by race and Figure 18.38 for the boxplot of discipline referrals by race.

Survey Analysis

Faculty Survey Summary of Data

The faculty survey revealed interesting perceptions on the level of knowledge of the code of conduct, training of the code of conduct, and other points of interest.

Race	GPA
Asian	3.47
Black	2.54
Hispanic	2.63
Indian	2.66
Mixed	2.67
White	2.97

FIGURE 18.33
GPA by race.

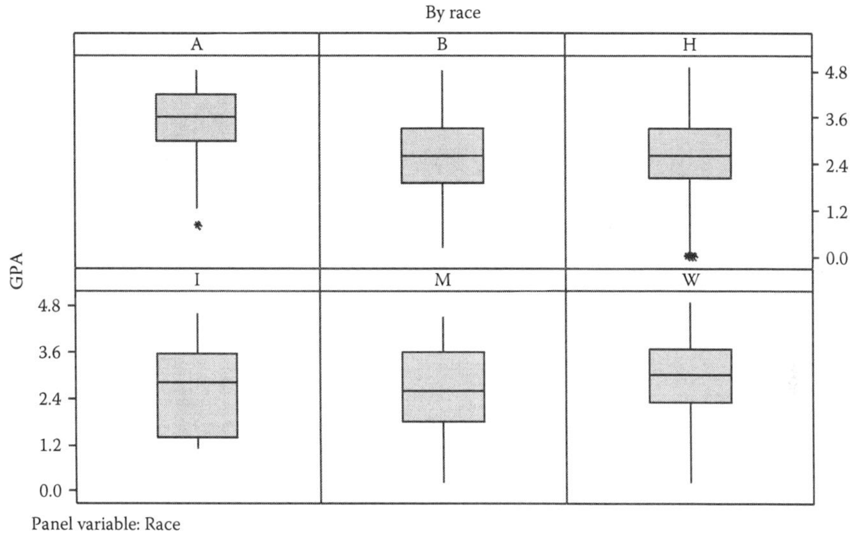

FIGURE 18.34
Boxplot of GPA versus race.

Grade	Average Number of Discipline Referrals
9	6.37
10	8.20
11	7.52
12	8.78

FIGURE 18.35
Average number of discipline referrals by grade.

Seventy-five percent of the respondents noted that the most important thing to them regarding the discipline program is the minimization of classroom discipline issues. This means they want a program that minimizes classroom disruption, disrespect, tardies, and other discipline issues. With regard to discipline in the classroom, the number one weakness that the

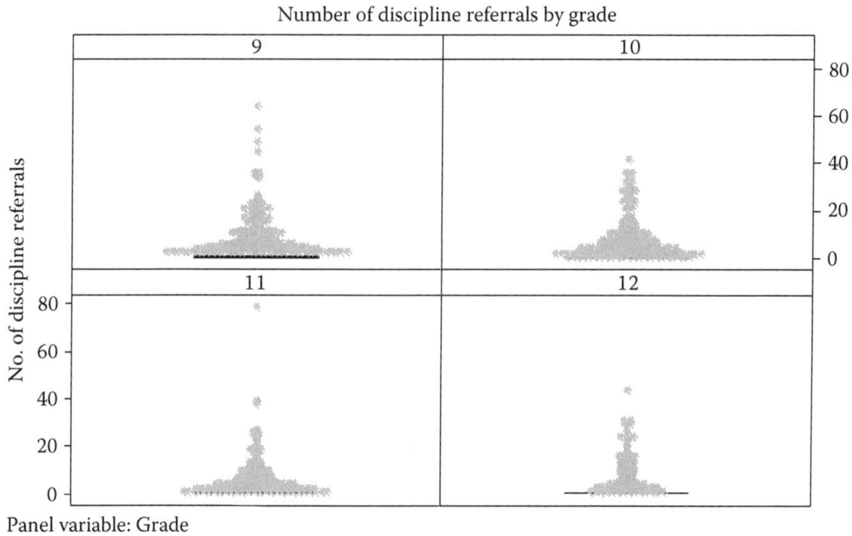

FIGURE 18.36
Boxplot of discipline referrals by grade.

Race	Number of Discipline Referrals
Asian	5.76
Black	7.04
Hispanic	7.30
Indian	6.50
Mixed	5.50
White	7.66

FIGURE 18.37
Average number of discipline referrals by race.

faculty identified was a lack of consistency across the board when it comes to classroom discipline policies and administration of policies. The faculty believes that some faculty members tend to let students slide when it comes to code of conduct infractions, thus making it harder to enforce these infractions in their classrooms. Additionally, the faculty believes that there is a lack of parent/teacher/counselor integration. They believe that these three areas are not linked in a fashion that enables the student to be highly successful.

The respondents generally ranked their personal level of knowledge of the code of conduct high, their peers' level of knowledge moderate, and the students' level of knowledge low. While at the same time, feeling that the level of training of the code of conduct is "not enough" and they admit that they spend "less than a period" instructing the students on this topic.

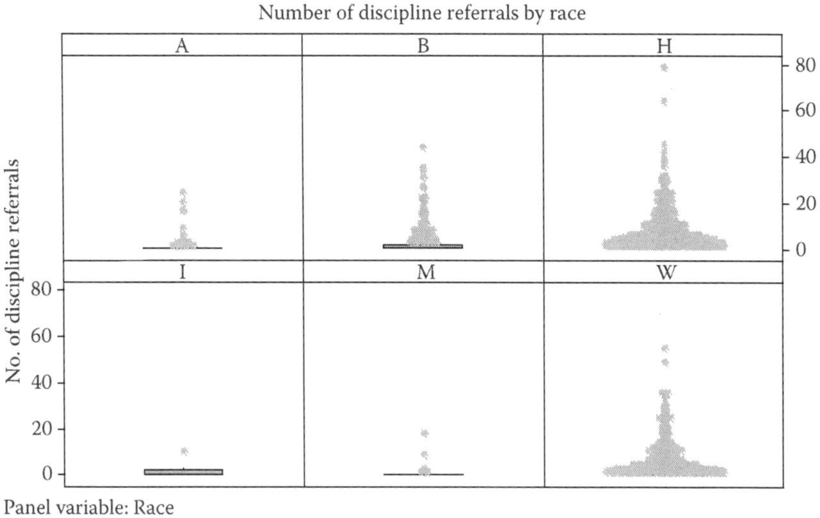

FIGURE 18.38
Boxplot of discipline referrals by race.

Several areas of concern were identified by responses from the faculty survey. First, 31% responded that it is "not very important" to prepare school work for students who are ISS. That would equate to almost a third of the faculty who do not understand the academic importance to those students who are serving in ISS.

Second, 36% of the respondents sometimes or never log tardies into the attendance system. This is a clear indicator of the problem of so many students remaining in the hallways after each period's bell rings. However, 87% of the respondents do think it is important to log tardies into the system. This data is backed up by the student survey in that 57% of the respondents state that three or less of their teachers count them tardy when late to class.

Several areas show favorable perceptions of the faculty. Seventy-four percent of the respondents are satisfied or very satisfied with classroom discipline at SHS. Seventy-six percent of the respondents are satisfied or very satisfied with the Dean's office discipline at SHS. Sixty-nine percent of the respondents are satisfied/very satisfied with the overall discipline at SHS. Complimentary of this data is the 69% response rate that the administration will back them when it comes to discipline-related issues.

Also positive to note is the favorable response for potentially implementing a Positive Behavior System (PBS) at SHS. The goal of the PBS is to create a baseline standard for classroom discipline as well as rewarding students positively or negatively for overall behavior. Seventy-three percent of the respondents stated they would support a PBS system at SHS.

Open-ended questions offered some constructive recommendations from faculty members. A few of the constructive comments provided by faculty members is as follows:

> Remove chronic disrupters into an 'at-risk' program
>
> More parent involvement and motivation is necessary!!! Most of the problems come from teachers who let their students talk while they are teaching, and then those students think they can do so in another teacher's class. Teachers need to learn to be the BOSS!!! ….
>
> Review Code of Conduct with all incoming Freshman and Parents. A required night/program they must attend before they enroll their child. This could be video that middle of the year transfers must watch with parents.
>
> Maybe semester trips for students without any form of discipline referrals. My school did it when I was in high school and it worked for us.
>
> Consistency would be the key ingredient that's necessary to improve the discipline process at SHS.
>
> CONSISTENCY … too many teachers allow students to wander hallways without passes, allow food/drink in class, let classes go early, don't count tardies, etc.

Demographics concluded that the respondents covered a wide spectrum of faculty members. The age range, length of teaching experience, and curriculum were fairly evenly distributed while type of classes massed on General Education and Combination. This indicated that the population of students that the respondents are exposed to is diverse.

Student Survey Summary of Data

The student survey that was used to gather the student VOC is included in the instructor's guide.

The student survey revealed interesting points of interest. Four freshmen teachers' classes and two upperclass teachers' classes participated in the survey. The result was 543 students who participated of which 71% were freshmen and 29% were upperclassmen. The response of the student survey skews toward the freshmen response, but coincidentally, the database data points toward the freshmen campus as the source of the majority of the discipline issues.

Sixty-nine percent of the student respondents feel the discipline imposed by their classroom teachers is "just right" and 52% feel the punishment imposed by the discipline deans is "just right."

Eighty-nine percent of the students stated that they only viewed three or less of their teachers contacting parents when it comes to classroom-related discipline issues. This is an interesting contradiction to 31% of the faculty respondents stating that the most effective form of in-class discipline is calling the student's parent! This issue is further complicated by faculty comments on issues with inaccurate parent contact information in the school database. Furthermore, it was estimated by one dean that the success rate for contacting a parent is less than 20%.

Of the students who took the survey and have had a referral, the students believed that 34% of their parents had "no opinion" of them getting in trouble. Furthermore, the students believed that 31% had "no opinion" of the punishment their child received at school. This possibly infers that up to a third of the parents of SHS students have no opinion or are disengaged from their children getting into trouble at SHS.

Also, another interesting note is when asked what discipline actions the students dislike the most, attendance contract, detention, and ISS were the actions most disliked. These actions cause an inconvenience to the students and/or take time away from them. These actions should prove to be the most effective form of punishment to deter repeat offenses.

DPPM/DPMO

The DPPM (Defective Parts Per Million) for the discipline time (see the section on "Process Capability") is 671,990, which equates to a little over 1.0 sigma.

Process Capability

We performed a process capability analysis for the average discipline time. The lower specification limit was identified as 10 minutes, and the upper specification limit was 30 minutes. The process is not capable related to the discipline time at the Dean's office, as shown is Figure 18.39.

Improve Phase

Following is the Improve phase for the SHS Discipline Process Improvement project, including the key deliverables developed as part of the prior exercises.

The Improve phase of the DMAIC process is designed to identify improvement recommendations, implement them, and then assess the improvement. The objectives of this phase in relation to the SHS Discipline Improvement Project are as follows:

- Identify the improvement recommendations.
- Develop the action plans for implementation.
- Pilot the improvement recommendations.
- Assess the improvement.

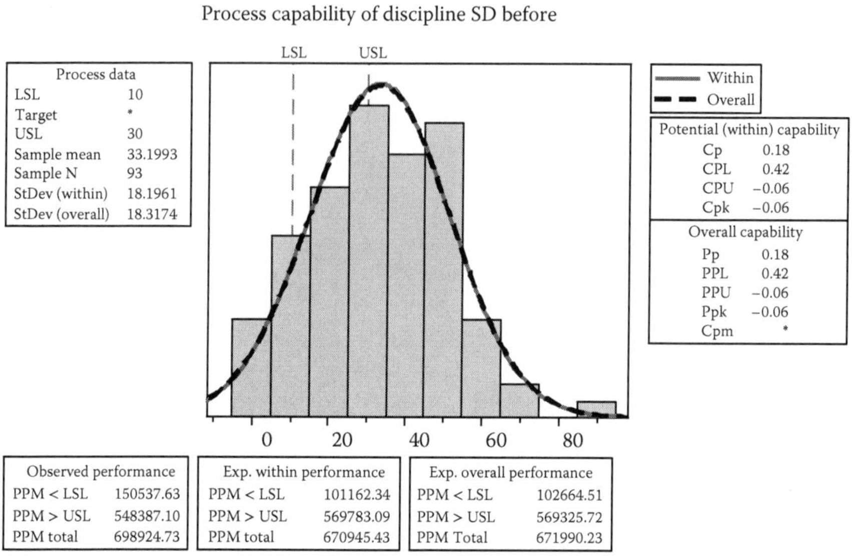

FIGURE 18.39
Process capability of discipline referrals before improvement.

Recommendations for Improvement

In the Improve phase, the analysis of the data measured is used to make improvements to the process or system. The LSS team put together multiple recommendations that would improve the overall discipline system and subsequently the academic environment as a whole. With the recommendations, suggested implementation plans are included as a guide for the SHS leadership team to follow when ultimately designing and implementing changes to the system. The recommendations that follow are based on data collected from the student database, teacher and faculty surveys, interviews with leadership team staff and faculty members, and benchmarking of best practices in like educational systems.

Recommendation 1

Create a unique and tailored SHS Discipline Program. This is a written publication that outlines the specific SHS policy on discipline and behavior. The publication would cover all aspects of discipline including training, faculty responsibilities, and student responsibilities. The guide would outline specific punishments that would result from specific offenses by the student, that is, clarifying the rules. Simply running the discipline program under the umbrella of the county code of conduct is not enough. Many schools have taken their county conduct policies a step further and published their own guide to discipline. This creates a baseline for classroom discipline that will help to create consistency among classrooms.

Recommendation 2

Create discipline dashboard for the SHS Principal. Develop control charts and post on a "dashboard" weekly discrete data that will be used to make decisions on the discipline/attendance programs. The dashboard is developed by the Principal with consultation from the discipline and attendance deans. The dashboard is given to the principal at the end of each week and is a visual snapshot of the current state of both programs. Control charts are statistical tools used to monitor processes and to ensure that they remain "In Control" or stable. Moreover, it helps to distinguish between process variation resulting from common causes and variation resulting from special causes.

Data Basis

A tool/mean is needed to assist the SHS administration to closely monitor the discipline process. A discipline dashboard is a very effective tool to communicate and decide on what type of action is needed if an out of control pattern is recognized.

Recommendation 3

Create a behavioral program specifically designed for the 9th grade campus. Consider implementing a Positive Behavior System (PBS) for the 9th grade campus. The PBS is a research-proven system that creates consistency in classroom discipline (i.e., creating baselines for acceptable behavior). PBS is data-driven and can be individually tailored for the specific school it is implemented in. The implementation of a PBS system is not simple, but the rewards for the work put forth are invaluable to the teaching environment as a whole. This team recommends a PBS team be formed as soon as possible to create the tailored PBS system for the freshmen campus. This team should be formed from the Discipline Deans' staff, teachers with strong discipline skills, counselors, or teachers with child development experience. The team should be trained in the specifics and history of PBS before they convene to create the SHS PBS system. The team should then concentrate their efforts to create a system by the next school year for implementation with next year's freshmen class.

Data Basis

Forty-two percent of the offender population at SHS is freshmen. Although the total number of referrals at SHS has dropped with the inception of the 9th grade campus, the percentage of freshmen offenders has remained the same. Additionally, incoming freshmen receive the same amount of training on the code of conduct as upperclassmen who have institutional knowledge and experience at the high school. The majority of the teacher respondents say they spend a period or less time per semester reviewing the content of the code of conduct while they also say that the amount of training spent on this training is not enough. The preponderance of the freshmen's first week at SHS should consist of orientations to the school, programs, and the discipline guide. Sometimes school leadership may be hesitant to take that much time away from classroom time, but in the long run, establishing solid discipline expectations will result in better classroom environments and more class time due to reduction in disruptive situations. Research has shown that punitive school and classroom environments, unclear rules and expectations, and inconsistent application of consequences contribute to increased levels of student antisocial behavior and truancy (Metzler et al., 2001). This is the situation currently within the freshmen ranks. Discipline is not reinforced in the parental home environments in today's society as it was years ago. Students are not as fearful of the consequences they face for inappropriate behavior. Because this team or SHS cannot impose positive parental support for the parents of 3600 students, the issue of teaching appropriate behavior must be attacked from a different angle. PBS allows the school to create a positive environment to teach and reinforce positive behavior. This system is perfect for the freshmen campus at SHS because the maturity level is very close to that at a middle school level where PBS has proven successful in multiple studies.

Recommendation 4

Identify high-risk freshmen before the school year and monitor their status the first quarter of the school year. This program should be done the month before the start of the school year when SHS has a fairly solid roster of the incoming freshmen class. A query can then be generated using FileMaker Pro, then sorted to identify freshmen who, in middle school, have multiple referrals, a low GPA, attendance problems, and low FCAT scores. This population is then considered the high-risk population for the incoming freshmen class. These are going to be the majority of students who fall within the 42% population of offenders. With this population generated, they can now be put into classrooms with teachers known for strong classroom management skills. This entire recommendation can be accomplished by discipline deans identifying the high-risk students and then having the guidance counselors ensure they are put into the proper classrooms where they are set up for success. For example, a high-risk student taking Biology should be put in Mr. V's class. A high-risk student taking English should be put into Mrs. B's class.

Data Basis

Students with low GPAs, low FCAT scores, and so on statistically have discipline issues. There is no special program in place that identifies these students before they commit offenses. With minimal effort, this can become a proactive system in lieu of the current reactive system.

Recommendation 5

Emphasize common offenses when training students on acceptable behavior at SHS. Concentrate training on the 14 common offenses that account for 80% of the offenses committed. When developing the SHS Discipline Guide, publish specific consequences for the common types of offenses. For example, publish a table that mirrors specific consequences for five or more tardies, 10 or more, and so on. When publishing this policy for the common offenses, all teachers will have a baseline of consequence to train the students.

Data Basis

Twenty-two percent of the offenses on the referral form (14 offenses) account for 80% of the offenses committed. The number one weakest area of classroom discipline as answered by the faculty survey was a lack of consistency among the teachers. Additionally, open-ended comments for this survey revealed the common theme that they want better consistency in discipline. The published policy with emphasis on the common offenses will provide this baseline for which consistency will result.

Recommendation 6

Impose on the faculty members the importance of logging tardies and their responsibility to monitor the hallways in between classes. Instead of administrative staff patrolling the hallways and pushing students to class, administrators should focus their efforts on managing the faculty to push the flow of students through their hallways. As part of teacher orientation, train the faculty on the importance of accurate tardy reporting and their responsibilities.

Data Basis

By contract, the teachers are required to be in the hallways greeting their students at every class. This is not happening at SHS and the preponderance of teachers do not get out in the hallways. This allows for student malingering on their way to class and increases the chances for hallway disruptions and probability of a high tardiness rate.

Recommendation 7

Establish an alternate consequence schedule for students who are in the lower 30% FCAT population. When a student is sent to the dean's office for a referral, one of the first actions by the dean is to check to see if the student is in the lower 30% FCAT population. If that student is in that population, then special consideration is given for the punishment of that student. Unless absolutely necessary, this student should not be given OSS.

Data Basis

Students who rank in the lower 30% statistically have lower GPAs. Students who are given out-of-school suspension have statistically lower GPAs than those who do not serve OSS. By giving a student who is in the lower 30% FCAT OSS is a detriment to the student's chance of improving. If possible, give students in the lower 30% ISS where they can be controlled and given additional reading assignments.

Recommendation 8

Create a faculty reward system for active discipline and classroom management skills.

Implementation

Create a small committee who can select a faculty member monthly who demonstrates outstanding classroom management skills and practices.

Data Basis

Currently, there is no positive environment award system in place to influence faculty members to actively promote good discipline at SHS. The level of classroom management skills varies across the faculty. If good teachers are rewarded for good practices, others will notice and learn from those teachers' tacit knowledge.

Recommendation 9

Create a parental involvement contract for repeat offenders. This is a written contract that will apply only to those students that are repeat offenders. This contract could be part of the code of conduct contract in the student enrollment process to acknowledge not only the parents of those repeat offenders but also the whole population. Parents of repeat offenders will acknowledge their written commitment to this contract and this can be performed at the beginning of each school year or semester. This contract can include a minimum number of community service hours or school involvement if their adolescent becomes a repeat offender. The primary goal of creating this parental involvement contract for those repeat offenders is to increase parental reaction when students perform a misconduct in the school.

This requirement could be initiated at the beginning of the school year and is part of the written acknowledgment of the SHS discipline policy guide and the county's code of conduct. The requirement of this parental contract would be included in the SHS discipline guide and would outline the consequences the parents must meet should their adolescent become a repeat offender at the school. Whether the stipulations are service at the school or mandatory parental counseling, the contract is designed to influence more parental involvement for the repeat offenders.

Data Basis

When looking at the repeat offender population survey respondents, 30% state their parents have no opinion at all when their son/daughter has a misconduct in school. Evidence shows that when schools work together with families to support learning, children tend to succeed not just in school but also throughout life. Recent research has shown that, particularly for students who have reached high school, the type of parental involvement that has the most impact on student performance requires their direct participation in school activities.

Recommendation 10

Create a knowledge-sharing program for classroom management best practices. This will include the involvement of teachers to discover best practices of effective classroom management. By creating this knowledge-sharing program, teachers will have the opportunity to share their strategies with respect to classroom management with other teachers. Teachers will learn from their colleagues how to work with students who have many types of special needs and apply a variety of management techniques to help students become self-regulated learners. This program could help the entire faculty population in learning how to increase student motivation, build student–teacher relationships, and increase home–school communication. The main purpose of creating this knowledge-sharing program is to enable teachers from SHS to learn from the experiences, methodologies, and achievements of colleagues. There exists a variety of information and communication technologies that may be used by the teachers to communicate and share their ideas and inputs on the topic. A knowledge management system for this area could be as simple as a best practices committee that publishes a bi-semester newsletter to a more complicated information technology design that stores best practices in a database.

Data Basis

The underlying theme to the respondents in the faculty survey is that they desire consistency among their peers in classroom discipline. The demographics among the teaching staff is very broad and the discipline management ability is just as broad. Many of the teachers who have weaker classroom management/discipline abilities could leverage from the experience that lies with many of the more experienced teachers. Currently, there is no knowledge-sharing system in place that gives teachers the opportunity to share their experiences, techniques, and methodology with respect to classroom management. By creating this system, teachers and students will benefit from such a positive improvement recommendation.

Recommendation 11

When a referral is issued and it is necessary to impose a disciplinary action, utilize lunch/after school detention (action code G) and ISS (action code P) as the primary actions for the most common offenses found. If a student repeats the offense, consider imposing an attendance/behavioral contract (action code I) to the student since it is the most undesirable action according to the student survey.

Data Basis

When looking at the most frequent actions taken during the study period, verbal reprimand (action code C), counseling and direction (code B), and parental contact (action code A) were the most frequent discipline consequences. However, the student survey revealed that the students consider the attendance/behavioral contract, lunch/after school detention, and ISS as the three most effective (undesirable) disciplinary actions, respectively. Therefore, if those actions are applied more often, the likelihood that the student will repeat the offense may be lower.

Recommendation 12

Consider imposing ISS rather than OSS *unless absolutely necessary.*

Data Basis

ISS ranked as the third most undesirable action imposed by the school. Moreover, it was statistically proven that students who are suspended tend to have a lower GPA in contrast to those who are not suspended. Consequently, OSS is one of the factors that affect the academic performance of a student, which may be mitigated by applying ISS.

Recommendation 13

Automate the referral process by developing a reliable program. This program should replace the current referral process. Seek funding to support a project for developing a program that is capable of automating and managing classroom referrals. To minimize the use of referral forms (paper-based), a replica of the form should be electronically developed. On top of that, limited access has to be given to the faculty members. Teachers could enter the necessary data while the student walks to the dean's office. A referral will be generated and queued under the new referral list. As soon as the student walks into the Dean's office, the Dean can pull that referral along with the student history just by one simple click. After deciding which consequences are assigned, the Dean can enter the required information into the program. Once this step is done, an automated e-mail goes to the teachers and to the parents explaining the nature of the misconduct and consequences.

To have a successful program, the essential requirements of the automated system are as follows:

- It should be capable of handling every task in the referral process.
- Deans/faculty inputs should be incorporated during the development cycle.
- The interface should be user-friendly.
- It should have the ability to retrieve/pull up historical data from different databases or systems. The system should be integrated with the existing programs.
- The system should be reliable and available.
- Some of the fields in the program have to be mandatory to prevent any type of error.

Data Basis

The referral system is antiquated considering the school and county has a fairly modern and extensive IT network.

Action Plan

A Pareto chart (Figure 18.40) shows the prioritized list of recommendations to identify which improvement recommendations should be implemented first.

Following in Figure 18.41 is a summary of the action plan with the recommended improvements and the time frame to implement them.

Future State Process Map

A revised process map incorporating the improvement recommendations is shown in Figure 18.42.

Training Plans, Procedures

Train incoming freshmen and parents during summer orientation on the new standardized discipline process. Train current students in assemblies. Train faculty during faculty meetings on the new standardized discipline process.

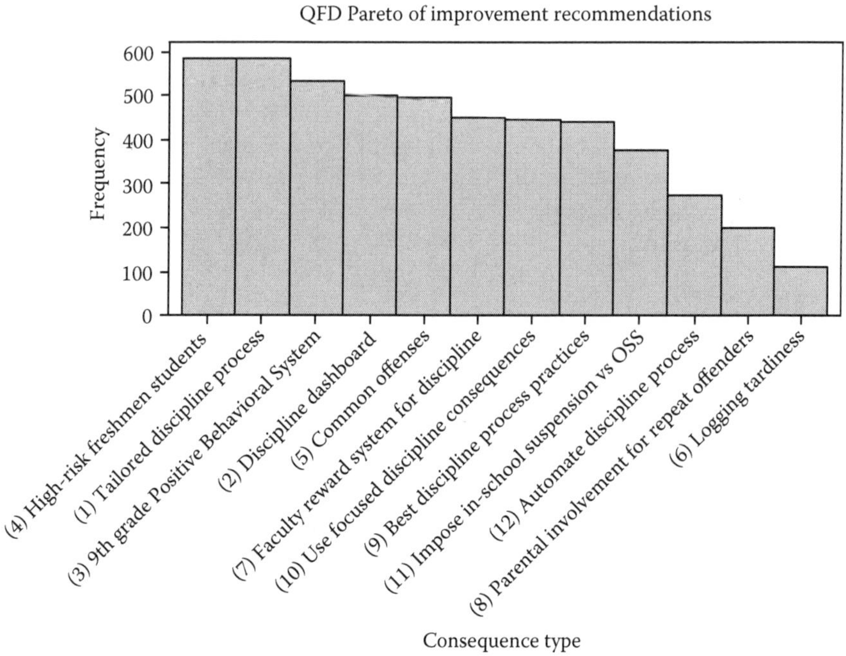

Frequency	588	586	534	504	496	450	448	441	376	276	198	110
Percent	11.7	11.7	10.7	10.1	9.9	9.0	8.9	8.8	7.5	5.5	4.0	2.2
Cum %	11.7	23.4	34.1	44.2	54.1	63.1	72.0	80.8	88.3	93.8	97.8	100.0

FIGURE 18.40
QFD Pareto of improvement recommendations.

Recommendations	Priority	Time frame
(4) High-risk freshmen students	588	Month 1
(1) Tailored discipline process	586	Month 3
(3) 9th grade (Positive) Behavioral System	534	Next Fall
(2) Discipline dashboard	504	Month 1
(5) Common offenses	496	Month 3
(7) Faculty reward system for discipline	450	Months 3 to 6
(10) Use focused discipline consequences	448	Month 3
(9) Best discipline process practices	441	Month 3
(11) Impose in-school suspension versus OSS	376	Month 3
(12) Automate discipline process	276	Long Term
(8) Parental involvement for repeat offenders	198	Month 12
(6) Logging tardiness	110	Month 12

FIGURE 18.41
Action plan.

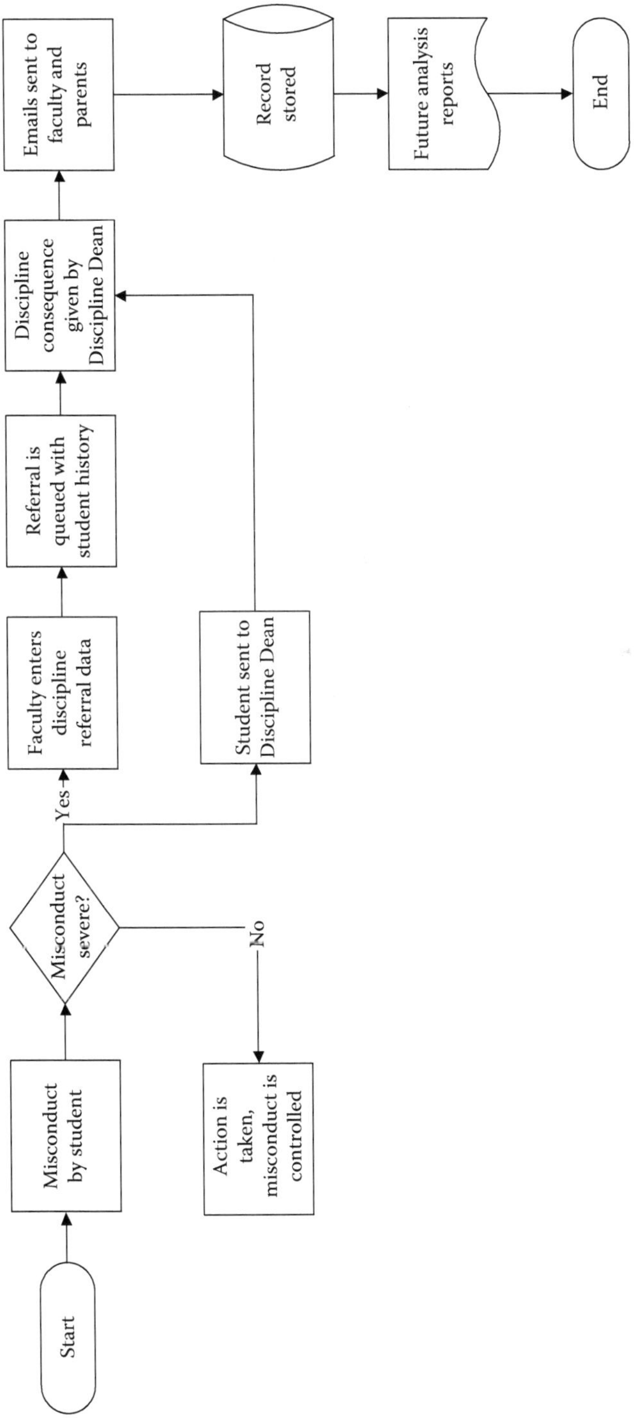

FIGURE 18.42
Future process map.

Control Phase

Following is the Control phase for the SHS Discipline Process Improvement project, including the key deliverables developed as part of the prior exercises.

The purpose of the Control phase of the DMAIC process is to design, develop, and incorporate controls into the improved processes. The activities of this phase are to

- Assess the gains that were realized by implementing the improvement recommendations in the Improve phase.
- Develop the control plan to maintain the gains.
- Standardize the process.
- Develop future plans for improvement.

Hypothesis Tests, Design of Experiments

We performed a two-proportion test on the total number of discipline referrals before the improvements (5430) and after the improvements (4698). The p-value was 0, so we reject the null hypothesis that there is no difference in the number of discipline referrals before and after the improvements. Therefore, we conclude that the total number of discipline referrals was *significantly* reduced after the improvement recommendations were implemented.

Mistake Proofing

Ideas for mistake proofing are as follows:

- Automate the discipline process so that the teacher can create the discipline referral on-line, on-time, or scan the document and send through e-mail to the office.
- Implement a process to verify student and parent contact information. Audit data by calling parents on a sampling basis once a month.

Control Plan

The Revised VOP matrix is shown in Figure 18.43. It provides the updated targets for each CTS and metric. The control plan for the recommendations is shown in Figure 18.44.

Process Capability, DPMO

The average discipline time before the improvements were implemented was 33.2 minutes and after was 29.50 minutes. We first tested for normality

CTS	Factors	Operational Definition	Metric	Target
Minimize Classroom discipline Issues	Freshmen training on code Clear guidelines	Training exists and is performed Clear guidelines exist	Number of disruptions	Reduce number of disruptions by 50%
Classroom discipline consistency	Guidelines Teacher training	Clear guidelines exist Teacher training each year	Guidelines Number of faculty trained	100% of faculty are trained within 3 months of hire or Jan. 1
Teacher/ parent/ counselor integration	Apathy of parents Data missing	Engaged parents assessed by parent survey	% of responses on survey for identified questions	Increase % of ratings in high categories by 10%
Adherence to code of conduct	Training Expectations	All students will be trained in code of conduct for 2 hours per semester Clear expectations conveyed	Number of students trained	100% of students are trained within first month of school or transfer
Classroom control	Teacher training	All teachers trained in classroom management Mentors for new teachers	Number of teachers trained Number of teachers with mentors Rating of mentoring program	100% of teachers trained 100% of new teachers have a mentor
Teacher/ parent contact	Apathy of parents Data missing	Engaged parents assessed by parent survey	% of responses on survey for identified questions	Increase % of ratings in high categories by 10%
Reduction of referrals	Freshmen No emphasis on common offenses	Guidelines	Freshmen and transfers trained in code of conduct and guidelines	100% of freshmen and transfers trained in code of conduct and guidelines

FIGURE 18.43
Revised VOP matrix.

and concluded that the discipline time appears to be a normal distribution. We checked for equal variances, found a p-value of 0, and concluded that the variances are not equal. We then did a t-test for unequal variances, where the p-value was 0.085. We failed to reject the null hypothesis and concluded that there is no significant difference in the discipline time before and after the improvements were implemented. There is still room for improving the process to reduce the discipline waiting time.

Recommendations	Control Plan
High-risk freshmen students	This recommendation can be controlled by evaluating at the mid-term and end of semester marks how many of these students encounter discipline actions.
Tailored discipline process	This recommendation would be controlled and evaluated after the fall semester when discipline data is measured against the prior year's data. Additionally, teacher feedback for the policies created within the guide should be solicited with revisions planned for future editions.
9th grade (positive) behavioral system	The use of the dashboard would assess whether the Positive Behavioral System would have a positive impact on reducing the number of discipline referrals school-wide.
Discipline dashboard	Implement control charts to monitor the number of discipline referrals. If the process appeared to be out of control, the Deans will know immediately as they are indicated on the control charts. The Principal will know weekly the status of the process as reported on his dashboard.
Common offenses	Provide a weekly report on the Principal's dashboard that shows the number of offenses by type for the previous week. The Principal can then react to that data to provide his/her guidance to the teachers to impart on their students.
Faculty reward system for discipline	Create a small committee who can select a faculty member monthly who demonstrates outstanding classroom management skills and practices.
Use focused discipline consequences	Monitor the number of referrals issued per month and compare the results against previous years to determine if the referrals have been reduced and to make sure the gain attained is sustained.
Best discipline process practices	Monitoring and evaluating the teacher's participation on this recommended system would be managed through the school established procedures of teacher reviews. This program can be performed at the end of each school year.
Impose in-school suspension versus out-of-school suspension	After the fall semester, compare the students who where out-of-school suspended and in-school suspended, and determine if those who received in-school suspension have less repeat offenses and better academic performance than those who received out-of-school suspension.
Automate discipline process	On a monthly/weekly basis, generate a report to highlight how many offenses were committed. Identify the time to receive referrals at the Dean's office after implementing the automated system. After analyzing that report, a proper action has to be taken to mitigate the situation if needed.
Parental involvement for repeat offenders	This recommendation would be controlled by the number of repeat offender parents that actually get involved.
Logging tardiness	Measure the number of teachers seen at their doors every period. This is always a reflection of the number of tardies reported and one of the items that is reported weekly to the Principal on his "dashboard."

FIGURE 18.44
Control plan.

Discipline Scorecard	
Week ending:	8/12/2007
Weekly Number of Discipline Referrals	42
Level 4	0
Level 3	0
Level 2	0
Level 1	42
9th Graders	20
10th Graders	13
Number of Behavior Contracts:	35
Number of Attendance Contracts:	56
Number of Days Missed Excused:	219
Number of Days Missed Unexcused:	1096

FIGURE 18.45
Discipline Scorecard sample.

Standard Work

The discipline process procedures were documented and standardized across the Discipline Deans. The best practice classroom management techniques were shared across the faculty.

Dashboards/Scorecards

The dashboard provides a systems view of the entire process and the critical metrics. Following is the proposed Principal's dashboard that includes tracking number of discipline referrals by week; the number of discipline referrals by severity level; the number of discipline referrals by 9th and 10th graders; the number of students on behavior and attendance contracts; and the numbers of days missed that were excused and unexcused across the student body. Target levels were identified that would identify the area to be green (OK), yellow (warning—carefully watch), and red (investigate the problem). A sample dashboard is presented in Figure 18.45.

Conclusions

The LSS tools and methodology applied were very successful in providing a systematic approach that was effective in attaining consistency in the student discipline process at SHS and in reducing the number of discipline referrals. This helped to reduce classroom disruptions that can negatively impact student learning both by those that act out inappropriately and by those that are in the classroom witnessing the infractions. By reducing the discipline referrals, the students, faculty, administration, parents and guardians, and society as a whole benefit by reducing the time taken up by disciplining students and focusing instead on learning.

References

Furterer, S. L. *Lean Six Sigma in Services: Applications and Case Studies*. Boca Raton, FL: CRC Press, 2009.

Metzler, C., Biglan, A., Rusby, J., and Sprague, J. (November 2001). Evaluation of a Comprehensive Behavior Management Program to Improve School-Wide Positive Behavior Support. *Education and Treatment of Children*, 24 (4), 448–479.

19

*Financial Services Improvement in a City Government: A Lean Six Sigma Case Study**

Sandra L. Furterer

CONTENTS

* The original case study was published in *Lean Six Sigma in Service: Applications and Case Studies*, CRC Press. It is adapted to focus on the Lean Six Sigma aspects applied for this case study (Furterer, 2009).

Financial Process Overview

Lean Six Sigma can improve the efficiency of processes and the quality of service to citizens and reduce the costs of providing these services. The author worked with a local government's financial administration department to implement Lean Six Sigma. The goal of the project was to streamline the processes and subsequently reduce the financial process cycle time. The city is a 7000-citizen municipality. It is a city manager form of government where the City Manager manages the city employees and implements policy defined by the mayor and City Council members. The Finance Director reports to the City Manager and is responsible for developing and managing financial budgets, financial processes, the mayor's court processes, income tax collection, and utility billing and collection processes.

The financial processes include payroll, purchasing and accounts payable, accounts receivable, monthly reconciliation, and budgeting. The Finance Clerk generates paychecks for administrative personnel, the Police Department, the Fire Department, the Public Works Department, and City Council. The International Association of Fire Fighters represents the firefighters, who require union dues to be held from the members' pay once a month to be submitted to the union. The processing also includes pension matching, making pension payments, and reporting. The Payroll Department also processes income tax payments, garnishments, child support, and other withholdings to the appropriate agencies. Employees receive paychecks every 2 weeks with pension reporting performed on a monthly basis. The customers of the payroll process are internal city employees and external agencies that receive withholding payments and reports. The Finance Director realizes that the current processes, with respect to the processes before the Lean Six Sigma program is implemented, are inefficient, error-prone, lengthy, and have an extensive number of non-value-added steps. The entire payroll, pension reporting, and withholding payment process take between 13 and 70 employee hours per pay period, depending on if information processing problems occur.

The purchasing and accounts payable processes enable city personnel to purchase materials, products, and services to run the city. Purchase requisitions are generated by personnel. The Finance Clerk generates the purchase order, which is then approved by the City Manager, the Finance Director, and City Council, if necessary. Invoices are received by the Finance Director and processed by the Finance Clerk, with the appropriate approvals and signatures. Payments to vendors are frequently late. Multiple invoices for the same payment are frequently received and must be reviewed to determine if they have been paid. The upfront purchasing process takes approximately 7–10 days to generate and approve the purchase orders after the approved purchase requisition is received. The purchase orders are then filed until the invoices are received. The entire accounts payable process takes approximately 2 weeks to process a batch from initial invoice receipt to vendor payment.

The Finance Clerk records revenue receipts and deposits revenue checks in the bank. In the current process, there is a lag between when the revenue checks are received in the Finance Department and when they are entered into the financial system and deposited in the bank, due to process inefficiencies and workload capacity issues.

In addition, the Finance Clerk is responsible for reconciling the financial records on a monthly basis. Reconciliation includes comparing the bank statements for the payroll account, a general account, and several investment accounts, to the financial system entries. Due to mainly either process inefficiencies or workload capacity issues, or both, monthly reconciliation currently is rarely performed in a timely manner. Sometimes the Finance Director reconciles the books and other times it is outsourced to an accountant.

The Finance Director is responsible for managing the budgeting process throughout the city. He receives budget requests from department managers, consolidates them into a city budget, prepares budget reports for state and county agencies, and makes budget journal entries into the financial information system (IS). The Finance Director is also responsible for ensuring that expenditures are within the approved budgets as well as providing budget information to city management. There are some training issues with respect to using the financial system for budgeting as well as duplicate data entry into multiple ISs. The financial IS is also limited with respect to a user-friendly ad hoc budget reporting system.

Define Phase

A successful implementation of the Lean Six Sigma problem-solving approach and quality and Lean tools will be measured by the reduction of process inefficiencies, the reduction of the time it takes to process the financial transactions, and the assignment of appropriate staffing levels to handle the workload. No quantitative or qualitative measures of process or quality characteristics existed before the project for any of the financial processes.

The Define–Measure–Analyze–Improve–Control (DMAIC) problem-solving methodology from the Six Sigma approach was used to improve the financial processes. The goal of the Define phase of the DMAIC Six Sigma problem-solving process is to define the need for improving the financial processes, develop the project charter, and perform stakeholder analysis.

Lean Six Sigma Project Charter

The Finance Director identified the need to streamline the financial processes. The Finance Clerk complained of needing additional staff and not being able to complete her work. She was responsible for the purchasing, accounts payable, accounts receivable, payroll, and monthly reconciliation and closing. The vendor payments were frequently late, resulting in vendors constantly calling the Finance Department requesting payment. The revenue receipts were frequently held in the Finance Department for over a week before processing and depositing. The estimated current payroll processing time ranged from 13 to 70 hours, with a mean time of 40 hours. Employees frequently complained about payroll paycheck errors. The monthly reconciliations were not performed on a regular basis. Adjustment journal entries were frequently made months after the error should have been discovered.

The Lean Six Sigma Quality Facilitator, the Process Analyst, and the Consulting Manager interviewed the finance personnel to understand the goals of the Finance Department, the project scope, and objectives. Figure 19.1 shows the project charter describing the problem, the goals and scope of the project, the customers and stakeholders, and what is important to their satisfaction (critical to satisfaction [CTS]), financial benefits, and potential project risks. The goal of the project is to streamline the financial processes, reduce cycle time, and improve quality and accuracy of the processes. The scope of the project is the financial processes including payroll, purchasing and accounts payable, accounts receivable, monthly reconciliation, and budgeting. Potential financial benefits are in the cost avoidance of not having to hire additional resources, and all work being performed by one person, instead of 1½ full-time equivalents, which could result in a fully loaded payroll cost of $66,000.

Stakeholder Analysis

The Lean Six Sigma team consisted of the Finance Clerk, who performed the accounts payable, accounts receivable, payroll and pension reporting, and monthly reconciliation processes within the finance department; the Finance Director, who managed the financial processes, the mayor's

Project Name: Financial Process Improvement

Problem Statement: The Finance Director identified the need to streamline the financial processes. The Finance Clerk complained of needing additional staff and not being able to complete her work. The vendor payments were frequently late, resulting in vendors constantly calling the Finance Department requesting payment. The revenue receipts were frequently held in the Finance Department for over a week before processing and depositing. The estimated current payroll processing time ranged from 13 to 70 hours, with a mean time of 40 hours. Employees frequently complained about payroll paycheck errors. The monthly reconciliations were not performed on a regular basis. Adjustment journal entries were frequently made months after the error should have been discovered.

Customer/Stakeholders: (Internal/External) Financial Departments, city departments, external vendors, governmental agencies (tax reporting, county and state, pension)

What Is Important to These Customers—CTS: Accuracy, timeliness.

Goal of the Project: To streamline financial processes, reduce cycle time, improve quality and accuracy.

Scope Statement: The financial processes include payroll, purchasing and accounts payable, accounts receivable, monthly reconciliation, and budgeting.

Financial and Other Benefit(s): Cost avoidance, not having to hire additional resources, and all work being done by 1 person, instead of 1 ½ FTEs. $66,000.

Potential Risks: Stakeholder buy-in; consulting resources not approved by City Manager.

FIGURE 19.1
Project charter.

court processes, income tax collection, and utility billing and collection, and also performed the budgeting preparation and tracking for the city; a Team Quality Facilitator, who developed the implementation plan and provided technical Quality and Lean principles and tools knowledge; a Process Analyst, who helped to collect and prepare process documentation; and a Consulting Manager, who provided business knowledge and direction and maintained the formal business relationship between the city and the consulting firm. The Team Quality Facilitator, the Process Analyst, and the Consulting Manager were hired from an external consulting firm. The team profiled the people and cultural state to understand the level of skills and training of the employees and their resistance or acceptance levels to change. At the start of the project, the Finance Clerk was very resistant to change. As the project progressed, she became very receptive to the improvement ideas because she saw how it would help her get her work done more quickly and with fewer errors. She also enjoyed receiving the attention related to the improvement effort. The Finance Director was very receptive to change and the improvement effort. He embraced the vision of improved and streamlined financial processes.

The stakeholders are defined in Figure 19.2, and the Stakeholder Commitment Scale is shown in Figure 19.3.

Stakeholders	Who Are They?	Potential Impact or Concerns	+/−
Finance Clerk	City employee who performs detailed financial processes including processing payroll, accounts payable and accounts receivable.	Standardized processes Fewer errors Reduction of time and work Resistance to change	+ + + −
Finance Director	Manager of the Finance and Administration Departments, including Finance, Mayor's Court, Utility Billing, and Income Tax.	Ensure accounting and finance standards and procedures are followed Citizen and council satisfaction Avoid hiring additional staff	+ + + −
Quality Facilitator and Process Analyst	Provides Black Belt expertise, identifies improvement recommendations, documents process, collects data, performs statistical analyses.	Reduce resistance to change with Finance Clerk Complete project on time and within budget Add value and improve processes	− + +
Consulting Manager	Manages client relationship for consulting company.	Client satisfaction Complete project on time and within budget	+ +

FIGURE 19.2
Stakeholder analysis definition.

Stakeholders	Strongly Against	Moderate Against	Neutral	Moderate Support	Strongly Support
Finance Clerk	X				O
Finance Director					XO
Quality Facilitator and Process Analyst					XO
Consulting Manager					XO
X = At start of project					O = By end of project

FIGURE 19.3
Stakeholder commitment scale.

Team Ground Rules and Roles

The consulting engagement "statement of work letter" described the roles and anticipated involvement of the Finance Clerk, the Finance Director, and the consultants. It was clearly identified that the consultants would work with the city to gather and analyze data and provide recommendations based on their best practice experience to help improve the financial processes. However, it was ultimately the Finance Department's responsibility to implement the processes and make change happen.

Project Plan and Responsibilities Matrix

The Quality Facilitator created a letter of understanding to document the roles and responsibilities of the team members. The team created a project plan with activities, a timeline, and resources, shown in Figure 19.4. Figure 19.5 identifies the team mission and team members' roles and responsibilities.

Activity Number	Phase/Activity	Duration	Predecessor	Resources
1.0	Define			
1.1	Define process improvement need	1 day		Quality Facilitator, Finance Director
1.2	Identify goals	2 days	1.1	Quality Facilitator
1.3	Form team	2 days	1.2	Finance Director, Consulting Manager
2.0	Measure		1.0	
2.1	Profile current state	14 days		Quality Facilitator, Process Analyst, Finance Clerk
2.2	Identify problems that contribute to process inefficiencies an errors	5 days	2.1	Quality Facilitator, Process Analyst, Finance Clerk
2.3	Identify root causes	5 days	2.2	Quality Facilitator, Process Analyst, Finance Clerk
3.0	Analyze		2.0	
3.1	Analyze gaps from best practice	5 days		Quality Facilitator
3.2	Identify improvement opportunities and develop an improvement plan	5 days	3.1	Quality Facilitator, Process Analyst, Finance Clerk
3.3	Perform cost and benefit analysis	5 days	3.2	Quality Facilitator
4.0	Improve		3.0	
4.1	Implement improvement solutions	20 days		Finance Clerk
4.2	Measure impact of the improvements	5 days	4.1	Quality Facilitator

FIGURE 19.4
Project plan.

Activity Number	Phase/Activity	Duration	Predecessor	Resources
4.3	Document procedures and train employees on the improved procedures	10 days	4.2	Quality Facilitator
5.0	Control		4.0	
5.1	Design and implement process performance measures	5 days		Quality Facilitator
5.2	Implement a continuous process improvement approach	Ongoing	5.2	Quality Facilitator
5.3	Celebrate success	½ day	5.3	Quality Facilitator

FIGURE 19.4 (***Continued***)
Project plan.

Team Mission	
Document the current financial processes to create desktop procedures and to identify and implement financial process improvements.	
Role	**Responsibility**
Finance Clerk as Process Owner	Provides process knowledge and identifies and implements improvement opportunities.
Finance Director as Project Champion	Establishes team mission and goals. Provides project team resources and support.
Team Quality Facilitator as Black Belt	Provides team facilitation. Provides technical Quality and Lean tool knowledge. Provides best practice for financial processes.
Process Analyst	Prepares documentation. Collects process data. Identifies improvement opportunities.
Consulting Manager	Provides business knowledge and direction. Manages consultants.

FIGURE 19.5
Team mission, roles, and responsibilities.

SIPOC

The Suppliers–Inputs–Process–Outputs–Customers (SIPOC) describes the scope of the financial process improvement project. The SIPOC is shown in Figure 19.6. It provides the stakeholders identified as the suppliers who provide input to the process (time sheets, data, payments, etc.) and the customers that receive the outputs from the processes (paychecks, invoice checks, etc.). The SIPOC also identifies the high-level process steps included in the scope of

Suppliers	Inputs	Process	Output	Customers
City employees	Time reports	Payroll	Checks, pension reports, taxes paid	City employees, taxing authorities, state, county
Vendors, city employees	Invoices, requests	Accounts payable	POs, checks	Vendors
State, county	Checks, direct deposits	Accounts receivable	Funds available or invested	City departments
City departments	Financial transactions, receipts, checks, invoices, bank statements	Monthly reconciliation	Balanced accounts, adjustments, financial reports	Finance Director, City Council
City departments	Budgeting needs	Budgeting	Budget, appropriations	City Council and citizens

FIGURE 19.6
SIPOC.

the project including accounts payable, accounts receivable, payroll, monthly reconciliation, and budgeting.

Measure Phase

The goal of the Measure phase of the DMAIC Six Sigma problem-solving process is to understand and document the current state of the processes to be improved and identify the process problems that are causing inefficiencies and errors and their root causes.

Process Maps

The team used process flowchart analysis to map the current state processes. These flowcharts identified the steps involved in the Finance Department activities related to budgeting/investments, purchasing/accounts payable, accounts receivable, monthly reconciliation, and payroll. Various system functions were identified in the process flows that were used to perform the financial processes. The process flows identified the written (of which few existed) and unwritten policies that governed the processes. It was decided that budgeting would not be included in the scope of the project after the initial process maps were developed; therefore, only the processes performed by the Financial Clerk would be in the project scope.

Operational Definitions

No process measures existed for the financial processes before the Lean Six Sigma project. The Finance Clerk estimated the average and range of the processing times based on her experience with the processes. The estimated processing times are displayed in Figure 19.7. The team also profiled the technology to determine if the financial system was meeting their needs. They had implemented the system about 6 months before the start of the project, and there were many training issues related to the software. There were also some inefficient IS flows required by the software applications. Ad hoc financial reporting capability was difficult, time consuming, and required extensive knowledge of data tables and query ability.

The operational definition for measuring the accounts payable process cycle time is defined as the time to process a batch once the batch is organized. It does not include the time waiting for the invoice to be matched once it is received in the mail. Accounts payable included the processes for receiving invoices from vendors who perform work for the city, or who the city purchased goods from, through to paying the vendors for the goods or services.

The operational definition for measuring the accounts receivable process cycle time is the time from when the revenue check or receipt is received in the finance office until it is deposited in the city's bank account. The accounts receivable processes include logging and depositing tax revenues received from governmental agencies for the city's operating revenues.

The operational definition for measuring the payroll process cycle time is the time from when the last time sheet is received from the city departments to when the payroll is complete and the paychecks or direct deposit information is printed. It does not include the pension reporting processing and printing time.

Process	Estimated Elapsed Processing Time Range	Estimated Average Elapsed Processing Time
Payroll and Pension Reporting	13–70 hours	60 hours
Purchasing/Accounts Payable	30–40 hours per batch (only about half of the due invoices are processed every other week)	40 hours
Accounts Receivable	40–80 hours (including delay due to workload capacity issues)	60 hours
Monthly Reconciliation	40–80 hours (if performed)	60 hours
Budgeting	No estimate available	No estimate available

FIGURE 19.7
Estimated processing times.

The operational definition for measuring the monthly reconciliation time is the time it takes to reconcile the books with the bank statements, make any appropriate adjustments, and print the appropriate financial reports. The monthly reconciliation process includes the review of the bank statements and ensuring that the bank transactions match the receipts, deposits, and transactions across the multiple city departments and that the "checkbook" balances.

The operational definitions for the defects for each process will be further defined after the defect types are collected using the check sheets discussed in the section "Data Collection Plan."

Data Collection Plan

Since there was no process measurement system in place to assess the CTS criteria related to cycle time, accuracy, and customer satisfaction, the data collection plan is a critical tool to help provide a way to measure the CTS. The data collection plan is shown in Figure 19.8. Cycle time and accuracy should be measured at a detailed process level for each subprocess, including accounts payable, accounts receivable, monthly reconciliation (Recon), and payroll.

Critical to Satisfaction (CTS)	Metric	Data collection mechanism (survey, interview, focus group, etc.)	Analysis mechanism (statistics, statistical tests, etc.)	Sampling plan (sample size, sample frequency)	Sampling instructions (who, where, when, how)
Cycle Time	AP: cycle time—vendor invoice received to paid	Track for 4 weeks	Mean, standard deviation, control charts	All invoices for one month	Process Analyst tracks date received to when paid
	AR: time to deposit funds in bank from when check received	Track for 4 weeks	Mean, standard deviation, control charts	All revenue receipts for one month	Process Analyst tracks date received to when paid
	Recon: time it takes to close	Track for 2 months	Mean, range	Time to close for 2 months	Process Analyst tracks time to close
	Payroll: time to process payroll	Track for 2 payroll cycles	Mean, range	Time for 2 payroll cycles	Process Analyst tracks time to close
Accuracy of the Process	AP: types and number of defects	Check sheet	Pareto chart	Defects for 1 month	Finance Clerk to track on check sheet
	AR: types and number of defects	Check sheet	Pareto chart	Defects for 1 month	Finance Clerk to track on check sheet
	Recon: types and number of defects	Check sheet	Pareto chart	Defects for 1 month	Finance Clerk to track on check sheet
	Payroll: type and number of defects	Check sheet	Pareto chart	Defects for 1 month	Finance Clerk to track on check sheet
Customer Satisfaction	Vendors	Survey	Statistical analysis	Survey 20 vendors	Quality Facilitator to create survey and collect survey data
	Internal customers	Survey	Statistical analysis	Survey internal city departments: police, fire, streets, administration	Quality Facilitator to create survey and collect survey data

FIGURE 19.8
Financial process data collection plan.

Pareto Chart

The Quality Facilitator and Process Analyst noticed that there was a large quantity of invoices for a city of this size. The Finance Clerk was constantly inundated with invoices coming in on a daily basis. As the team further investigated, asking why several times, it became evident that there was no centralized purchasing. Although all of the purchase requisitions came to the Finance Clerk to be approved by the Finance Director and City Manager, each city department decided what they would purchase and who they would purchase it from. Each department ordered their own office supplies from their favorite office store supplier. There was no preferred or certified vendor list for purchases under $10,000.

The team decided to analyze the accounts payable data for the year-to-date and identify the number of vendors by the dollar value that was purchased by each vendor within the first 8 months of the year. The resulting Pareto chart (Figure 19.9) shows over 250 vendors with a total purchase activity for an 8-month period under $500. Each invoice requires a purchase requisition to be completed and approved, a purchase order to be created, printed, and approved (in duplicate), an invoice to be received and matched with the shipping or receiving paperwork, the invoice to be entered and processed, and a check to be printed and signed (in duplicate), as well as the resultant monthly reconciliation of all of these transactions. An opportunity for consolidating the purchasing activity and eliminating the large number of non-value-added activities is identified for the Analyze phase.

The Finance Clerk was constantly overwhelmed when she ran into a problem with the IS. She claimed that the financial system was wrought with problems and just did not work. She said she constantly had to call the financial system vendor's IS help desk to have them fix a problem. The way that the Finance Clerk dealt with an IS problem was she would call the

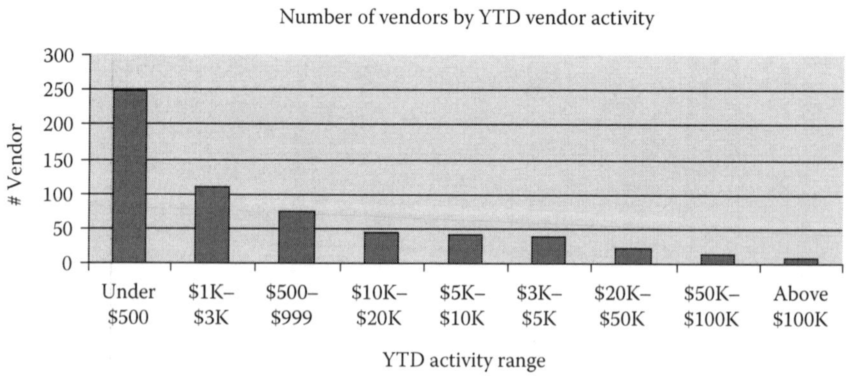

FIGURE 19.9
Pareto chart of year-to-date (YTD) vendor activity.

vendor's IS help desk, report the problem, and then sit at her desk waiting (not working on another task) for a call back, which could take 2 hours or more.

The team decided to investigate the causes of the IS problems and understand if the financial system was broken or perhaps if it was a training issue. The team collected data from the IS vendor's help desk system on the problems reported by the city's Finance Department, which included problem type, time to resolve, and resolution category. The Pareto chart in Figure 19.10 shows that 54% of the "problems" reported to the IS help desk were related to training (or lack of training) issues, not the "perceived software problems."

Voice of the Process Matrix

The voice of the process (VOP) matrix helped to link the CTS criteria to the metrics, targets, and potential process factors that affect the CTS. The VOP matrix was used to summarize the VOP, shown in Figure 19.11. The CTS criteria were defined as cycle time, accuracy of the process, and customer satisfaction. The factors that potentially impact cycle time were having standard procedures, streamlined processes, training, and the volume of the invoices and paychecks. The cycle time for each process was defined to be measured. The accuracy of the processes would be potentially impacted by training in procedures and the financial software and would be measured by assessing the number and types of defects in each process. Customer satisfaction could be impacted by whether there was a repeatable process and whether the city would collect and measure voice of the customer (VOC) information. The VOC could be measured through surveys. The proposed target for each of the metrics is also included in the matrix.

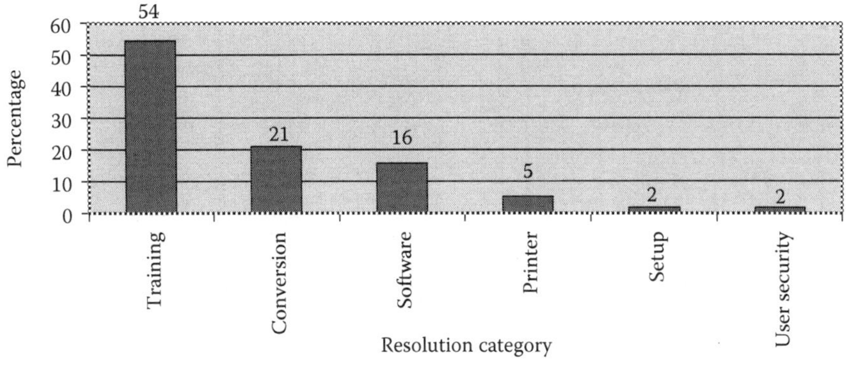

FIGURE 19.10
Information system problems broken down by resolution category.

CTS	Process Factors	Operational Definition	Metric	Target
Cycle Time	• Standard procedures exist • Streamlined processes • Training • Volume of invoices and paychecks	Measure each process time	AP: cycle time—vendor invoice received to paid AR: time to deposit funds in bank Recon: time it takes to close Payroll: paid on time per schedule	AP: 10 business days AR: 2 days Recon: 10 days Payroll: paid on time
Accuracy of the Process	• Training in procedures and software	Measure each process and defect types	Defects by process and type	95% accuracy
Customer Satisfaction	• Repeatable process • Collect and assess VOC	Measure customer satisfaction through customer and vendor surveys	% of positive responses for identified survey questions	80% of responses are rated 4 or 5 for identified questions

FIGURE 19.11
VOP matrix.

Cost of Poor Quality

The following are potential costs of poor quality elements for the financial processes related to the following categories:

- Prevention:
    - Training on the processes
    - Training on the IS
    - Developing a quality management system
    - Developing a vendor certification program
    - Developing a measurement system
    - Implementing a continuous process improvement program
- Appraisal:
    - Certifying vendors
    - Assessing the measurement system
    - Assessing the accuracy and quality of the processes
    - Assessing customer satisfaction
    - Assessing the process cycle times
    - Internal failure

    - Process defects in each of the financial subprocesses found before they reach the internal customers in other departments or external vendors
    - Accounting adjustments during or after monthly reconciliations
- External failure:
    - Process defects that reach the vendors or internal customers in other departments
    - Process defects that reach external taxing authorities or state or county agencies
    - Incorrect or missing garnishments
    - Financial errors or adjustments that City Council or the Financial Auditor discovers
    - Lack of citizen good will due to financial errors or adjustments; this means that the citizens may not be satisfied or have faith in the city government due to errors and adjustments in financial data

Analyze Phase

The goal of the Analyze phase is to analyze the problems and process inefficiencies. Also part of the Analyze phase is a cost and benefit analysis to understand whether the improvements are too costly compared to the estimated benefits to improve productivity and quality.

Cause and Effect Diagram

The project team used the process flowcharts and several Lean tools including waste identification and elimination, standardization of operations to identify and eliminate non-value-added activities, and good housekeeping (part of the 5S) to identify process problems such as inefficient sorting and filing of purchase orders and invoices. The team used brainstorming techniques to identify problems.

The team used cause and effect analysis to identify root causes related to people (such as lack of training and skills), methods (lack of standardized procedures), information technology (IS human factors and confusing and inefficient processing flow was), and hardware (broken and inefficient printers). A cause and effect diagram is presented in Figure 19.12.

The team identified gaps comparing the current state processes to best practice financial processes. The Team Quality Facilitator and the Process Analyst used their understanding of financial processes and the concepts of Lean principles and the process flowcharts to identify non-value-added

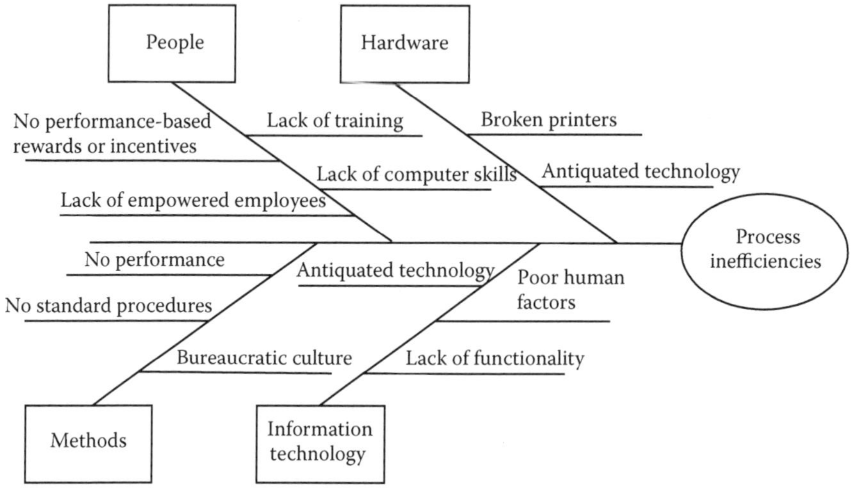

FIGURE 19.12
Cause and effect diagram.

activities, especially related to unnecessary work and rework. The team used the concept of implementing improvements that would prevent problems and rework due to printer jams and inefficient use of the technology to reduce the financial processing time. The Team Quality Facilitator performed an analysis of reported financial IS problems using Pareto analysis and statistical process control charts, across the finance and administration departments. The purpose was to identify employee training and knowledge gaps with respect to the financial and administrative IS.

Cause and Effect Matrix

A cause and effect matrix (Figure 19.13) was created to understand similar causes that produced defects in each of the financial processes. The bureaucratic culture contributed the most to the process defects. Next was lack of training, then antiquated technology. Lack of standard procedures and then lack of functionality were the last two root causes in priority order contributing to process defects.

Why–Why Diagram

A why–why diagram (Figure 19.14) was created to identify the root causes for why payroll processing time takes so long. The root causes are similar to what was already identified when generating the cause and effect diagram and matrix. Some of the root causes are lack of training, lack of procedures, no focus on process improvement, bureaucratic culture, and focus on price versus value.

	Effects					
	AP Defects	AR Defects	Payroll Defects	Recon Defects	Total	Relative Weighting
Causes/Importance:	8	4	10	6		
Lack of training	9	9	9	3	216	2
Lack of standard procedures	3	3	3	9	120	4
Antiquated technology	3	3	9	9	180	3
Lack of functionality		9	3	3	84	5
Bureaucratic culture	9	9	9	9	252	1

FIGURE 19.13
Cause and effect matrix.

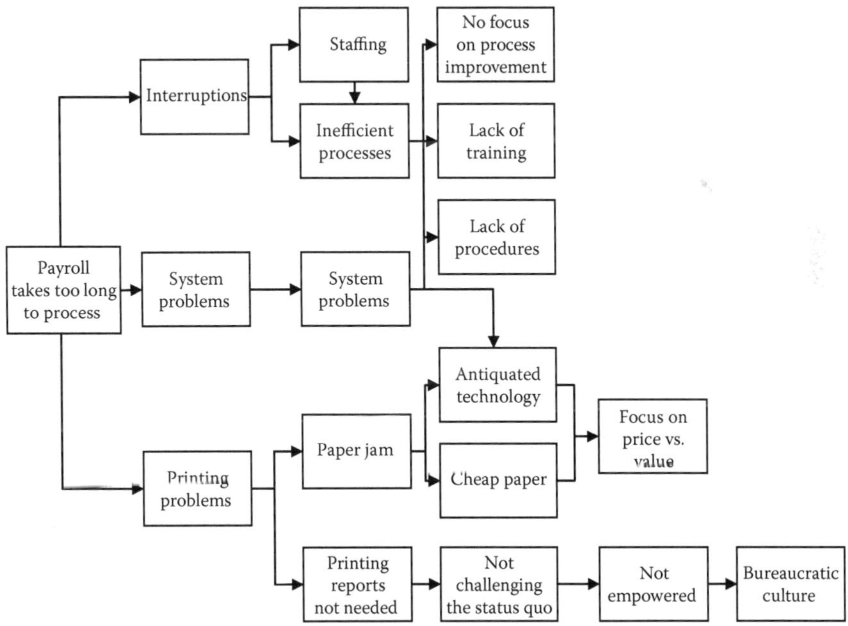

FIGURE 19.14
Why–why diagram.

Process Analysis

The process analysis was performed for the accounts payable (AP), accounts receivable (AR), monthly reconciliation, and accounts receivable processes. The number of value-added versus non-value-added activities was compared for each process. The monthly reconciliation process had the highest percentage (93%) of non-value-added activities, followed by the accounts receivable process with 86% non-value-added activities. The payroll process had 83% of the activities identified as non-value-added, while the accounts payable process had 61% non-valued-added activities. Balancing the books in the monthly reconciliation process is necessary from a financial audit and controls perspective, but the defects and inefficiencies from all of the upfront processes, such as payroll, AP, and AR, flow into the downstream reconciliation process and cause the balancing problems. The focus in the Improve phase should be to improve the upfront processes to reduce reconciliation problems. Figure 19.15 shows the summary of value-added and non-value-added percentages in each process. Figures 19.16 and 19.17 show the actual activities that are identified as either adding value or not adding value to the processes.

Process	Value-Added Percentage of Activities	Non-Value-Added Percentage of Activities
Accounts Payable	39%	61%
Accounts Receivable	14%	86%
Monthly Reconciliation	7%	93%
Payroll	17%	83%

FIGURE 19.15
Financial process value analysis summary table.

Process	Value-Added Activities	Non-Value-Added Activities	Value-Added % of Activities
Accounts Payable	• Perform bidding process • Council approves • Enter new vendor • Enter PO in FSS • Approve PO • Fire Dept. calls with amount • Treasurer gives amount needed • Fill out requisition form • Pay PO	• Obtain PO number • Fill out requisition form • Print PO • Verify premium invoice number • Verify PO exists • Store PO • Send invoice to supervisor • Total invoices on calculator • Print report • Verify total • Fix problems • Print checks • Send checks • File copy	39%
Accounts Receivable	• Post receipt in system • Deposit at bank	• Make copy of check • Total on calculator • Staple deposit slip and copy • Print report • Verify total • Match report to receipts • Fix problems • File receipts • Stamp back of checks • Fill out deposit slip • Staple to report • Store	14%

FIGURE 19.16
Financial process value analysis: accounts payable and accounts receivable.

Process	Value-Added Activities	Non-Value-Added Activities	Value-Added % of Activities
Monthly Reconciliation	• Reset month in system	• Print reports • Compare report totals to bank statements • Call help desk for help • Fix problem • Reconcile bank statements • List outstanding checks • Compare totals • Verify items • Review check register • Review wire transfers • Make adjustments • Bank make adjustments • File bank statements • File reports	7%
Payroll	• Enter time sheets in system • Print checks • Print deduction checks • Print direct deposit vouchers • Perform direct deposit transfer	• Verify time sheets • Create manual hours sheet • Print reports • Compare hours totals • Fix hours in system • Print reports • Compare totals • Fix hours • Print payroll reports • Fix printer problems • Redo payroll in system • Void printed checks • Reprint checks • Change printer paper • Print successful bank report and send to bank • Fix direct deposit problems • Fix paycheck problems • Write manual check • Bank fixes problem • Fix problem in direct deposit • Bank calls with problem • Write check for general fund • Deposit in bank • File copies of report	17%

FIGURE 19.17
Financial process value analysis: monthly reconciliation and payroll.

Figures 19.18 and 19.19 show the individuals and moving range control charts of the time (in hours) that it took the software vendor to resolve reported IS problems for the city. This showed that problems with the system contributed to out of control conditions and, therefore, process inefficiencies. The out of control conditions were assigned to a cause related to a computer program archiving process that was extremely difficult to identify because it only happened during a monthly archiving process.

Waste Analysis

The Lean Six Sigma team performed a waste analysis for the following processes: accounts payable, accounts receivable, monthly reconciliation, and payroll. There were multiple instances of each of the eight types of waste across all of the processes, as shown in Figure 19.20.

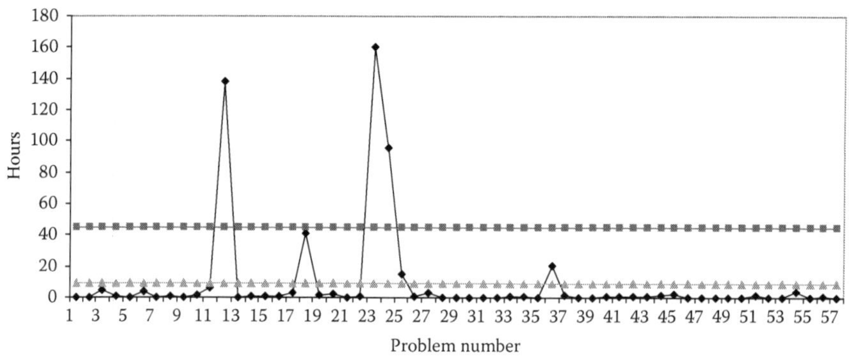

FIGURE 19.18
Mean time to resolve problems control chart.

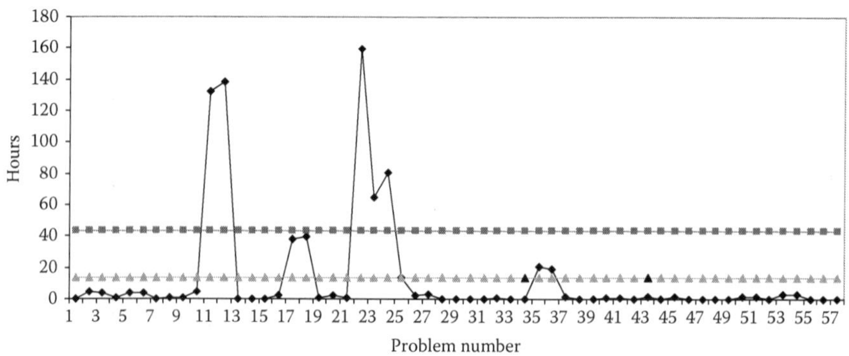

FIGURE 19.19
Moving range time to resolve problems control chart.

Waste Type	Process	Waste Element
Transportation	AP, AR, Payroll	Moving manual checks, moving funds manually, not using direct deposit so moving paychecks
Overproduction	AP, AR, Payroll, Monthly Reconciliation	Printing reports that are not used
Motion	AP, Payroll	Walking to printer in other room
Defects	AP, AR, Payroll, Monthly Reconciliation	Matching totals, process defects (wire transfers, direct deposit errors, information system process errors, printer problems), paycheck errors, time sheet errors
Delay	AP, AR, Payroll, Monthly Reconciliation	Waiting for AP processing, waiting to deposit AR checks, not getting to Monthly Reconciliation process, paying outside accountant to balance books, payroll late
Inventory	AP, AR, Payroll, Monthly Reconciliation	Filing/storing reports, purchase requisitions, purchase orders, invoices, time sheets
Processing	AP, AR, Payroll, Monthly Reconciliation	Matching and balancing, not using direct deposit (printing checks), not moving funds automatically at bank
People	AP, AR, Payroll, Monthly Reconciliation	No focus on process improvement, not using people's ideas

FIGURE 19.20
Waste analysis.

Improve Phase

The goal of the Improve phase is to implement the improvements, measure the impact of the improvements, document procedures, and train employees on the improved procedures.

Recommendations for Improvement

The team identified improvement opportunities that were grouped in the following Lean categories: standardized processes and procedures, good housekeeping, kanban and visual control, waste identification and elimination, and one-piece flow.

Standardized Processes and Procedures

The team suggested that the Finance Department develop standardized desktop procedures. No written procedures existed in the current state.

The Finance Clerk would keep handwritten notes, but this did not lend to standardization and repeatability.

Another improvement area was to use an Excel spreadsheet to standardize batch calculations for matching and dividing repeating invoice amounts across different account numbers.

The Fire Department had converted from an association to a city department during the improvement effort. The team encouraged the Finance Department to integrate the Fire Department into the standardized payroll and accounts payable procedures.

The team recommended that the employees who used the financial system get training from the software vendor tailored specifically to their streamlined financial processes. Initially, when the city implemented the new financial system, the software vendor would train a generic process that encouraged printing of lengthy reports that the city did not need to print. The software vendor was able to provide additional understanding on the more extensive software functionality and tailor the processes better to the city's needs.

The team recommended that the city standardize the time sheets across all of the departments to help reduce payroll data entry errors and the time to enter the time sheets. The team also recommended that the Finance Clerk use time sheets in Excel spreadsheets to calculate the total time sheet hours by department, to compare to the payroll reports, instead of using a calculator.

Kanban and Visual Control

The team created a kanban and used visual control for the accounts payable processing. A kanban is a Lean tool that is used as a signal to pull work. The kanban was designed as a file hanging system that was easily visible to the Finance Clerk and the Finance Director. In the current process, the invoices, purchase orders, and requisitions that needed to be either assigned account numbers or approved by the Finance Director were frequently lost in the piles of work. The kanban was organized in the order of the process steps. The documents that needed to be assigned account numbers were placed in a red folder in the first slot of the filing system. The purchase orders that needed approvals were placed in the next slot, so that the Finance Director would easily see them and quickly process them. The appropriate documents for each step were placed in the bin, so that the Finance Clerk and the Finance Director would have visual cues for the work that needed to be done. This greatly reduced the purchasing and accounts payable processing times. Figure 19.21 graphically depicts the purchasing and accounts payable kanban system.

Waste Identification and Elimination

The team identified unnecessary steps in the processes such as printing lengthy reports that were never used. The team encouraged either eliminating the printing of unnecessary reports or printing them to an electronic file, which took seconds instead of hours.

PO Requisition
Assign Account Numbers
Enter POs and Print
Approve POs
File POs
Store POs, for invoice
Enter invoice, sign checks

FIGURE 19.21
Purchasing and accounts payable Kanban system.

The team encouraged the use of new accounts receivables technology that automatically transferred journal entries, instead of requiring redundant data entry.

The team identified direct deposit as an improvement opportunity to eliminate printing of payroll checks. They suggested having a payroll direct deposit contest between departments to encourage use of direct deposit. This was after identifying and eliminating problems with the direct deposit process.

The team recommended extensive information technology improvements that further streamlined the processes and eliminated redundant data entry.

One-Piece Flow

One-piece flow is a Lean concept that tries to reduce the batch processing size to one or very small to flow work through the process more quickly. Another improvement idea that the team identified was to reduce the batch sizes of the accounts payable and accounts receivable batches. This would help to move closer to one-piece flow and enable vendors to get their payments quicker by processing smaller batches more frequently. This was also dependent on other improvements for both of the processes, so the batches could be processed more quickly. The team recommended the accounts receivable (revenue) batches be processed daily, instead of holding them for 1–2 weeks. This would increase the potential revenue from interest received by depositing the checks more quickly at the bank.

The team used the vendor Pareto analysis to identify duplicate vendors and recommended the number of vendors be reduced. The duplicate vendors were mainly due to each department choosing their own vendors for similar purchases across the city. This would also help the accounts payable processing to move closer to one-piece flow, or smaller batch sizes, by reducing the number of vendors and invoices.

Action Plan

The team implemented the initial financial process improvements to the payroll and pension reporting, purchasing, and accounts payable processes across a 4-month period. They implemented improvements to accounts receivable and monthly reconciliation throughout the next year, as time and resources permitted. They did not implement budgeting process improvements, because the Finance Director wanted to focus only on the processes performed by the Finance Clerk. The team first collected further information to validate the feasibility of the process improvement ideas presented in the Analyze phase. They created an implementation plan for any improvements that would take more than 1 week to implement or that required significant expenditures and defined the associated costs and benefits at a finer detail than in the Analyze phase. The team gained approval from the Finance Director to proceed with the implementation of the improvement opportunities. The team implemented the improvements and redesigned the appropriate processes to incorporate the improvements. As part of the project management of the implementation, the Team Quality Facilitator provided weekly status reports to the team that included the tasks that were completed and the status and estimated completion date. The Team Quality Facilitator documented any outstanding unresolved issues on the items for resolution (IFR) form. The IFR form included a description of the issue, the owner who was responsible for ensuring that the issue was resolved, the estimated resolution date, the priority of the issue, the status, the date the issue was opened and resolved, the impact of the issue to the project, and a description of the resolution.

Cost and Benefit Analysis

The Team Quality Facilitator and Process Analyst identified potential costs and proposed benefits of each proposed improvement to determine if the estimated benefits are greater than the costs to implement. They also provided advantages and disadvantages to each solution, so that the Finance Director could make an informed and data-oriented decision. Most of the costs were related to training and resources needed to implement and document the standardized procedures. The largest costs were related to consulting fees and obtaining laser printers for check printing. Four alternatives were identified that could automate the time sheet payroll hours entry and verification activities. The first solution—Alternative 1—was to create a Microsoft Access program that would allow entry of time sheet data and perform automated verification and summing of hours by department. Alternative 2 was to implement an existing module from the financial IS vendor to automate the time sheet data and allow remote entry by each department and automated integration of the time sheet data into the payroll system. The next alternative—Alternative 3—was to implement custom design and development of scanning and optical character recognition software to enable scanning or input from Microsoft Excel time sheet data. Alternative 3 would require the highest cost, the longest implementation time, and the highest level of technology skills from the department employees.

Alternative 4 was to develop Excel time sheets that would enable automated entry of time sheet data within each department and allow automated verification and summing of the time sheet data. The entered payroll hour's data could then be compared to payroll hour reports to ensure payroll data accuracy. Alternative 4 required the lowest cost, the shortest implementation time, and a lower level of technology skills from department employees. An economic analysis was performed to determine which alternative was the most economically attractive alternative.

The net present worth of the costs and benefits over a 5-year project life for the projects were

Alternative 1, Net Present Worth: –$15,349

Alternative 2, Net Present Worth: –$12,542

Alternative 3, Net Present Worth: –$74,961

Alternative 4, Net Present Worth: $7,289

Only Alternative 4 had a positive net present worth, or a benefit and cost ratio greater than 1. The internal rates of return for Alternatives 1, 2, and 3 were all negative. The internal rate of return for Alternative 4 was 48%. The payback period for Alternative 4 was 2.02 years.

The city implemented Alternative 4, which reduced the time needed by the Finance Clerk to enter and verify time sheet data. It also pushed accountability of time sheet data to the originating department, who had the most knowledge about whether the data was accurate. Alternative 4 also eliminated the cumbersome, time-intensive, off-line calculator-based payroll hours verification step. This alternative also standardized the time sheet format and process across all of the city departments. The time sheet errors and payroll processing time were reduced by automating and standardizing the payroll time sheet entry and verification process.

Future State Process Map

The team revised the process maps to include the improvement recommendations. Many of the non-value-added activities were removed by focusing on removing the wasteful activities.

Dashboards/Scorecards

Since there were no process measures in place before the Lean Six Sigma project, the team developed detailed process measures and a metrics guide document. The process measures are shown in Figure 19.22. The metrics guide document includes a detailed description of each metric and how to measure it, including the data collection mechanism. The metrics can be arranged in a dashboard, as shown in Figure 19.23.

Proposed Process Measure	Data Collection Mechanism
Payroll and Pension Reporting	
Number and type of payroll problems encountered per number of employees	• Payroll check sheet • Payroll metric log • Moving range and individual control chart of problems per employee
Payroll processing time by payroll period	• Payroll check sheet • Moving range and individual control chart of payroll processing time
Purchasing and Accounts Payable	
Number of problems per invoice	• Accounts payable check sheet • Moving range and individual control chart of AP problems per invoice
Time per invoice	• Accounts payable check sheet • Moving range and individual control chart of time per invoice
Percent invoices without purchase orders	• Accounts payable check sheet
Percent invoices paid within discount period	• Accounts payable check sheet
Accounts Receivable	
Time per receipt	• Accounts receivable check sheet • Accounts receivable metrics log • Moving range and individual control chart of time per receipt
Number of problems per receipt	• Accounts receivable check sheet • Accounts receivable metrics log • Moving range and individual control chart of problems per receipt
Monthly Reconciliation	
Number of problems by type	• Monthly reconciliation check sheet • Monthly reconciliation problem Pareto chart

FIGURE 19.22
Proposed process measures for scorecard.

Revised VOP Matrix

The revised VOP Matrix is shown in Figure 19.24, including incorporating a control chart to track defects in the accounts payable and payroll processes, with a more realistic target for tracking control of the process. The percentage for ratings in the positive categories (4—Agree and 5—Strongly Agree) was also revised to be more realistically aligned to the results of the surveys.

Training Plans, Procedures

The Lean Six Sigma team created detailed desktop procedures for each of the financial processes and trained the Finance Clerk in the procedures. The

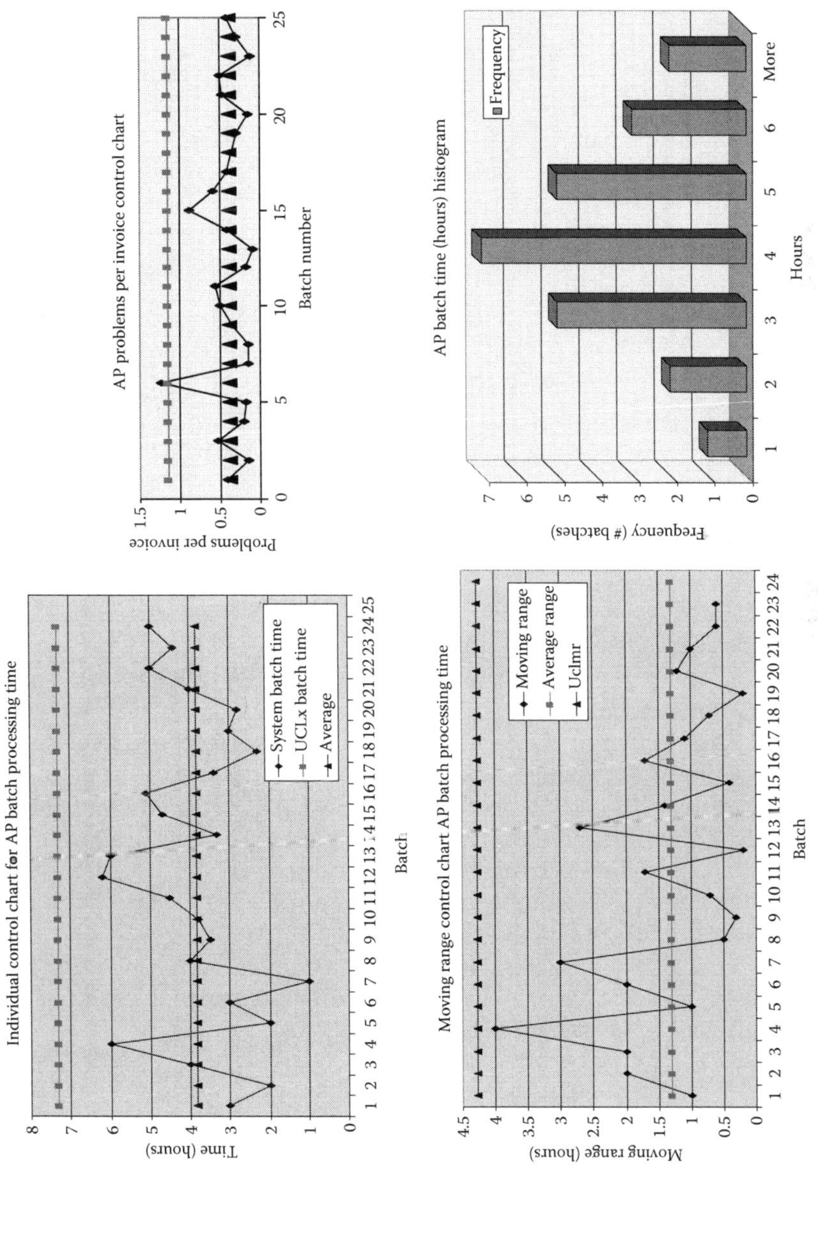

FIGURE 19.23
Dashboard example.

CTS	Process Factors	Operational Definition	Metric	Target
Cycle Time	• Standard procedures exist • Streamlined processes • Training • Volume of invoices	Measure each process time	AP: cycle time—vendor invoice received to paid AR: time to deposit funds in bank Recon: time takes to close Payroll: paid on time per schedule	AP: 10 business days AR: 2 days Recon: 10 days Payroll: paid on time
Accuracy of the Process	• Training in procedures and software	Measure each process and defect type	Defects per invoice (or paycheck) by process and type	No out of control points where assignable cause cannot be found
Customer Satisfaction	• Repeatable process • Collect and assess VOC	Measure customer satisfaction through customer and vendor surveys	% of positive responses for identified survey questions	60% of responses are rated 4 or 5 for identified questions

FIGURE 19.24
Revised VOP matrix.

procedures were extremely detailed and even included screenshots populated with sample data and step-by-step instructions. The procedures were very successful in helping to train the Finance Clerk, remove resistance to change, and eliminate problems reported to the help desk.

The procedures were developed based on our detailed knowledge of the financial IS acquired during the project. The desktop procedures were so thorough that on several occasions when the Finance Clerk was not available, the Finance Director and the Income Tax Clerk were able to perform the payroll process with limited advanced training.

The Finance Clerk was trained on all of the improved processes using the detailed desktop procedures. She also received process-specific training on the financial IS from the software vendor.

Control Phase

The goal of the Control phase is to implement performance measures and other methods to control and continuously improve the processes.

Assessments of Improved Processes

The team measured the impact of the improvements after the majority of the improvement opportunities were implemented for each financial process. The payroll processing time was reduced by approximately 60%. Although the errors were not measured before implementation of the improvements, no paycheck errors were found while migrating the Fire Department into the Finance Department procedures and financial systems, using the revised and improved payroll processes.

The purchasing and accounts payable processing time was reduced by approximately 40% and all the vendors started getting paid on a consistent and timely basis. The accounts payable improvements also completely eliminated some of the non-value-added processing steps such as no longer having to verify that duplicate invoices had been paid, due to paying invoices on time.

The accounts receivable processing time was reduced by approximately 90%. Revenue checks were getting deposited in the bank daily. The monthly reconciliation processing time was reduced by approximately 87%. Additionally, the monthly reconciliation process was performed on a consistent monthly basis, due to providing more capacity for the Finance Clerk. The increased capacity was a result of the elimination of non-value-added tasks and reducing the payroll, accounts payable, and accounts receivable processing times.

The financial processes were able to be performed by one person working 40 hours per week, instead of one-and-a-half employees before the Lean Six Sigma implementation.

Another significant improvement related to the improved processes and subsequent training was that the number of financial system problems reported to the software vendor greatly decreased from an average of 13 problems reported per month by the Finance Clerk to an average of 6 per month.

Figure 19.25 summarizes the estimated prior processing times, the estimated processing times after the improvements, and the percentage reduction

Process	Average Estimated Processing Time Prior to Improvements	Average Estimated Processing Time after Improvements	Percentage Reduction of Processing Times
Payroll and Pension Reporting	60 hours	24 hours	60%
Purchasing/Accounts Payable	40 hours	24 hours	40%
Accounts Receivable	60 hours	6 hours	90%
Monthly Reconciliation	60 hours	8 hours	87%

FIGURE 19.25
Improved financial processing times.

of processing times. More specific performance measures to measure actual cycle times per batch and quality of the processes were recommended to the city, but were not implemented before the end of the initial project.

The consultants encouraged the Finance Department to implement a continuous improvement process to continue to improve both the productivity and the quality of the financial processes. This would be especially important if turnover occurred, so that the culture would change to one that continually and always improved. The good news related to changing the culture was that an upstream process in the billing department saw the value of the improvements by the reduction in the number of the billing reconciliation problems when they had to send their journal entries to finance. Some time after the project, the Finance Clerk left the position, and the Utility Billing Clerk was able to step into the finance position and improve the financial close process to 1 day.

Mistake Proofing

As much as possible, the IS functionality was used for mistake proofing, by automating the steps that were possible to automate. This eliminated many of the manual or calculator-based activities for balancing that led to many mistakes.

Control Plan

One of the last, but very important steps of the Control phase is to take the time to celebrate the improvement effort, even if it is something as simple as going out to lunch to celebrate, which the team did. The Finance Department had not yet changed their reward and recognition system to accommodate continuous improvement and performance-based metrics.

The entire Lean Six Sigma implementation in the Finance Department took about one-and-a-half calendar years. The Define phase took 3 months, and the Measure and Analyze phases took 2 months each. The Improve and Control phases took about 1 year together.

Through implementing a Lean Six Sigma program, the city's Finance Department was able to significantly reduce the time to process payroll, purchasing and accounts payable, accounts receivable, and monthly reconciliation. Payroll processing time was reduced by 60%. Purchasing and accounts payable processing time was reduced by 40%. Accounts receivable processing time was reduced by 90%. Monthly reconciliation processing time was reduced by 87%.

The detailed metrics guide, summarized by the performance measures in Figure 19.22, was used as the control plan.

Process Capability, Defects per Million Opportunities

A formal process capability analysis was not performed due to the sample size being low at the time of the project close.

Control Charts

Several types of control charts are suggested in the process metrics guide, including the following:

Payroll: Moving range and individual control charts of number of problems per employee; moving range and individual control charts of payroll processing time

Accounts Payable: Moving range and individual control charts of number of AP problems per invoice; moving range and individual control charts of time per invoice

Accounts Receivable: Moving range and individual control charts of time per receipt; moving range and individual control charts of number of problems per receipt

Replication Opportunities

The Finance Department migrated the Fire Department into the city's standardized and improved financial processes and systems when they became a city department. The migration was seamless. No paycheck errors occurred during the first pay period when the Fire Department's payroll was processed by the Finance Department using the improved procedures.

Standard Work, Kaizen

The monthly reconciliation process was performed on a consistent monthly basis, due to providing more capacity for the Finance Clerk. Through development of the desktop procedures, the process activities were able to be performed in a standard way each time. Kaizen improvement sessions were held with the Finance Clerk to generate and implement many of the improvement ideas.

Combining the principles and tools of Lean Enterprise and Six Sigma provides an excellent way to improve the productivity and quality of providing financial services in a local government. Although the majority of Lean Six Sigma applications have been in private industry, focusing mostly on manufacturing applications, this case study is an excellent example of how Lean Six Sigma tools can be applied in a service-oriented, transaction-based entity such as a local government.

References

Furterer, S. L., and Elshennawy, A. K. (2005). Implementation of TQM and Lean Six Sigma Tools in Local Government: A Framework and a Case Study. *Total Quality Management and Business Excellence Journal*, 16 (10), 1179–1191.

Furterer, S.L. (2009). *Lean Six Sigma in Service: Applications and Case Studies.* Boca Raton, FL: CRC Press.

20

Application of Lean Tools in a Hospital Pharmacy

Seth Langston, Jason Park, and Raj Vemulapally

CONTENTS

Executive Summary

The purpose of this project is to perform an analysis using Lean techniques and tools and recommend Lean improvements that can help to streamline the prescription-filling processes at a hospital pharmacy. By utilizing Lean techniques and tools, there is a potential to decrease wait times by 40%, decrease the number of inventories conducted by 100%, and create over 8 person-hours of extra capacity for the staff to use for other tasks.

Customers and stakeholders express dissatisfaction at the amount of time they have to wait to pick up medications at the hospital pharmacy. Utilizing the Define–Measure–Analyze–Improve–Control (DMAIC) outline, this project focused on improving efficiency while eliminating waste with Lean tools.

During the Define phase, the voice of the customer was captured, as well as the current state business processes understood. The process baselines were measured in detail, data were collected, waste was identified, and opportunities where efficiency can be improved were proposed. The current value stream map was key as it helped to identify the process in detail by mapping out how value flows. A detailed analysis was conducted based on the data collected, looking at all areas of the pharmacy from auto-filling robots to the hand filling of prescriptions. From the analysis, a number of Lean tools were utilized to eliminate waste. The improvement tools that were proposed include the kanban system, cellular manufacturing layout, poka-yoke, 5S, single-minute exchange of dies (SMED), and total productive maintenance (TPM). A future state map was created, showing the desired end state. A solution implementation plan was introduced, as well as control measures that can help to deploy the proposed tools in an effective manner.

This project demonstrated that, although different from a normal manufacturing process, Lean tools can be applied to healthcare to reduce waste, thereby improving efficiency and ultimately providing more satisfaction to the customer. In this project, the cellular manufacturing process proved to be the most beneficial Lean tool in terms of significantly impacting the wait times for the customer. Tools such as the kanban system reduced the need for inventories and provided additional capacity to conduct other work. More importantly, the freed-up capacity from implementing Lean improvement could allow for extra time to apply additional Lean tools without disrupting the flow of work, with the potential for the pharmacy to improve efficiency even further. This project shows how Lean tools can be applied continuously in an attempt to achieve significant improvements in efficiency.

Introduction

Overview of Lean Thinking

According to Womack and Jones (2003), Lean thinking is a way to "specify value, line up value-creating actions in the best sequence, conduct these activities without interruption whenever someone requests them, and perform them more and more effectively." There are numerous types and kinds of waste, or *muda*, that exist across industries, remain unchanged, and cause excess amounts of inefficiency across the workplace. The goal of using Lean tools is to identify and eliminate waste and ultimately create a workplace that ensures the customer is fully satisfied every time.

Project Overview

Pharmacy customers process their prescriptions for medication through the Tri-Care Pharmacy system. There are several different ways to fill a new or existing prescription. This project focuses on the retail sector of filling prescriptions where the customer physically goes to the retail pharmacy at a military installation to process and receive their medication.

This project can be classified as a combination of Lean production and Lean healthcare. Although the pharmaceutical industry falls into the healthcare industry, many of the Lean production tools can be applied to create a system that focuses on efficiency and customer satisfaction. The motivation for this project came from customers who felt the frustrations of picking up prescriptions at the pharmacy due to excessive waiting. The pharmacists felt that the inefficiencies wasted time not only for them but also for the customers.

DMAIC

This project will utilize the DMAIC outline for improving and analyzing process along with Lean tools in the pharmacy. The steps, taken from Six Sigma, are separated into the following (iSixSigma, 2012):

1. Define the customers, their critical to quality issues, and the core business process involved.
2. Measure the performance of the core business process involved.
3. Analyze the data collected and process map to determine the root causes of defects and opportunities for improvement.
4. Improve the target process by designing creative solutions to fix and prevent problems.
5. Control the improvements to keep the process on the new course.

Define

Problem Statement/Scope

The team observed that the queue for the process of filling a prescription is inefficient, and a large amount of muda exists. The wait time from start to finish is extremely long and many times frustrating for the customer. The result of this is dissatisfied customers who want to receive their medication in a timely manner. The objective of this project is to utilize Lean thinking to improve efficiency and reduce wait times for the customer. The scope or boundaries of our project are limited to the retail portion of the pharmacy located at the main hospital.

Problem Objectives and Goals

When planning the approach to the project, the team decided to use the SMART (SMART stands for specific, measureable, attainable, relevant, and time bound) goals tool to create a set of objectives that were specific, measurable, attainable (but challenging), relevant, and time bound. This way, the team had quantitative objectives that were not only feasible but also able to show whether or not the process was successful. The team looked at Lean tools and decided which ones would be most applicable to the project. The main objectives of the project were to

1. Reduce wait times for the customer by a minimum of 30%.
2. Reduce time for inventory by 50%.

Other objectives included applying poka-yoke or mistake proofing to reduce pharmacy verification time, improve robot efficiency using SMED, decrease downtime through TPM on robots using preventative maintenance schedules, and create a future state map that would show the ideal process. These were accomplished by using Lean tools and combined tools to help achieve our main objectives.

Stakeholder Analysis

To capture the true voice of the customers to understand whether they were indeed dissatisfied with the current prescription refill process, the team collected the voice of the customer through a customer survey. The stakeholders involved were typical pharmaceutical customers. A survey with a sample size of 25 customers was conducted in which the results showed 76% of the sample population expressing dissatisfaction (dissatisfied or extremely dissatisfied) with the wait times at the pharmacy, as shown in Figure 20.1. As shown in Figure 20.1, no participants responded that they were extremely satisfied with the wait times.

Process Overview

With results that indicated customers were dissatisfied with the wait times, the team had to determine why the process was inefficient. The team decided to look at the process from the customer's point of view. What does the customer see? The team developed a process flow diagram illustrating the pharmacy process of picking up a prescription from the customer's point of view. As shown in Figure 20.2, the process is very straightforward from the customer's viewpoint, making it understandable that frustration would be present in an otherwise straightforward process due to several steps where the customer must wait (waiting to process their identification [ID] card and verifying the prescription, and waiting to pick up the medication).

The team interviewed key process owners to understand how the pharmacy prescription fill process worked. Therefore, a current state analysis was conducted on-site with the interim head of the pharmacy. During this investigation, it was discovered that the goal of the pharmacy is to fill approximately 600 prescriptions per day including refills. The workday consists of operating hours from 0700 to 1900, while work is being

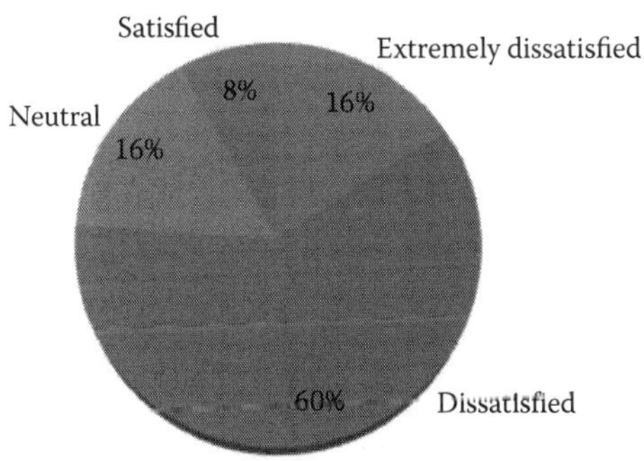

FIGURE 20.1
Customer satisfaction survey (wait times) results.

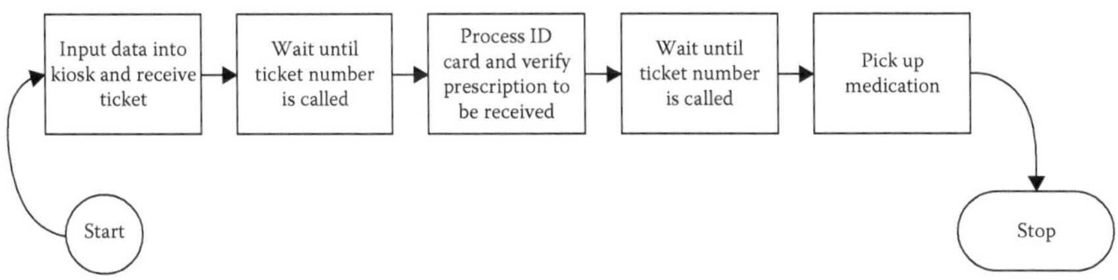

FIGURE 20.2
Process flow diagram from the customer's point of view.

conducted internally from 0600 to 2000. Refills and inventories are conducted from 0600 to 0700 and from 1900 to 2000. Refills must be called in at least 48 hours in advance and are completed in a batch and queue method, where the refills stack up and are all filled in bulk in the morning or evening. At any one point, there can be between three and five pharmacy technicians (four on average) and two pharmacists working within the system.

In terms of serving the customer, the order in which the customers arrive is the order in which they are served and receive medication. This first-in-first-out method means that there is no triage system where a certain type of customer gets priority over the other. Additionally, the arrival rates in terms of customers follow a uniform distribution with the peak demand between 1100 and 1400.

There are three courses of action when a customer is in-process to receive medication. The fastest course of action is the one in which the refilling robots, called R2D2, process the medication. There are two that process medications throughout the day and have a processing time of 35 seconds. The robots fill the medications that are most frequently prescribed and can hold about 100–150 medications at one time. One alternative course of action is for the pharmacy technician to fill the prescription by hand because it is not in the robot or because it is a gel or liquid, which the robot cannot process. The processing time for this course of action is 64 seconds. The final alternative, which takes the longest, is if the prescription is a controlled substance and must be separately filled. The processing time for this course of action is an average of 161 seconds. Figure 20.3 shows the time to fill the prescriptions based on the prescription type.

Inventories of the medications are conducted at 0600 and 1900, twice daily. During the day, if a bin is turned over it signals that it is time to order that medication. A pharmacy technician will write the needed medication on the whiteboard so that at the end of the day it can be ordered. During inventories,

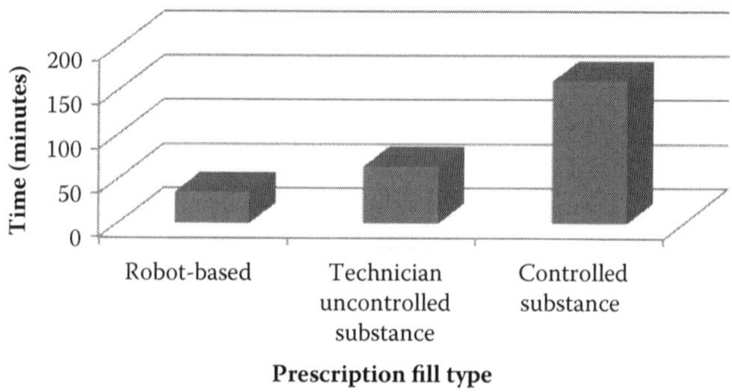

FIGURE 20.3
Average time to fill the prescriptions based on the prescription type.

pharmacy technicians physically count out each bin to determine if more of that medication needs to be ordered. If a medication needs to be ordered, the pharmacy technician will use a scanner gun to scan the medication and process it at one time at the end of the day for the next shipment to arrive overnight and be delivered at 0600 the next day. Additionally, if there are any expired medications they are turned into the supplier for proper disposal. The process map is shown in Figure 20.4.

The main Lean metrics used were the number of inventories conducted daily; lead time in seconds, which measured the time from which a customer was in-process to the time the prescription filling began; and processing time in seconds, which measured the time the medication was actually being filled. Additional metrics included value-added time, which measured the time that the customer requirements were being completed, and non-value-added time, which is the time for those activities that did not meet customer requirements.

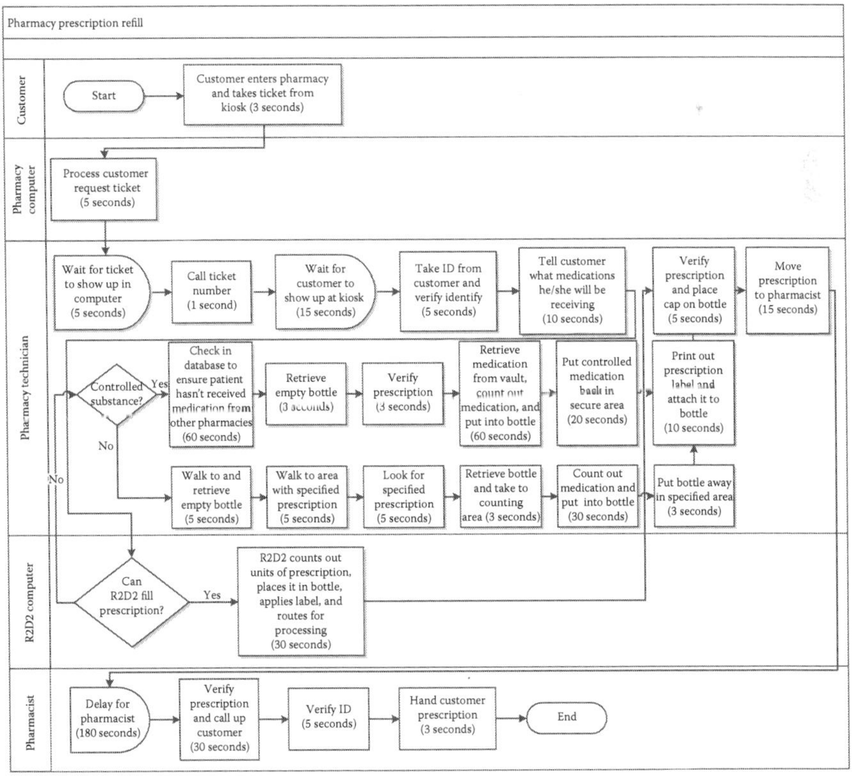

FIGURE 20.4
Process map of pharmacy prescription fill process.

Measure

Baseline

To measure the process, a baseline had to be defined. There were a number of ways the team could measure the baseline. When a customer is in-process, if there are two or more courses of action in terms of filling the prescription, the processes occur simultaneously, making the longest processing time the one that defines how long the total process takes. After looking at different ways to measure the baseline, the team decided to use the worst-case scenario approach, meaning that the longest processing time was the time recorded to perform the process. Other methods that could have been implemented include using the average of times based on the percentage utilized for that certain process. For example, robots fill about 50%–60% of prescriptions, hand fills about 15%–20%, and controlled substances about 15%–20%. Another method that could have been used was to create a weighted average of the percentages by multiplying by the processing times. However, for the purpose of showing the most immediate and significant improvement, the team decided to use the worst-case scenario method.

Data Collection Plan

To collect the data needed, the team met with the interim head of the pharmacy. After discussing the process in detail, the team determined the average times needed to process a prescription step by step. Next, the team conducted a walk-through of the pharmacy where the team looked at how the process worked in person and the current layout of the pharmacy. Due to privacy restrictions, the team was unable to take pictures or get detailed information on the medications; however, the team was able to acquire sufficient data to conduct the Lean analysis.

Current State Value Stream Map

Because this was not a normal manufacturing process, the team divided the current value stream maps into two different maps. The team utilized the gathered data as well as the process map to create two value stream maps. The first value stream map consisted of the overall system, or daily value stream map, where inventories and how the process worked step by step could be seen, as shown in Figure 20.5.

However, the team realized that the real time savings for the customer would not come from the daily operations. The team had to look at the value stream map of steps that involved the customer and wait times. Therefore, the team created an additional value stream map that laid out

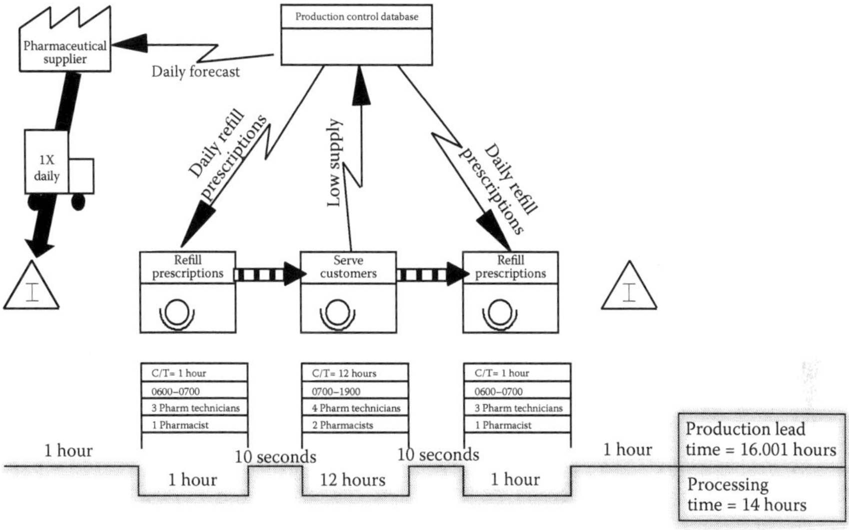

FIGURE 20.5
Current value stream map: daily operations.

the process from the time the customer arrived to the time the customer left, which was the most important aspect of the project. From the process map (see Figure 20.4), the team created a value stream that depicted the flow of processes from when a customer arrived to when the customer left (Figure 20.6).

The team now had a more detailed value stream map that could be used to create a future state with the goal to diminish the wait time for the customer. From this, the team determined a processing time of 597 seconds and a lead time of 25 seconds. The team could now identify the non-value-added processes and streamline the process as a whole by implementing Lean principles.

Analyze

Many Lean tools were used to analyze the pharmacy. The team began by simply using the five Lean principles, which are specify value, map the value stream, create flow, establish pull, and seek perfection (Womack and Jones, 2003). Initially, value was specified as the customer obtaining a prescription. The non-value-added time would include the amount of time a customer (patient) would have to wait to get his or her prescription. The customer wants to spend as little time as possible waiting to get his or her

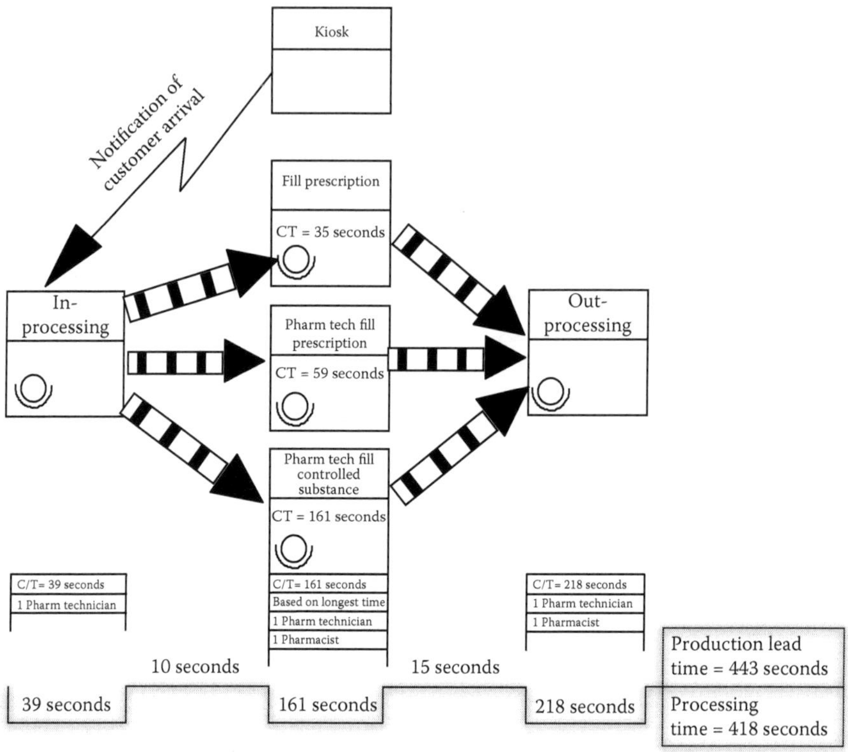

FIGURE 20.6
Current value stream map: prescription-filling process.

medications. The team also specified value as the amount of money saved on the part of the pharmacy as a whole. This dollar amount consisted of time and inventory. The time aspect of value for the pharmacy is made up of the amount of time it takes the pharmacy to fill, verify, and distribute a medication to a customer once the order comes in. The inventory aspect of value is the amount of medications and supplies taking up space inside the pharmacy. Ultimately, the team concluded that the overall value could be significantly improved by reducing the amount of time on both the customer and the pharmacy portions of the entire value stream. There are also significant improvements that can be made with respect to the inventory process on the pharmacy's part of the value stream extending to the supplier. The overall value is both the time and the money saved by both the patient and the pharmacy.

Once the team established the definition of value from a customer and pharmacy perspective, the next step was to identify the value stream. This would allow the team to see the areas where the majority of the waste was actually occurring (see Figure 20.4). The team began with the customer checking in at the pharmacy kiosk. This first step started the whole process of the prescription being filled, unless the patients called in the prescription beforehand, which means they would only have to wait until someone brought their medications to the window and called them forward. Going up to the next level of the value stream, we get to the actual pharmacy itself. In the pharmacy, the stream diverges into three directions (see Figure 20.6). The first, being the most common, is the part of the stream that involves the two robots. The second is the stream that involves pharmaceutical technicians who manually fill a prescription that the robots cannot. Such prescriptions include gels, lotions, and other medications that cannot be dispensed in the pill form. The third stream involves narcotic and controlled-substance prescriptions. This stream requires a pharmaceutical technician to get the medication from a vault and manually fill the prescription. All three of these streams converge on the pharmacist who has to check and verify every prescription before it is issued to the patient at the window. The next part of the stream goes from the pharmacy to the supplier. In the current state, the inventory is checked twice a day, once before the pharmacy opens and once after the pharmacy closes. Shipments of medications come from the supplier once a day in the morning. With these elements of the value stream defined, the team was able to identify areas for improvement. These areas included the multiserver queue setup on the patient end, the malfunctioning of robots, the layout of the pharmacy, the inventory process, and the bottleneck step with the pharmacist verifying all prescriptions.

The next step was to determine how to make the medications move from the pharmacy to the patient in single-piece flow. This was already occurring with the patients who signed in at the kiosk. The majority of the waste, however, comes from the refills that are prefilled based on the patient calling them in or requesting them online. A significant amount of space in the pharmacy is used to hold these refills, and they can stay there for up to 2 weeks. It can be argued that this saves time because the prescriptions can be filled while the pharmacy is closed, but nonetheless a significant number of prescriptions remain on the shelf well into the second week. A simple fix would be to reduce the inventory window from 2 weeks to 2 or 3 days or to have a separate area or location for the refills to be picked up.

After the team analyzed the pharmacy and looked for ways to improve the flow, the team could determine how the prescriptions should be pulled by the patient. The pharmacy currently does a good job pulling the prescriptions. A prescription is never filled until it is pulled from a patient whether it is personally from the kiosk or online or through calling ahead. The primary area with respect to pull that can be improved involves the process of receiving daily shipments based on forecasts and inventories. There is certainly room for improvement at some level with this process because, despite the daily inventories, forecasting, and deliveries, waste still exists. This waste is primarily in the form of extra inventory that eventually expires and has to be disposed of.

The final principle the team used when analyzing the pharmacy, which ties to the next part of this chapter, was continuous improvement. Some of the tools that the team applied include 5S, TPM, SMED, kanban cards, poka-yoke, and cellular manufacturing layout. By implementing a concrete plan in the pharmacy that utilized all of these tools, value can be significantly increased. The hardest component, however, is ensuring that the plan is known and adhered to by everyone involved. Even if this is accomplished, a culture of continuous improvement must be implemented as well. These tools must be continuously reevaluated and areas of improvement constantly identified and acted upon to make the pharmacy a truly Lean organization.

Improve

Implementation of 5S

The first tool implemented to reduce waste was 5S. Through the use of a 5S work sheet, the pharmacy could act on reducing waste on day one. The 5S stands for *seiri, seiton, seiso, seiketsu, and shitsuke* (Womack and Jones, 2003). This roughly translates to sort, straighten, shine, standardize, and sustain. We developed a 5S work sheet, see Appendix 1, for the pharmacy based on what was observed during our visit.

Beginning with sort, the team noticed some areas that could be acted upon immediately. This included ensuring that only necessary materials were present at each workstation, expired medication and defective products were segregated, and standard work guidelines were posted for pharmaceutical technicians. Next, the team looked at straighten, including making sure walkways and areas were clearly marked, location indicators were present for equipment, and equipment was arranged in the order of use. With regards to shine, the team evaluated the pharmacy based on overall cleanliness. With regards to standardize and sustain, there was no apparent system of ensuring that everything was sorted, straightened, and cleaned on a regular basis. This is the obvious reason why 5S was a tool that was selected first as a way to quickly improve the flow and organization of the pharmacy.

TPM and SMED Implementation

The next two tools that the team looked at were TPM and SMED. These tools would be applicable to the two robots that the pharmacy relied heavily upon. Based on an interview with the pharmacy commander, the two machines broke down on average approximately three to four times a month.

Fortunately, these breakdowns were not simultaneous for both robots. Despite this fact, having even one machine down for any period of time can have a detrimental effect on the entire operation of the pharmacy with regard to time and efficiency. The team began by looking at the current way the robots were operated. The team discovered that the biggest area for improvement regarding time was the "firefighter" mentality when a machine malfunctioned, and the traditional approach to maintenance. Only the contractor was able to fix and perform maintenance on the two robots. If a robot broke down, the contractor would then be called. The amount of time it took for the contractor to arrive and for the robot to be put back into operation varied, sometimes lasting as long as a couple of days. A lot of time could be saved if personnel within the pharmacy were trained on how to conduct preventive and routine maintenance on the robots. The team could improve this by using the four-phased approach to TPM. The first phase would involve inspecting the robots for problem areas, such as conveyor belts, robot claws, power cables, and pill bin lids. All abnormalities would then be tagged and documented using a tag log. In phase two, the technician would prioritize the abnormalities in the tag log and identify the causes, such as dirt in the conveyor belt or poorly trained pharmaceutical technicians not closing the pill bins properly. Phase three would include developing preventive maintenance standards by using an inspection standards work sheet to inspect the robots. Next, an inspection frequency would be developed and the schedule published along with the inspection standards work sheet. The final and fourth phase would consist of training all required personnel on the maintenance and repair of the robots. Over the course of a month, the overall equipment effectiveness (OEE) of the robots could be significantly improved by applying TPM.

SMED will also make the robots more efficient by saving time on the routine tasks of refilling the pill banks on each robot. The robots are currently set up to only allow for refilling of pill banks while the robots are not in operation. By changing this task from internal to external, approximately 5 minutes for each time this task is conducted can be saved (see Appendix 2).

Kanban Implementation

The final three tools used to improve the pharmacy were a kanban system, poka-yoke, and cellular layout. The goal of implementing the kanban system was to reduce inventories and the time used to conduct them. The current state involves taking inventory twice a day, once in the morning and once in the evening. By transitioning to a kanban system, the need for these inventory checks would be eliminated. This would ultimately result in the savings of 2 person-hours of time that could be applied elsewhere. This would also lower the amount of expired medications in the pharmacy and reduce inventory.

Poka-Yoke

Poka-yoke would be one of the more difficult and resource-consuming but beneficial tools to implement. A pharmacist takes approximately 3 minutes per prescription to verify. One method of mistake proofing the verification process is to create a system in which the pharmacist checks prescriptions at the beginning of a process or pulling from the value stream. Pharmacists must verify that the medication the patient is receiving is compatible with the other medications that he or she is currently taking. If there was software that could compare medicines in the database and output a color scheme, this would ensure that the pharmacist makes less mistakes and shorten verification time. For example, a red color would indicate that the medicine is definitely not compatible with the current medication that the patient is on. The pharmacist could automatically halt the prescription-filling process. A yellow color would indicate that verification is needed by the pharmacist to make a decision because the medication may or may not be compatible. A green color would indicate that the pharmacist does not need to verify it in the database because it ensures that the medication is compatible. This could lessen the verification time by up to 66% (from 180 seconds to 60 seconds).

Cellular Manufacturing Layout

After analyzing the process flow, the team determined that a cellular manufacturing layout would be the best solution in terms of providing the most significant savings and the maximum removal of waste. With the current layout of the pharmacy, considerable time is wasted while the customer is waiting and the technician is walking back and forth to grab bottles, verify prescriptions, and find medication. By creating a U-shaped layout and splitting the process up into two main production lines, a substantial amount of time is saved through efficiency. The old layout compared to the new layout is shown in Figure 20.7.

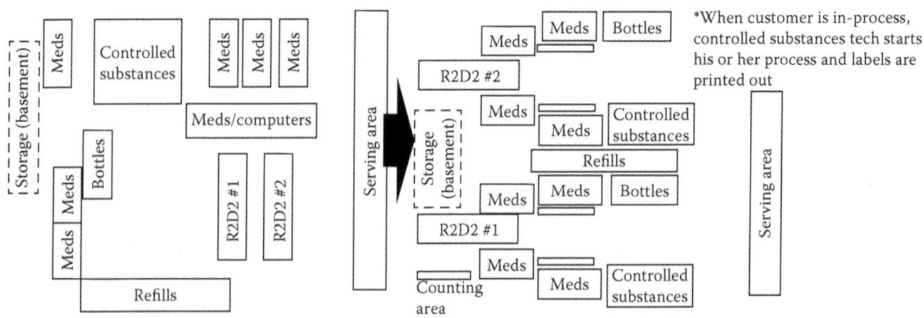

FIGURE 20.7
Old layout versus new layout.

More specifically, after applying the cellular manufacturing layout the team discovered a savings of 15 seconds per prescription with the new layout. Additionally, 53 seconds could be saved in filling controlled substances. There are approximately 120 controlled substance prescriptions a day. We estimate saving an average of 53 seconds per controlled substance prescription due to the new layout. This results in 106 minutes saved per day, during the time that a customer is waiting.

Solution Implementation Plan

The first step to begin implementing the proposed improvement ideas would be to conduct a 2-day executive planning session with the executive management of the pharmacy and a sensei, a Lean expert who is familiar with the healthcare industry. The purpose of this would be to gain executive buy-in, establish strategic direction, and assess cultural change readiness. Additionally, the pharmacy could request funding for contractor support, personnel hours for improvement workshops, and material budget for equipment needs (County of Ventura, 2012).

The next step would be to conduct a value stream analysis in which further waste along the value stream is identified. This should be done not only with management but also the pharmacy technicians and pharmacists to understand the process from their perspective. Once waste is identified, they could schedule events off of a possible, implement, challenge, and kill (PICK) chart where they could determine which Lean tools would have the biggest payoff and be the easiest to implement. By ranking and ordering the Lean tools based on the chart, they could then create a schedule in which they could estimate which ones to implement first, which ones would take longer, which would need more resources, and so on. The PICK chart is shown in Figure 20.8.

A possible implementation schedule is taking the first 2 weeks to implement the kanban system during the time it takes to perform inventories and the 5S work sheet to minimize clutter. These would require little resources and could be done relatively easily. The second 2 weeks would see the more challenging tool implementations such as SMED and TPM. Finally, as time allows and resources are approved, the team could work on the cellular manufacturing layout and poka-yoke, which will take a significant amount of time and resources but provide big payoffs.

	Big payoff	Small payoff
Easy to implement	Implement	Possible
Hard to implement	Challenge	Kill

FIGURE 20.8
Possible, implement, challenge, and kill chart.

The final step would be to conduct quarterly organizational assessments in which they could assess the return on investment for the Kaizen events. By this time, the staff would have seen how effective the Lean tools are and would be able to identify further continuous improvement opportunities. In this step, they could see sustainment as well as record lessons learned or identify course corrections. Recording lessons learned is critical as this will be the next step to try and implement future changes to the pharmacy.

Future State Map

The final step was to create a future state value stream map that showed the ideal process. The future state map showed how the process improvements would result in time saved for the customer by implementing Lean tools. The team decided to create one where there would be more continuous flow by pulling information as well as showing the benefits achieved as far as time saved. The two future state maps or the ideal states are shown in Figures 20.9 and 20.10.

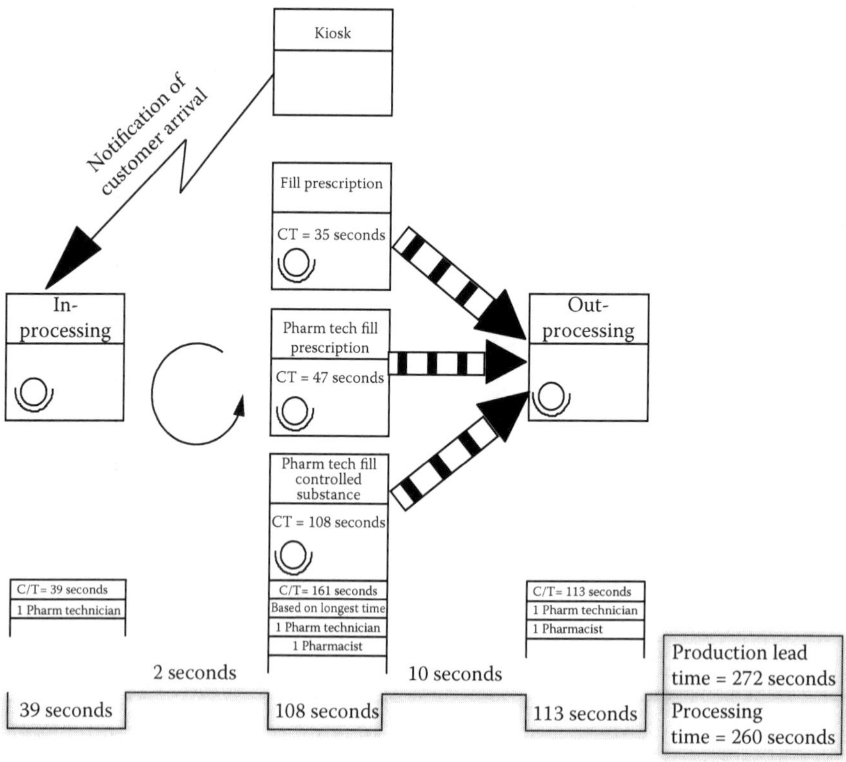

FIGURE 20.9
Future state map 1.

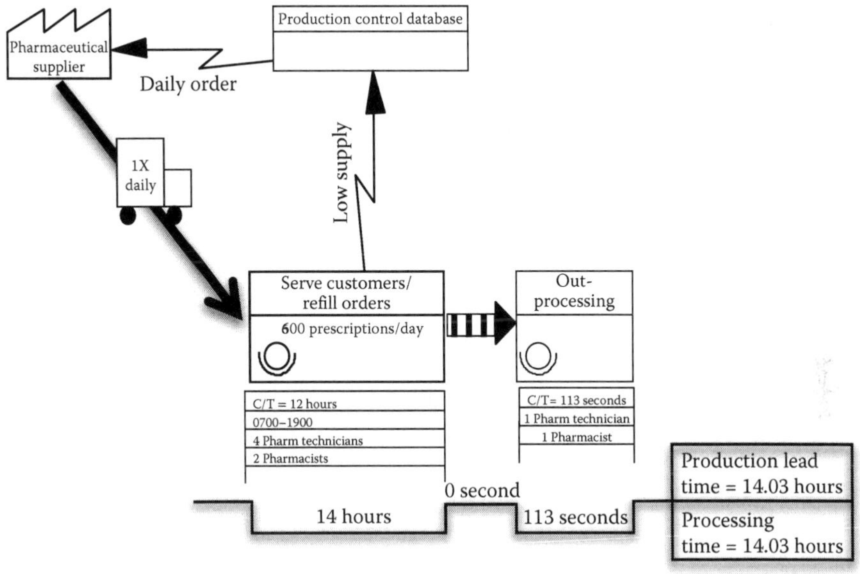

FIGURE 20.10
Future state map 2.

Control

Now that the ideal state is determined, it is critical that all the improvements are documented and continuous training conducted. In the beginning of the process, it was important to show how easily low-cost change could be implemented and the benefits achieved. Once the initial Kaizen is conducted, change will become more difficult but very possible. To control the improvements, management must institute control plans and standard work to ensure that continuous improvement is achieved. Pharmacy technicians should have a daily checklist of items to check and procedures to conduct that are specifically detailed step by step. This is mistake proofing the work that the pharmacy has achieved thus far. There should be a handbook with a list of procedures that must be conducted daily, weekly, monthly, and so on. This ensures that continuity is maintained as there is employee rotation. A brand new pharmacy technician should be able to walk into his or her first day of work, follow the standard operating procedure, and have minimal questions as to what needs to be performed in what order and by what time. At this point, policies should be developed concerning how to further implement change and encourage feedback. Control will be challenging as human nature will continue to resist change. This is why it is critical to set up a system that will encourage pharmacy technicians and pharmacists to not fall behind and continue to move forward.

Conclusions/Results

Due to the scope of the effort and the time available to the Lean team members, the proposed improvements were not implemented during this analysis. However, based on the analysis, the team provided potential estimated time savings based on their proposed Lean design. The estimated time that can be saved by implementing kanban cards is 120 min/day. Estimated time saved by using SMED is 5 min/day. Estimated time saved from using the cellular manufacturing layout is 406 min/day. Estimated time saved by continuously pulling refills is 406 min/day. This could create approximately 8.85 extra person-hours per day that the pharmacy could use to perform other work. More importantly, the extra person-hours could allow Lean tools to be implemented during this time without disrupting the flow of daily activities for the initial Lean improvement implementation steps. By calculating the average time it takes to process one prescription per customer, the team was able to achieve the main goal of reducing the customer wait time by about 40%. Table 20.1 shows the benefits achieved through implementing Lean tools.

It is evident that the application of Lean tools to a pharmacy scenario can result in comparable success as when applied in a manufacturing setting. The focus of this project does not cover the span of time needed to see these tools implemented and carried out over the next few months. Consequentially, we must rely on our forecasted data in Table 20.1 as evidence that significant reduction in waste can be achieved. We can also point to an example in which a similar project, over the course of a year, reduced waste by $289,256 (Hintzen et al., 2009). It describes the project and results of applying Lean techniques to the inpatient pharmacy at the University of Minnesota Medical Center (UMMC) in Fairview, Minnesota. The goals that this project sought to achieve included improving workflow, reducing waste, and achieving substantial cost savings. The main Lean tools Hintzen et al. (2009) used in the project included 5S and value

TABLE 20.1

Benefits Achieved

	Current State	Future State	Improvement (%)
Total lead time	25 seconds per prescription	12 seconds per prescription	52
Inventory turns	2	0	100
Customer wait time	7.22 minutes per customer	4.33 minutes per customer	40
Additional capacity time for pharmacy	0	8.85 person-hours per day	

stream mapping. By mapping out the value stream of the entire pharmacy, the project team identified the sterile products area (SPA) and the inventory area as places along the value stream that had the most non-value-added time and waste. Process improvement goals for the SPA included the reduction of missing doses, errors, and patient-specific waste by 30%, 50%, and 30%, respectively. They also expected a significant reduction in pharmaceutical inventory and returns due to outdating. The benefits of applying Lean concepts included an annual cost saving of $289,256 due to waste reduction, improvements in workflow, and decreased staffing requirements. This is an excellent example where Lean was successfully applied and resulted in substantial waste reduction. Therefore, there is no reason that these principles and tools would not work in our pharmacy project. This ultimately shows that improving the processes of a pharmacy is in the best interests of customers whether it is the patient, hospital, or taxpayer.

References

County of Ventura, 2008. "Lean Continuous Process Improvement." Accessed April 20, 2012. http://www.ventura.org/rma/service_excell/pdf/RMAImplementationPlan.pdf.

Hintzen, Barbara L., Scott J. Knoer, Christie J. Van Dyke, and Brian S. Milavitz. "Effect of lean process improvement techniques on a university hospital pharmacy." *American Journal of Health System Pharmacists*. 2009, 66(22), 2042–2047.

iSixSigma, 2005. "DMAIC." Accessed April 20, 2012. http://www.isixsigma.com/dictionary/dmaic/.

Womack, James P., and Daniel T. Jones. *Lean Thinking: Banish Waste and Create Wealth in your Corporation*. New York: Free Press, 2003.

Appendix 1 5S Work Sheet

	Jan		Feb		Mar		Apr		May		Jun	
Pharmacy	Y	N	Y	N	Y	N	Y	N	Y	N	Y	N
1S Sort (seiri)—organize needed from unneeded; remove the unneeded.												
Are only necessary materials present? (bins, bottles, lids, paperwork, bags, medications)												
Are defective materials identified and segregated? (expired medication, defective products)												
Are unused machines, equipment, or parts marked with a red tag? (medication not picked up)												
Are red-tagged items in the red tag area?												
Is standard work present and current? (pharmacy techs adhering to the current standard of work)												
2S Straighten (seiton)—arrange neatly, identify for ease of use.												
Are areas and walkways clearly marked?												
Are materials visually controlled location indicators and at the point of use?												
Are maximum and minimum allowable quantities indicated?												
Is nonconforming material segregated, identified, and documented?												
Is equipment clearly identified and arranged in order of use?												

	Jan		Feb		Mar		Apr		May		Jun	
Pharmacy	Y	N	Y	N	Y	N	Y	N	Y	N	Y	N
3S Shine (seiso)—conduct a cleanup campaign to polish and shine the area.												
Are floors and areas clean and free of trash, liquids, and debris?												
Is equipment inspection combined with equipment maintenance?												
Is the work area lighting sufficient, ventilated, and free of dust and odors?												
Are cleaning operations assigned and visually managed?												
Are cleaning supplies easily available and stored in the marked area?												
4S Standardize (seiketsu)—continue the habits frequently and often.												
Are standard procedures clear, documented, and actively used?												
Are there records of defective material entries with corrective action?												
Are equipment TPM procedures in place and current?												
Are waste and recyclable material receptacles emptied regularly?												
Are scrap and product nonconformance areas cleared at a set interval?												
5S Sustain (shitsuke)—institute the 5S practices as a part of the culture.												
The fifth S is awarded when three consecutive months of 1S, 2S, 3S, and 4S have been sustained.												

Appendix 2 SMED Work Sheet

Setup Analysis Chart | **Machine** 1 | **From:** A | **To:** B

Step #	Changeover Element	Time Element (seconds)	Changeover Categories: Internal	External	Waste	Improvement Plan (Design Changes)	Goal of Improvement Plan: Eliminate	Internal to External	Reduce
1	Turn off Script Pro machine	1	1			Design an inventory bank door in the rear of the machine where bins can be accessed while the machine is still operating while avoiding empty bins	×		
2	Open inventory bank door	15	15		×	Door now opens from the rear to not interfere with the machine		×	
3	Flip open bin release switch	5	5		×	Release switch is now on the rear		×	
4	Remove empty bin	5	5		×	Empty bin is removed from the rear door while machine is operating		×	

5	Fill empty bin with needed pills	120	120		×	Pills are counted into the bin based on type of medication and forecasts (so pills do not sit in bins and expire before use) specific number of pills, not just "fill the bin." This can be reduced by having pills measure out before operation for typical types of meds		×	×
6	Place filled bin back in inventory bank	5	5		×	Filled bin is replaced while machine continues to operate		×	
7	Close bin release switch	5	5		×	Close bin release switch now on the rear			
8	Close inventory bank door	15	15		×	Close inventory bank door now on the rear		×	
9	Run Script Pro diagnostic	120	120		×	No diagnostic needed since machine was never turned off, no need to reboot	×		
10	Place machine back in operation	5	5		×	Machine was never taken out of operation	×		
Totals (seconds)		296	296	0		Totals (seconds)	126 Eliminated	170 External	
							5 minutes of waste eliminated from the value stream		

21

Using Value Stream Mapping to Identify Performance Gaps for Hoshin Planning

Elizabeth A. Cudney

CONTENTS

Executive Summary

Carjo Manufacturing represents a composite company that produces axle components for the automotive industry. The company's main products include shafts, tubes, and gear housing assemblies. The company's sales two years ago were $18 million, but dropped to $16 million. Sales this year are expected to be down again. To be competitive in the market, Carjo must implement Lean to drive out wastes and improve its profit margins. In addition, Carjo must develop a strategic plan to ensure the organization focuses on the appropriate process improvement opportunities.

Organization's Current State

The executive management at Carjo Manufacturing understands the necessary change in the manufacturing environment and sees what they need to do for the organization to survive. In addition, Carjo's senior leadership has heard about the success of Lean and has driven some Lean implementation activity. The problem has been that the company has seen only modest gains from implementing Lean. The senior leadership team expected to achieve more dramatic gains from the Kaizen events that have been held.

Carjo has a strategic vision at the executive leadership level, but this vision is not cascaded down through all levels of the organization. The executive team realizes that this is a hurdle for the organization and that Carjo can achieve more significant results by linking its improvement efforts to its strategic long-term objectives. The executive team has determined that they must cascade down the long-term vision of the organization to prioritize their continuous improvement (CI) efforts. In terms of CI, the employees have been exposed to Lean techniques through minimal training. Only a select group of employees have been trained on Lean and only as it applies to their specific job function. There is no formal program for mentoring or coaching employees with respect to Lean, Kaizen, or Plan–Do–Check–Act. It is currently an ad hoc approach of employing these tools where they are needed if employees are aware of them. A broader training of the entire organization on Lean is necessary to achieve the gains Carjo needs.

Lean and Hoshin Kanri Integration

Executive management formed a cross-functional, CI team that was familiar with Lean and provided them formal training. Executive management tasked this team to lead the effort of determining the current state and future state of the organization. Concurrently, executive management focused on the strategic planning efforts of the organization while the CI team worked on value stream mapping.

The CI team began by creating an overall current state map of how Carjo currently creates value for their customer for the gear housing assembly. Then, the team created an enhanced future state map, which incorporated best practices into their processes through research and benchmarking. The goal is to holistically optimize the entire process of value flow by eliminating waste and controlling variation.

Achieving full implementation of the enhanced future state value stream map is far more complex than developing it. The executive team decided to use Hoshin Kanri to begin their Lean journey by linking the long-term strategic plan of the organization to process improvement efforts. Typically, organizations select their Kaizen events and process improvement projects based on anecdotal data. It is critical for the success of the organizational effort to develop skills and know-how of process improvement. Decisions must be based on data. Organizations must focus on developing the skills of their employees by providing appropriate training on process improvement tools, data analysis, and decision-making.

Lean drives CI and as a stand-alone tool is effective. However, linking Lean initiatives to the strategic vision of the organization considerably enhances the effectiveness of Lean. The leadership team at Carjo has identified a disconnect between the long-term strategy of the business and the improvement efforts that have taken place. The company leaders have decided to use Hoshin Kanri to flow down the long-term vision of the organization and use it as the strategy for business process excellence. Therefore, Carjo utilized a five-phase methodology (described in Chapter 10) to implement the enhanced future state value stream map using Hoshin Kanri to expedite the rate of improvements.

Lean and Hoshin Kanri Implementation

The approach developed by executive management uses the strategic long-term vision of the organization to prioritize CI efforts. A value stream map provides a picture of all of the activities required to produce a product. The Hoshin strategic plan summary also provides a picture of the overall strategy of an organization and how the strategy cascades throughout all levels of the organization. The linkage is clear on how each strategic goal is measured and who has the ultimate responsibility.

The senior leadership team at Carjo Manufacturing scheduled a leadership retreat to develop the long-term vision of the organization. The team developed four strategic goals for the organization including the following:

1. Develop world-class new business/marketing.
2. Expand product offerings.
3. Implement cost reduction projects.
4. Implement Kaizen projects.

The senior leadership team then used the strategic goals to develop three breakthrough goals for the organization including the following:

1. Improve financial returns by 5% in 3 years.
2. Grow sales by $3 million by in 3 years.
3. Achieve world-class supplier status in 3 years.

The leaders then reviewed the relationships between their strategic goals and their breakthrough goals to ensure all of their strategic goals were being addressed properly. Next, the team identified the appropriate metrics that would tie directly to their breakthrough goals. The metrics were quantitative to indicate whether the process improvements have an impact on the overall organization and are trending in the right direction. Finally, the senior leaders assigned ownership of the breakthrough goals to specific members of their team. Figure 21.1 provides Carjo's Hoshin strategic plan summary.

The leadership team next focused on developing the Hoshin plan summary. As shown in Figure 21.2, the team carried down the strategic goals and owners from the company's Hoshin strategic plan summary. The short-term and long-term goals tie back to the measures outlined in the Hoshin strategic plan summary. The leadership team at Carjo decided to have their short-term focus on improvements in the next year and the long-term goals 3 years out.

The next big decision for Carjo's leaders was to determine what their implementation strategies would be based on related to their strategic goals. For example, to develop world-class new business/marketing, the leaders determined their implementation strategy would be to grow sales. Then, on the

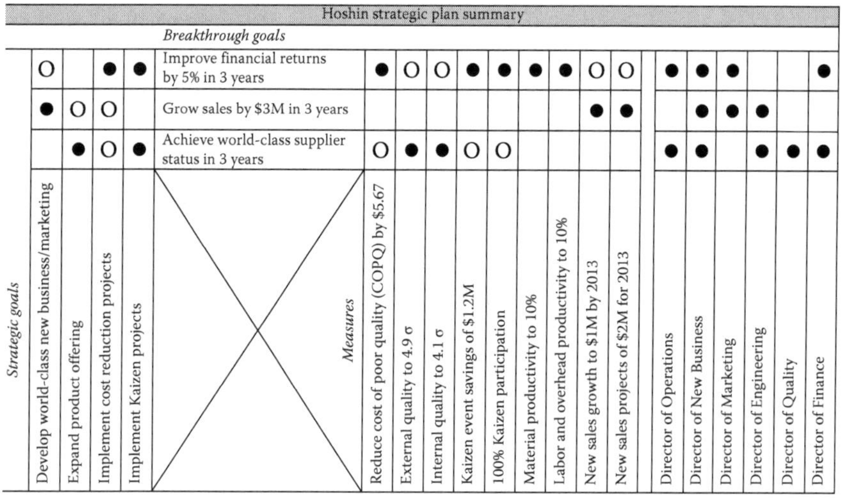

FIGURE 21.1
Carjo's Hoshin strategic plan summary.

Hoshin plan summary								
Strategic goals	Mgt. owner	Goals		Implementation strategies	Improvement focus			
		Short term	Long term		Safety	Quality	Delivery	Cost
Develop world-class new business/ marketing	Director of New Business	$1M by next year	$3M in 3 years	Grow sales				●
Expand product offerings	Director of Engineering	$2M for next year	$5M in 3 years	New sales projects		O		●
Implement cost reduction projects	Director of Operations	$1.2M annual savings	$4M annual savings	Improve financial returns	O	●	●	●
Implement Kaizen projects	Director of Operations	100% participation	100% participation	Achieve world-class supplier status	●	●	O	●

FIGURE 21.2
Carjo's Hoshin plan summary.

basis of the implementation strategy, the leadership team determined which improvement focus area was impacted. Using four improvement focus areas (safety, quality, delivery, and cost), the leadership team at Carjo outlined the impact of each strategic goal. Using this information, the leadership team was able to clearly see that two strategic goals impacted all four improvement focus areas. These two strategic goals are implement cost reduction projects and implement Kaizen projects because they show a relationship with each improvement focus.

This helped Carjo's senior leaders prioritize their implementation strategies. Since these two implementation strategies have a significant impact, the leadership team determined these should be the company's highest priority.

Using the information from the Hoshin strategic plan summary, Hoshin plan summary, and implementation strategy, each department in Carjo began using their current state value stream maps to identify opportunities relating to the implementation strategies. This linked the department-level value stream map of the gear housing to the high-level value stream map developed by the CI team.

Before starting the mapping process, the process improvement team first sought to understand the customers' requirements. For Carjo's main product, the gear housing assembly, the daily customer demand is 775 units. The plant runs a two-shift operation. Therefore, the takt time is calculated as shown in Figure 21.3.

Takt time is 67 seconds per gear housing assembly. This information is critical to the process improvement team for the current state value stream map since all process operations should be less than takt time. Figure 21.4 shows the current state maps for the gear housing assembly.

Based on the current state map and customer demand, the process improvement team at Carjo can identify the improvement opportunities.

Calculating Takt Time			
8	Hours =	480	Minutes (based on standard work shift)
		−30	Minutes (break time)
		−10	Minutes (wash time)
		−0	Minutes (cleanup)
		−5	Minutes (team meetings)
	Total	435	Available minutes per shift
435	Minutes available × 60 = 26,100 seconds per shift		
26,100	Seconds per shift × 2 shifts = 52,200 seconds per day		
52,200	Seconds divided by 775	pcs/day = 67 seconds	
Takt time = 67 seconds per piece			

FIGURE 21.3
Takt time.

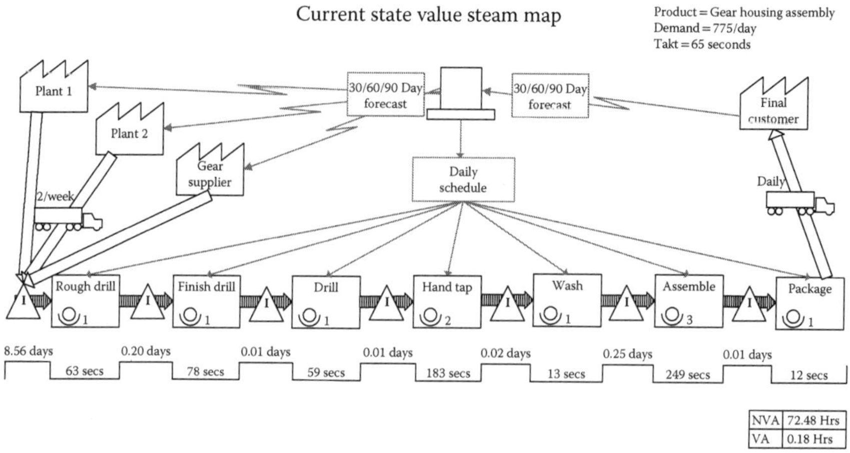

FIGURE 21.4
Current state value steam map.

The team uses the current state map and takt time to pinpoint bottlenecks that hinder flow and cause wastes.

There are several issues with the gear housing assembly process that lead to excessive waste. Waste is any process or operation that adds cost or time and does not add value. There are seven original wastes that are described in Figure 21.5. An eighth waste of unused creativity should also be considered because it is critical to focus on our people and tap into their intellectual creativity.

The process improvement team used this information to identify waste within the current state map for the gear housing assembly. Using the value steam map in Figure 21.4, the process improvement team added the wastes to the current state map. Figure 21.6 shows the current state map for the gear housing assembly with the waste in the processes identified using Kaizen bursts.

There are several key points from analyzing the information in Figure 21.6. First, the value-added time is 656 seconds. The total lead time is 7.2 days. This calculates to 0.18% value-added time. In several instances, operators are assigned to a machine that is operating considerably below takt time, which is causing inventory to build up. This leads to the waste of overproduction and waste from idle/waiting time.

Given the takt time of 67 seconds, there are three operations in the gear housing assembly process that cannot be completed within this time. These are bottleneck operations that cause other operations to wait. In addition, the long cycle times lead to overtime to produce enough products for shipments, which increases production costs.

Wastes	Description
Overproduction	Producing more than what the immediate internal or external customer needs. Overproduction requires additional space, material handling, and storage that otherwise would not be needed.
Inventory	Inventory not immediately needed by the customer. This is typically caused by push scheduling.
Waiting	Time spent waiting for materials. This is typically caused by unbalanced production lines.
Transportation	Transportation does not add value since it does not contribute to transforming the final product. Point-of-use techniques can help minimize transportation waste.
Processing	Overprocessing waste can be caused by poor tool or product design.
Motion	Wasted motion involves double handling, reaching for parts, and stacking parts to name a few. Point-of-use techniques can help eliminate wasted motion as well.
Defects	Poor quality of products requires production of additional products (causing overproduction) to replace the defective parts and creates an inventory of unusable products. Poke-yoke techniques can help prevent defects from moving down the production line.

FIGURE 21.5
Seven original forms of waste.

There is also considerable inventory within the facility that exists mainly at the beginning and end of the process, because of infrequent deliveries. There is also inventory built up as work-in-process throughout each facility caused by the lack of balance between operations.

Carjo's previous method for process improvement was for each individual team to determine their process improvement activities. The Kaizen event selection was up to each team leader. No formal method existed to select projects. The current state value stream maps were used to drive the improvement project identification, but there was no clear method for prioritization of the projects. This results in some improvement gains, but not a big impact on the organization. However, by linking the process improvement efforts to Hoshin Planning, Carjo can now prioritize their process improvement activities to realize a significant impact to the organization.

Using the information from the Hoshin Plans and current state value stream maps, the Carjo executive team led the effort to drive down the strategic goals into its daily management and action plans. One of the key strategic goals was implementing cost reduction projects because of the goal's impact on safety, quality, delivery, and cost. The department managers reviewed the current state map with their teams to identify improvement opportunities for cost reduction projects. A key success factor for this project was the

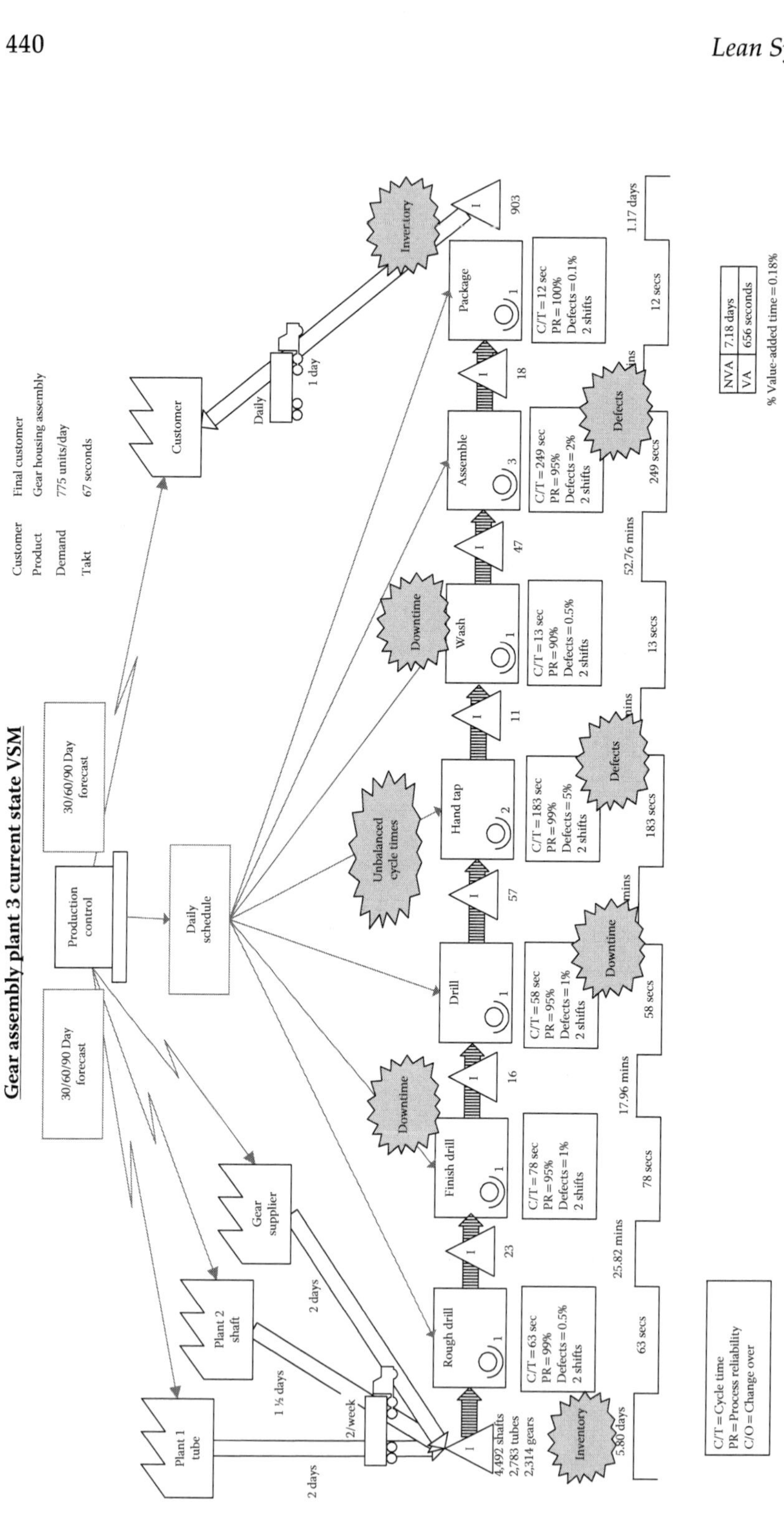

FIGURE 21.6
Gear housing value stream map with wastes.

training and coaching the teams were provided. The executive management team designed the training and coaching into the Hoshin Kanri and Lean rollout as part of the five-phase methodology.

In addition, the managers noted any high internal defect rates. The hand tap process currently accounts for 5% of the total product defects. This defect had a current internal quality defect rate of 56,704 parts per million (PPM). Therefore, the department manager selected this project as part of the company's cost reduction strategy. Using the current baseline, the team developed short- and long-term goals for the project. Figure 21.7 shows the Hoshin action plan for this project. As a team, each department discussed each breakthrough goal to identify implementation strategies that would impact the overall organization. The departments developed a Hoshin action plan for each of these strategies.

Next, the executive team at Carjo cascaded these strategies down to the Hoshin implementation plan, also called a bowling chart. To determine the target improvements, the team revisits the current state maps for this process to determine the baseline for the improvement strategy. By understanding the current performance level, the team determines the target improvement. Using the monthly improvement targets, the team manages the projects and monitors trends using a bowling chart as shown in Figure 21.8. The shaded values are the actual values for that month that did not meet the goal for that month. This allows the team to manage by exception.

HOSHIN ACTION PLAN		
Breakthrough Goals: Improve financial returns by 5% by 2015	**Team:** Caroline, Josh, Wes	
Management Owner: Brian C.	**Date:** 9/14	
Department: Gear-housing machining	**Next Review:** 10/15	
Situation Summary: Internal defects are currently 5% of total product cost		
Objective(s): Implement cost reduction projects to improve financial returns		
Short-Term Goal: Internal quality to 32,798 PPM **Long-Term Goal:** Internal quality to 28,154 PPM	**Strategy:** Six Sigma project on hand tap	**Targets and Milestones:** Conduct measurement systems analysis 9/21 Process failure modes and effects analysis 10/2

FIGURE 21.7
Hoshin action plan.

HOSHIN IMPLEMENTATION PLAN														
Breakthrough Goal: Implement cost reduction projects **Strategy Owner:** John F. **Date:** 9/14														
Strategy	**Performance**		**Schedule and Milestones**											
			J	**F**	**M**	**A**	**M**	**J**	**J**	**A**	**S**	**O**	**N**	**D**
Six Sigma project on hand tap	Target	28,154 PPM	57,304	54,654	52,004	49,354	46,704	44,054	41,404	38,754	36,104	33,454	30,804	28,154
	Actual	31,652 PPM	54,007	54,900	51,583	48,128	42,860	43,957	38,100	37,905				
Kaizen participation	Target	100%	10%	20%	30%	40%	50%	60%	70%	80%	90%	100%	100%	100%
	Actual	70%	12%	23%	31%	42%	55%	58%	74%	82%				

FIGURE 21.8
Hoshin implementation plan.

Conclusion

Using an approach to link Lean into the strategic vision of the organization enables organizations to realize the full benefits of Lean. Lean projects that are selected based on their impact on the entire organization have the most effective results. For Lean efforts to be successful, employees must be adequately trained and coached to develop their skills. Effective mentoring is essential for employees to understand and implement new techniques. Therefore, the approach linking Hoshin Kanri and Lean must address training and coaching to develop the skills of employees at all levels of the organization.

It is critical to ensure that your strategic vision cascades down throughout your organization into the daily activities of all employees. This clear linkage enables an organization to move in a common direction with common goals. When employees understand the direction of the organization, they can make the appropriate improvements that will enable long-term success. By using the strategic vision, an organization can employ Lean techniques to eliminate waste and improve flow.

Section IV

Planning and Implementation Strategies for Lean Initiatives

22

Planning and Executing a Kaizen Workshop

David M. Dietrich

CONTENTS

Planning the Lean Initiative Workshop

As the Lean practitioner engages in the application of the tools and techniques described in previous chapters, a specific sequence of events needs to occur to make sure the candidate initiative is successful. As with most Lean activities, a practitioner would hope to achieve improvements in safety, cycle time, cost, quality, and customer satisfaction by envisioning possibilities, energetically pursuing opportunities, and rapidly implementing process changes. The following sequence of workshop events would ensure that the company engaging in Lean would be successful in its endeavor.

Before beginning any Lean initiative, inevitably, the participants involved in a Lean workshop will ask, "What is in it for me?" At the same time, the Lean practitioner should be asking, "What is in it for the customer?" Because Lean efforts affect the entire company, it should be made clear that participation in the workshop will yield process improvements that will enhance the stability of the company and will increase job satisfaction among the participants because results can happen fairly quickly. In addition, management support incentives may be offered to team members. As with all Lean activities, the customer should be the sole focus of the candidate Lean initiative.

The Lean workshop leader should work with management to determine the value of the workshop.

Roles and Responsibilities of Team Members

Participants in the workshop each have specific roles and responsibilities that should be documented and reinforced by the team leader.

Management team: The process sponsor, process owner, and management who support the area of focus for the workshop have the following responsibilities:

- Down-select a process target for Lean improvement.
- Assist with the preparation of workshop objectives that fulfill business needs.
- Provide uninterrupted team member participation that frees the team members from daily routines.
- Empower the team to make rapid changes.
- Remove impediments that discourage participation.
- Host an end-of-event celebration that focuses on rewarding the participants.
- Provide follow-up at regular intervals to ensure that gains are held and action items completed.

Workshop participants have the following responsibilities:

- Provide process understanding that encompasses product or service value and flow.
- Use facts and data to make decisions.
- Present unbridled fresh ideas in an aggressive manner.
- Do not say "we can't." If impediments exist, suggest alternatives. Identifying impediments without suggesting alternative ideas is considered complaining.
- Be an active participant, not a passive one.

Critical to Success Factors

For the workshop to be successful, there are a number of factors that must exist prior to the workshop being conducted. For example, people's issues must be addressed before the workshop week: work schedule conflicts, chain of command challenges, mitigation of office politics, and nonparticipant

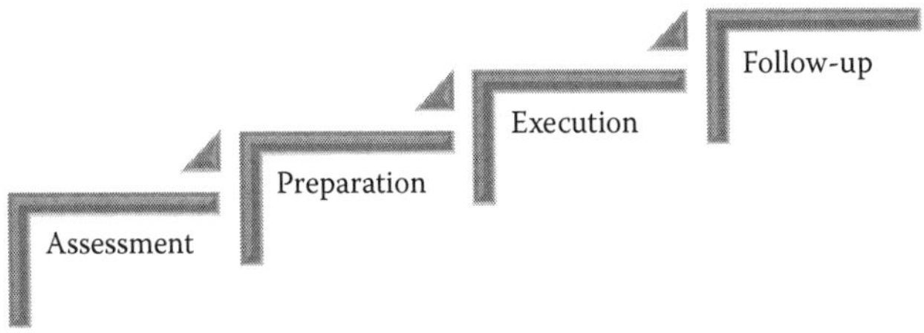

FIGURE 22.1
Workshop phases.

communication. In addition, the following list considers other items to keep in mind for the workshop:

- The workshop will not solve everything that is wrong with the facility, process, or company. However, periodic controls should be put in place to ensure the successes in the workshop.
- Additional communication effort is required for area nonparticipants.
- Targeted improvements should be clearly defined to show how they integrate together to fulfill company objectives.
- Decisions should be based on all the consensual data gathered by the team and not on preconceived notions or opinions.
- You cannot pick Lean tools à la carte. You must follow the events sequentially to build on tools and successes.

There are four phases of the workshop with each time-dependent phase defining what should occur. Figure 22.1 describes the phased layout of the workshop.

Phase 1: Assessment

The purpose of this phase is to provide management with an evaluation of the organization's readiness potential for improvement from a workshop. This effort starts with a meeting to review business analysis materials with the person responsible for the process area. During this phase, potential workshop topics and priorities will be discussed. Identification of key stakeholders and a schedule for completion of the plan will also be determined. This is performed by the workshop leader, and he or she looks at opportunities around the following areas:

- Process flow impediments
- Troublesome equipment identified by total productive maintenance efforts

- Disorganization of office and/or production areas
- Fluctuating customer demands creating excess work-in-process (WIP) inventory
- Constrained processes

The workshop leader also evaluates readiness for the organization in terms of

- Responsiveness of company changes
- Rapid need for improvement
- Enthusiasm/spirit
- Management support

This phase's exit criterion ends with a report out to management to review the findings and recommendations and a plan for the next steps. There may be several reasons to defer a workshop. These reasons may include topics such as low enthusiasm among the workers or management, little workplace organization (e.g., needs a 5S first), deprived corporate culture for various improvement sustainment, lack of commitment by managers, disagreement in location, scope and timing, and so on. If any of these reasons apply, delaying the workshop is appropriate until team members have been educated about the benefits of Lean, some progress is made with 5S, or an improvement plan exists for the area in question.

The following time line is recommended based on past successful workshops. As general education, data availability, and standard work become institutionalized, this time line may be reduced. Initially, every first-time workshop leader should follow this checklist and time line while going through the certification criteria process.

By the end of the assessment phase, the Lean workshop leader should conduct the following:

- Hold the first meeting; review the value state map.
- Conduct a process walk and area evaluation with managers.
- Define and document roles and responsibilities.
- Brief the management team on the environment scan.
- Coordinate items required for workshops with local management. Items may include logistical supplies, food, training material, conference rooms, and so on.
- Identify potential process owners.
- Schedule initial meeting dates/times/places with management.
- Schedule initial area evaluation.
- Coordinate notification letter submittal to participants (date/time/place).
- Send e-mail to the management team responsible for the workshop area.

Phase 2: Preparation

To achieve a successful workshop, a large amount of preparation is required. The more effort put into this phase, the more successful phase 3 becomes. The purpose of the preparation phase is to develop a charter for the Lean workshop that is tailored to the area based on facts and data. The person responsible for this effort is the Lean workshop leader with help as required from the organization's management team and other key individuals.

During this phase, a definition of a clear charter to focus the workshop is established. Plan the meeting logistics (rooms, etc.), and select team members who are knowledgeable about the topics covered in the workshop. It is critical to define the workshop objectives and the approach to be used in the workshop to achieve those objectives. If the workshop is reacting to a known problem or condition, observe the process, collect data, and document per the tools taught in this book such as Pareto diagrams, histograms, and time observations. Identify notional root causes of problems to be resolved in the project charter and the appropriate tools to be used to resolve the problems. A simple way to gather data items for a root cause analysis is a KNOT chart. The KNOT chart simply defines what is known (K), what is not known (N), what is opinion (O), and what we think (T) we know (see Figure 22.2).

Once a notional root cause has been defined, a charter may be developed. If you are starting with a new process development or greenfield site, use past

	Specific Data Item	Know	Need to Know	Opinion	Think We Know	Action
1	Two workers clocked in at 5:00PM on December 4	X				
2	Conveyor on the machine appeared to be faster than usual			X		Locate and interview other witnesses
3	Operator heard of conveyor issues from the same machine model at previous company			X		Locate and interview other witnesses
4	Motor speed was set at 1200 rpm				X	Verify by review of machine logs
5	Loose screws on the floor?		X			Interview cleaning crew
6	Thermocouple registered 65° F at 11:00 PM	X				

FIGURE 22.2
Example of a KNOT chart.

experiences at other facilities to assist with process improvement charter development. The general output from this effort includes the following:

- Objectives and boundary definitions: the objectives should be SMART (specific, measureable, attainable, relevant, and time sensitive) bounded.
- Data collection and documentation: from gemba walks and data-gathering tools.
- Workshop logistics: meeting locations, reoccurrence frequencies, and so on.
- Team member selection: try to make it as cross-functional as possible with a keen eye to mix experience with fresh ideas.
- Charter development.

The management team is responsible for assisting with charter development and making certain the workshop goals and objectives are well understood. In addition, the process owner works with the workshop leader to determine the scope of the activity, clarify the boundaries, and define the workshop deliverables.

The charter document is used to shape your workshop. This document must clearly state what you want accomplished because it defines what the team will do. The management team must consider this document carefully so that their expectations for the workshop are met. The Lean team leader must provide a copy of the charter, as a daily reference, to each team member and have a poster-size copy of it to display in the area.

The team members must cover the complete work statement. Managers must provide workshop team members who are qualified to accomplish the desired deliverables. Management must also remove any roadblocks to success and is tasked with holding the gains after the workshop week. It is also recommended that management participate in the kickoff meeting and report-outs, as well as post-workshop follow-up meetings.

Selecting workshop team participants is critical, as they must include representatives from the focus, supplier, and customer areas who have experience, knowledge, and credibility. In a multishift process, representatives from all shifts should be on the workshop team. Remember to include resource representatives from support groups. They are vital when information and action are needed to get things done (such as information technology people to link workstations or forklift operators to move equipment). Facilities and computing support, for example, might be on-call participants. If they have participated in workshops and understand Lean principles, they do not have to attend the entire week. Make sure they are invited to the kickoff, report-outs, and education day to introduce them to the workshop team and keep them informed of the workshop's progress. If possible, invite representatives from the customer and supplier processes as key stakeholders.

A responsibility matrix is a clean way to document a team member's roles and responsibilities. In the matrix, "R" indicates who is responsible for the task, "C" shows who is to consult that task, "S" illustrates who is supporting the task, and "A" determines who is approving the task. See Figure 22.3 for an example of a responsibility matrix.

The Lean workshop leader should conduct a large amount of tasks that require attention across many themes. By the end of the preparation phase, the following should be completed across the following themes:

Human resources:

- Identify team members and on-call participants.
- Begin backfilling resources for workshop participants.

Management communication:

- Identify safety, health, or process restrictions. Obtain the necessary training if needed.
- Define management educational needs.
- Ensure that on-call participants and their managers are notified.
- Schedule management briefing to confirm charter and targets.

Technical analysis:

- Perform gemba walk with process owners.
- Draft the charter documents.
- Identify current standard work packages associated with the area.
- Collect baseline data: time studies and observations, takt time, and process flow diagrams.
- If necessary, obtain facilities layout.
- Complete baseline data collection and gather the data together in an organized folder.
- Conduct a workshop-planning meeting with core team members. This shall include a review of the workshop charter, agenda, and data collection.

Activity	Frank	Raj	Chaz	Kelly	Jorge	Ron
Manufacturing layout redesign	S	A	S	C	C	R
IT development	S	A	C			R
Time studies	R	A	S			C
Signature approvals	R	A	S	C	C	C

FIGURE 22.3
Responsibility matrix.

Logistics:

- Schedule meeting rooms for the workshop.
- Determine date/time/location of the workshop.
- Send out invitations for training and workshop kickoffs to participants.
- Ensure that the workshop packet, printed and electronic, materials are complete.

Phase 3: Execution

The execution phase includes the actual workshop session. The exit criterion for this phase is the appropriate completion of the workshop. The workshop may be broken into two segments:

- Segment 1: Identify and test proposed solutions
    - Cause and effect diagrams
    - Affinity diagrams based on cause and effect diagrams
    - Failure mode and effect analysis (FMEA)
    - Prioritization matrices
- Segment 2: Implement the solutions
    - Ordering of materials
    - Communication plan
    - Risk probability and impact assessment
    - The physical moving, removal, and installation of items/equipment
    - Construction of kanban cards
    - Installation of visual management tools
    - Project schedule tracking via PERT or Gantt charts

To make the workshop as productive as possible, it is preferable to have data on hand and ready for analysis for review by the workshop team members. When data are available, segment 1 becomes easier to complete. For example, data on hand make cause and effect diagrams more meaningful, affinity diagrams easier to construct, and FMEA easier to develop.

Segment 2 is the physical transformation of a work area or process. This transformation should be accomplished with all members of the team and management as well. It is critical to receive management commitment and political ramifications of work process disturbance rectified prior to beginning segment 2. It is a powerful statement to the entire workforce to see management physically moving heavy items, labeling and organizing work areas, and sweeping or cleaning. This often empowers other members of the workforce to adopt Lean principles as well. Remember that Lean activities should be suggested by the work team and come from a grassroots movement, not assigned by management to the work team.

By the end of the execution phase, the Lean workshop leader should conduct the following:

- Complete all action items defined in segment 1.
- Complete all action items defined in segment 2.
- Brief the management team on the completion of the Lean workshop.
- Release each Lean workshop participant with a list of potential Lean activities in other areas.

Phase 4: Follow-Up

During phase 4, practitioners should verify implementation and sustain the gains of the improvement by developing a sustainment strategy plan. The exit criterion for this phase is the creation of a process control sustainment plan that verifies workshop success.

To verify the implementation of the workshop, a chart should be developed that identifies the key processing steps improved on, what the defined target was, the process owner, what is being measured, when it was measured, where it was measured, and what is the resultant action taken. In addition, a project closure report should be completed that describes the before and after states of the project with measurable successes. In addition, it should contain follow-on actions for future improvement.

By the end of the follow-up phase, the Lean workshop leader should conduct the following:

- Review success sustainment with the process owner.
- Write the sustainment plan that controls the success from the workshop.
- Document Lean workshop cost, implementation cost, and time span of implementation.
- Document dollars and percentages of savings over time relative to the investment costs.
- Brief the management team on the return on investment of the Lean workshop.
- Disseminate success to other departments internally or outside suppliers to market an interest in future Lean projects.

Conclusion

Planning a Kaizen event should be carefully considered. More up-front planning in the first two phases results in a more fruitful workshop week. Be wary of resource constraints, office politics, cultural adoption for change, and executive sponsorship before any planning is conducted. Remember to create transparent communication before, during, and after the workshop to facilitate buy-in from various stakeholders benefiting from the Lean exercise. In addition, adopt Lean best practices, found in this book, to efficiently Define, Measure, Analyze, Improve, and Control the Lean project.

Homework Problems

1. Name the four phases of a Kaizen workshop. Briefly describe the actions taken during each phase.
2. Which of the following is a SMART objective:
   a. Develop a manufacturing process by the third quarter of 2016.
   b. Improve the assembly line oven by 10% energy savings by August 13, 2016.
   c. Reduce WIP inventory in June.
   d. Increase throughput of the assembly line.
3. Name the phase where each of the following tasks belong (assessment, preparation, execution, and follow-up):
   a. Giving a presentation to the management team on the workshop's success.
   b. Defining SMART objectives.
   c. Identifying troublesome equipment or processes.
   d. Documenting the savings incurred from the workshop.
   e. Establishing an effective communications plan to disseminate information to team members.
4. In your own words, describe three critical success factors and how they are influential in the development of Lean workshops.

23

Prioritizing the Lean Initiatives

Sandra L. Furterer

CONTENTS

Introduction

This chapter describes four main approaches to prioritizing an organization's Lean initiatives. The four methods increase in complexity and time to apply the prioritization technique. The first is a simple priority impact and effort matrix; the next is applying a Pugh initiative selection matrix; the third is applying a quality function deployment (QFD) house of quality tool; and the last, most complex, approach is applying a systems engineering approach from the business architecture body of knowledge. The advantages and disadvantages of each approach are shown in Figure 23.1.

Method 1: Priority Impact and Effort Matrix

The priority impact matrix can be used to prioritize the organization's Lean initiatives based on their potential impact and effort. The following are the steps to apply this approach:

1. Brainstorm potential Lean initiatives. Figure 23.2 shows an example of a hospital's Lean initiatives. An affinity diagram can be used to organize the potential Lean initiatives into logical Lean themes or project

Prioritization technique	Priority impact and effort matrix	Pugh initiative selection matrix	QFD house of quality	Strategic business process architecture
Advantages	Simple, easy to use	Compares competing projects with each other	Well-known technique	Better aligns priorities with strategic goals
	Quick to use	Provides multiple prioritization criteria	Provides quantification of priorities	Helps to align projects with operational processes
Disadvantages	May not tie to strategic initiatives	May not tie to strategic initiatives	More difficult to use	More difficult to use
			May take more time to apply	May take more time to apply

FIGURE 23.1
Advantages and disadvantages of the Lean initiative prioritization approaches.

5S Operating Room (OR)
Improve emergency department throughput
Improve operating room turnaround
Streamline OR instrument tracking
Pre-op patient readiness
Improve lab turnaround
5S pathology lab
Streamline outpatient rehab processes
5S nursing stations
Reduce bottlenecks in the orthopedic institute
Improve women's center intake processes
Improve women's center value stream throughput
Improve inpatient bed assignment
Streamline endoscopy throughput
Reduce outpatient registration wait time
Improve outpatient imaging diagnostic time
Standardize medical practice coding process
Standardize emergency department sepsis protocols
Standardize congestive heart failure treatment protocols
Standardize stroke treatment protocols

FIGURE 23.2
Potential Lean initiatives for a hospital.

types, shown in Figure 23.3. The themes or project types would be generated based on the logical groupings of the brainstormed initiatives. Additionally, several Lean themes could first be generated, and potential project initiatives could be brainstormed within each of the themes. Some potential Lean themes are as follows:

Implement 5S organization	Improve throughput and value stream	Improve turnaround	Standardize work	Reduce waste; improve processes
–5S OR –5S Pathology lab –5S Nursing stations	–Improve emergency department throughput –Reduce bottlenecks in the orthopedic institute –Improve women's center value stream throughput –Streamline endoscopy throughput	–Improve OR turnaround –Improve lab turnaround	–Improve pre-op patient readiness –Standardize medical practice coding process –Standardize emergency department sepsis protocols –Standardize congestive heart failure treatment protocols –Standardize stroke treatment protocols	–Streamline outpatient rehab processes –Streamline OR instrument tracking –Improve women's center intake processes –Improve inpatient bed assignment –Reduce outpatient registration wait time –Improve outpatient imaging diagnostic time

FIGURE 23.3
Lean initiatives affinity diagram.

- Value stream throughput improvement
- Waste reduction and process analysis
- Standardization of work activities
- Reduce turnaround time or changeovers
- 5S organization
- Improve flow
- Mistake proof processes
- Process redesign

2. Assess the impact and effort for each initiative. For impact, use a 1–5 rating scale, with 1 representing a low impact and 5 representing a high impact if the project is successful. For effort, use a 1–5 rating scale, with a 1 being a low effort and a 5 being a high effort. Then multiply the impact and effort ratings to get a priority level. Rate the priorities as a low for an impact effort product rating of 1–8, a medium for a rating of 9–17, and a high priority for 18–25. You will then have relative priorities for the Lean initiatives. An example is shown in Figure 23.4.
3. Gain consensus from the Lean steering committee on the Lean priorities.

This method is relatively simple to apply, and can probably be completed in a two-hour session with the Lean planning committee.

	Impact	Effort	Impact X Effort	Priority
Potential Lean initiatives	**1 = Low to 5 = High**	**1 = Low to 5 = High**		**High = 18–25 Med = 9–17 Low = 1–8**
5 S				
5S OR	5	2	10	Med
5S Pathology lab	3	3	9	Low
5S Nursing stations	5	2	10	Med
Throughput				
Improve emergency department throughput	5	1	5	Low
Reduce bottlenecks in the orthopedic institute	2	5	10	Med
Improve women's center value stream throughput	4	3	12	Med
Streamline endoscopy throughput	1	5	5	Low
Turnaround				
Improve OR turnaround	5	1	5	Low
Improve lab turnaround	3	3	9	Low
Standard work				
Improve pre-op patient readiness	3	3	9	Low
Standardize medical practice coding process	4	5	20	High
Standardize emergency department sepsis protocols	5	4	20	High
Standardize congestive heart failure treatment protocols	5	2	10	Med
Standardize stroke treatment protocols	5	3	15	Med
Waste elimination/process analysis				
Streamline outpatient rehab processes	4	4	16	Low
Streamline OR instrument tracking	3	2	6	Med
Improve women's center intake processes	2	5	10	Med
Improve inpatient bed assignment	5	3	15	Med
Reduce outpatient registration wait time	3	5	15	Med
Improve outpatient imaging diagnostic time	3	4	12	Med

FIGURE 23.4
Priority impact and effort matrix.

Method 2: Pugh Initiative Selection Matrix

The Pugh initiative selection technique is a technique for evaluating and selecting design concepts in new product development and Design for Six Sigma. This technique can be applied to prioritize your Lean initiatives. The following steps would be used to derive your Lean priorities:

1. First, brainstorm the potential Lean initiatives and create an affinity diagram grouping them by the type of Lean project as performed in method 1. Then generate the criteria on which to compare the initiatives. Some potential criteria could be the following:
   - Resource effort
   - Quality impact
   - Productivity impact
   - Cost impact
   - Risk impact
2. The next step is to select one of the Lean initiatives as the "candidate" initiative. A candidate initiative could be one that was already implemented in the past or one that you already know as a high-priority initiative. You then compare each of the other initiatives (new) to the candidate for each comparison criterion. If a new initiative is better than the candidate for those criteria, you would place a plus sign (+) in the cell where the new initiative intersects the criteria. If the new initiative is worse than the candidate initiative for the criteria, a minus sign (–) is placed in the cell. If the new initiative is the same as the candidate on those criteria, a zero (0) or S for same is placed in the cell. Figure 23.5 shows a generic Pugh selection matrix.

	Potential Lean initiatives						
Criteria	**1**	**2**	**3**	**4**	**5**	**6**	**7**
A	–	–	–	0		0	–
B	–	0	–	–		0	–
C	+	+	–	–		–	–
D	+	–	–	+		–	+
E	+	+	–	–	Candidate initiative	–	–
Pluses	3	2	0	1		0	1
Minuses	2	2	5	3		3	4
Zeros	0	1	0	1		2	0
Priority	1	2	6	3		5	4

FIGURE 23.5
Pugh initiative selection technique.

3. The highest-priority initiative is the one with the most pluses and the fewest minuses. The next priority initiative is the one with the next highest pluses and fewest minuses, and so forth.

The sample Pugh initiative selection matrix for a hospital's Lean initiatives is shown in Figure 23.6.

Method 3: Quality Function Deployment and House of Quality

QFD and the house of quality matrix is an excellent tool to help prioritize the project initiatives with the prioritization criteria. Figure 23.7 shows the format for the house of quality. The customer requirements in the typical house of quality are replaced by the prioritization criteria, and the technical requirements are replaced by the potential Lean initiatives.

The steps for creating a house of quality are (Evans and Lindsey, 2007) as follows:

1. Brainstorm the prioritization criteria. We will use the same criteria that we generated in the Pugh initiative selection technique. The team will then provide an importance rating for each prioritization criterion on a scale of 1–10, with 1 being of low importance and 10 being of high importance.
2. Brainstorm the potential Lean initiatives with the organization's Lean committee. We will use the same list as in the prior two prioritization techniques for the QFD house of quality example.
3. Develop the relationship correlation matrix by identifying the strength of the relationship between each prioritization criterion and each potential Lean initiative. Typically, a numerical scale of 9 (high strength of relationship), 3 (medium strength of relationship), 1 (low strength of relationship), and blank (no relationship) is used. For example, if a prioritization criterion is reducing cost, then if the potential Lean initiative has a high impact on reducing cost the 9 rating would be used; if it has a medium impact, a 3 would be used; if there is a low impact to reducing cost, a 1 would be used; and if the initiative does nothing to reduce cost, then no rating is given.
4. Develop the trade-offs or relationships between the potential Lean initiatives in the roof of the house of quality. You can identify a positive (+) relationship between pairs of Lean initiatives; as one initiative is implemented and the goals are achieved, it would have a positive impact or relationship on another Lean initiative. If there is no relationship between a pair of initiatives, the intersecting

	5S			Throughput				Turnaround		Standard work					Waste elimination/process analysis					
Criteria	5S OR	5S Pathology lab	5S Nursing stations	Improve emergency department throughout	Reduce bottlenecks in the orthopedic institute	Improve women's center value stream throughout	Streamline endoscopy throughout	Improve OR turnaround	Improve lab turnaround	Improved pre-op patient readiness	Standardize medical practice coding process	Standardize emergency department sepsis protocols	Standardize congestive heart failure treatment protocol	Standardize stroke treatment protocols	Streamline outpatient rehab processes	Streamline OR instrument tracking	Improve women's center intake processes	Improve inpatient bed assignment	Reduced outpatient registration wait time	Improve outpatient imaging diagnostic time
Resource effort	+	+	+	C	+	0	+	–	0	+	–	+	+	+	+	0	0	–	+	+
Quality impact	+	+	+	A	0	0	0	+	+	–	–	+	+	+	–	+	–	–	–	–
Productivity impact	–	–	–	N	–	0	–	0	0	+	–	–	–	–	–	+	–	+	–	–
Cost impact	–	–	–	D	–	–	–	+	–	–	–	+	+	+	–	–	–	–	–	–
Risk impact	+	–	–	I	–	0	–	+	+	–	–	+	+	+	–	+	–	–	–	–
Pluses	3	2	2	D	1	0	1	3	2	2	0	4	4	4	1	3	0	1	1	1
Minuses	2	3	3	A	3	1	3	1	1	3	5	1	1	1	4	1	4	4	4	4
Same	0	0	0	T	1	3	1	1	2	0	0	0	0	0		1	1	0	0	0
Priority	High	Med	Med	E	Med	Low	Med	High	Med	Med	Low	High	High	High	Low	High	Low	Low	Low	Low

FIGURE 23.6
Sample Pugh initiative selection technique for a hospital's potential Lean initiatives.

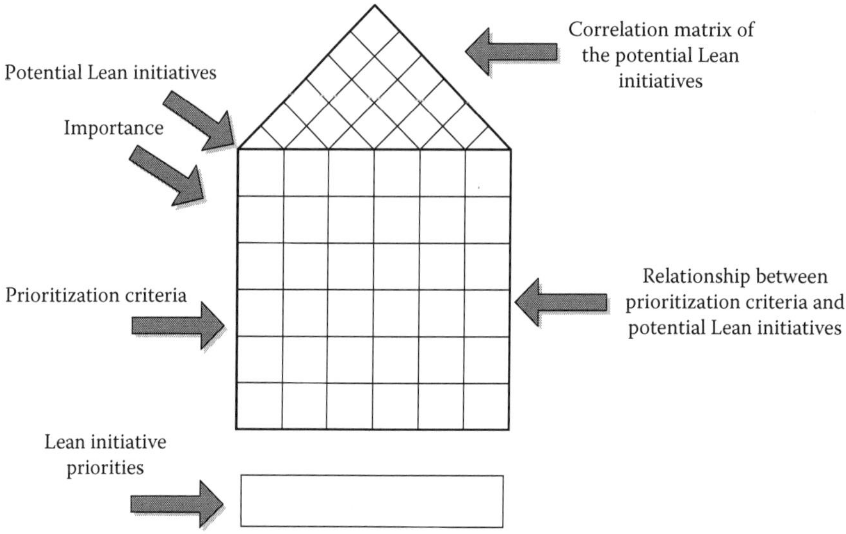

FIGURE 23.7
Quality function deployment house of quality matrix.

diamond is left blank; if there is a negative relationship between a pair of initiatives, such as when one project is worked on or implemented, the other project is negatively impacted, then a minus sign is placed in the intersecting diamond. An example of a positive relationship can be illustrated by the 5S project in the operating room (OR), which would complement and further enhance the OR turnaround project since a more organized environment will reduce the time searching for supplies, equipment, and instruments. A negative relationship can be illustrated by implementing congestive heart standard protocols at the same time as sepsis protocols in the emergency department as this could overtax the staff and be less productive than if only one project was implemented at a time. No relationship would be represented by two projects that do not impact each other at all, such as improving the OR instrument tracking may have no relationship to improving the women's center intake processes.

5. The priorities of the Lean initiatives can be summarized by multiplying the importance weightings of the prioritization criteria by the strength of the relationships in the correlation matrix. This helps to identify which of the Lean initiatives should be started first.

A sample QFD house of quality matrix is shown in Figure 23.8 for a hospital's Lean program.

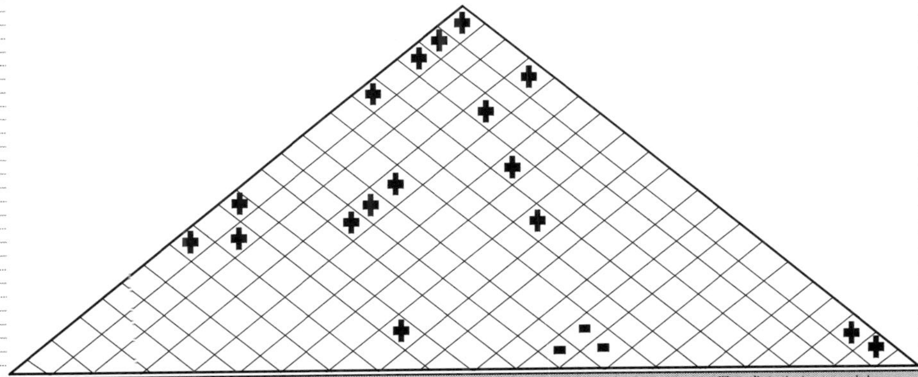

Criteria	Importance	5S			Throughput				Turnaround			Standard work				Waste elimination/process analysis					
		5S OR	5S Pathology lab	5S Nursing stations	Improve emergency department throughput	Reduce bottlenecks in the orthopedic institute	Improve women's center value stream throughput	Streamline endoscopy throughput	Improve OR turnaround	Improve lab turnaround	Improve pre-op patient readiness	Standardize medical practice coding process	Standardize emergency department sepsis protocols	Standardize congestive heart failure treatment protocols	Standardize stroke treatment protocols	Streamline outpatient rehab processes	Streamline OR instrument tracking	Improve women's center intake processes	Improve inpatient bed assignment	Reduce outpatient registration wait time	Improve outpatient imaging diagnostic time
Quality impact	10	3	1	3	9	3	9	3	9	3	9	9	9	9	9	3	9	3	3	1	3
Productivity impact	6	9	9	9	9	9	9	3	9	1	9	1	1	1	1	3	9	9	9	3	9
Cost impact	8	9	1	3	9	3	9	3	9	3	9	3	3	3	3	1	3	3	9	3	9
Risk impact	9	9	9	1	9	1	3	3	9	3	3	9	9	9	9	1	9	1	3	1	3
Absolute priority		237	153	117	297	117	243	99	297	87	243	201	201	201	201	65	249	117	183	61	183
Relative priority		6	13	14	1	14	4	17	1	18	4	7	7	7	7	19	3	14	11	20	11

FIGURE 23.8
Sample quality function deployment Lean initiative prioritization.

Method 4: Strategic Business Process Architecture Prioritization

The systems engineering business architecture–based approach that the author developed can be used to prioritize the Lean initiatives and ensure that they are aligned to the organization's key business processes and business strategies. Resources can be applied to these projects, from a central governing steering committee, to ensure alignment between the business strategic initiatives and the Lean projects and to assess the improvement projects' success and financial savings. An enterprise view can help an organization to elevate its Lean program to an enterprise level, provide a prioritization of Lean improvement efforts to align with the business strategy, and provide greater cross-functional process improvement to provide greater cost reduction. Providing an enterprise view can also help to align the organization's Lean program with key business strategies and core processes. This approach helps to understand the organization from a cross-functional enterprise view so that the Lean program can be aligned with the strategic and operational plans as well as focused on key bottleneck processes within the organization.

There are several critical success factors to a successful improvement program, including the following:

- Basing the Lean project selection on key business priorities (Furterer, 2008).
- Focusing on key business areas (Furterer, 2008).
- Aligning the program with strategic initiatives and prioritizing the projects based on business strategy (Furterer, 2008).
- Defining a vision of what the organization will look like after implementing Lean and embracing the change (Sureshchandar, Chandrasekharan, and Anantharaman, 2001).
- Top management support through leadership: this factor includes management setting goals and providing leadership and direction (Hoffman and Mehra, 1999).
- Strategy that is complementary and integrated with existing policies and strategies (Dale, 1994).
- Clear links between strategic goals and the change strategies (Newman, 1994).

These critical success factors point to the need to align an organization's Lean program with the organization's strategic plan and the key business processes that enable the enterprise to meet the customers' needs and expectations. It also includes the need to understand the Lean program's vision. Many times, an organization's Lean projects are identified from a bottom-up approach, not one that is top-down, and aligned with the enterprise goals and strategic plans.

Although these projects often achieve improvement and financial savings, there is no direct visibility or traceability to the organization's strategic goals.

The author developed strategic business process architecture (SBPA) models from the systems engineering business architecture body of knowledge. These provide an enterprise-wide understanding of the business. The business processes enable the extraction of key business elements that support required capabilities of the business to meet customer needs.

SBPA modeling techniques and methods can be used to provide prioritized alignment with the Lean program goals and the enterprise's Lean projects and improvement initiatives. This can provide alignment between the business strategies and goals and the organization's improvement plans.

Business architecture provides models that describe business entities (business processes and relevant business information) and the relationships, dynamics, and rules that govern their interaction to achieve enterprise-wide objectives.

Strategic Business Process Architecture Definition and Elements

SBPA definition:

Our definition of SBPA focuses on modeling the key concepts and elements of an enterprise by aligning business strategies and enabling processes to optimize the key groups of activities (components) that enable optimized information technology and business process improvement initiatives.

A simplified view of the key elements of SBPA is shown in Figure 23.9 (Furterer, 2009a).

The elements of SBPA describe the business system enterprise. The SBPA includes understanding the customers and their expectations. Another important element of business architecture is documenting the business strategies and goals, as well as the external and internal influencers of the business. The goals should relate to the capabilities that the business needs

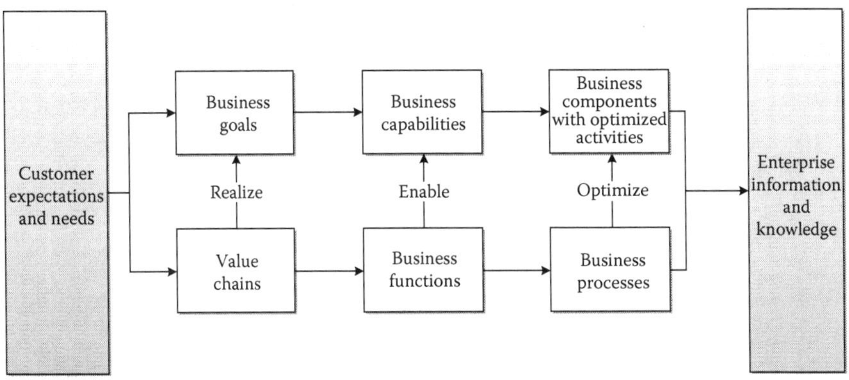

FIGURE 23.9
Key elements of strategic business process architecture.

to meet its business goals. A value chain is a chain of activities that provide value to your customer (Porter, 1985). Value chains are decomposed into the business functions. The functional decomposition diagram itself does not depict process flows, but rather the hierarchical organization of functions and the processes that they include. Each value chain and its subsequent business functions will be used to further decompose the processes. This ensures traceability to value chains that provide customer value.

The business capabilities enable the business functions. Although the way in which a business implements its processes is likely to change frequently, the basic capabilities of a business tend to remain constant. The advantage of a model that is based on the most stable elements of the business is its longevity. The business processes and their activities describe the sequence of activities that enable the business to meet the customer's expectations and provide value through value chains.

Optimized business components consist of the optimized activities that support the business. These components consist of the activities that require similar people, processes, and technology. They allow the standardization of business processes by componentizing the activities that can be used in multiple areas of the business, across many business units and markets (IBM Corporation, 2005). The enterprise information and knowledge describe the information and roles that exist in the business, which are part of the business processes.

Our complete SBPA models are more expansive than the simplified view shown in this chapter. However, the simplified elements can be used to illustrate how they can be used to prioritize Lean initiatives that align to core business processes and business strategies.

Figure 23.10 shows the enterprise prioritization elements that can be used to prioritize process improvement across the enterprise.

There are three levels to align when prioritizing enterprise Lean initiatives:

- Enterprise level—links the organization's strategic goals, improvement initiatives, and core processes.
- Operational level—describes and decomposes the key value chains, business functions, processes, and activities that provide value to customers.
- Performance level—provides an understanding of the critical to satisfaction (CTS) criteria that are most important to satisfying customers' needs, to the processes, and to the metrics that measure performance improvement.

The following matrices, shown in Figure 23.11, describe the interconnectivity and relationships between the enterprise, operational elements, and performance elements that help to ensure alignment between the strategic goals, processes, and improvement projects in the organization.

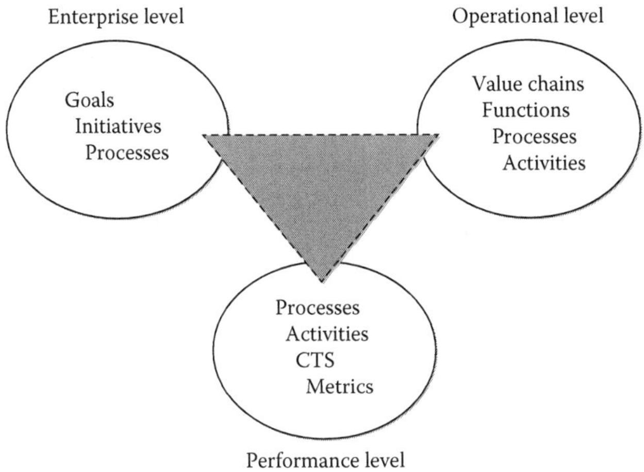

FIGURE 23.10
Enterprise prioritization elements.

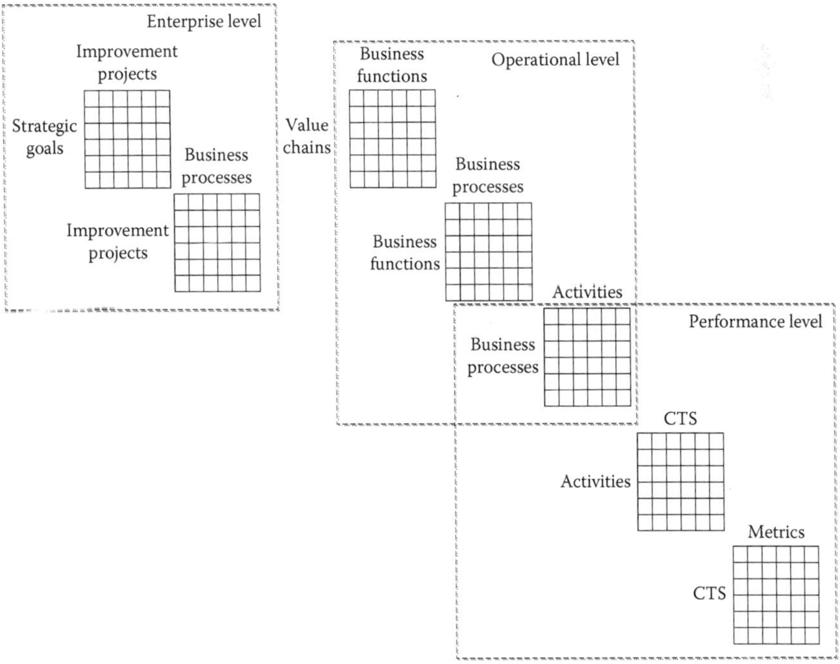

FIGURE 23.11
Enterprise prioritization relationship matrices.

Steps for Project Prioritization

Strategic Level

- The steps to prioritizing the Lean initiatives begin with listing the key strategic goals down the left side of the enterprise-level goals and improvement projects matrix and then listing the proposed Lean initiatives across the top of the matrix. We define the strength of the relationship between each goal and initiative. We use a scale of 9 to designate a high relationship for the initiative helping to meet the goal, a 3 to designate a medium relationship, a 1 to designate a low relationship, and a blank to designate no relationship. We define the importance of each strategic goal on a scale of 1–10, 10 being the most important and 1 being the least important goal. We then multiply the strength of each relationship (goals and initiatives) by its importance to get an absolute ranking of each initiative. We then define the relative ranking of each initiative to identify which is the most important initiative to focus on to meet our most important goals based on the point values (highest point equals priority 1, etc.).

A sample of a goal and an improvement project mapping is shown in Figure 23.12.

Operational Level

Operational-level matrices can be used to decompose the value chains that provide value to the customers into business functions (the value chains and business functions matrix, and the business functions and business processes matrix). Once the business processes are generated at the operational level, the project initiatives can be mapped to the processes.

Performance Level

The performance level helps us to map the work activities to the CTS criteria from a customer perspective. The CTS criteria are then mapped to the metrics that are used to measure the CTS criteria. The performance-level matrices are typically developed during the Lean initiatives deployment phase.

We used SBPA models to prioritize the Lean initiatives that align to the organization's business strategies. We then used operational models to derive the business processes that are improved by the Lean initiatives. These process improvement initiatives can help to enable enterprise performance excellence, because they are derived from the value chains that relate to the key business processes across the entire enterprise. This prioritization methodology allows us to ensure alignment throughout our business both vertically throughout all levels of the organization and horizontally across departmental silos to provide cross-functional process integration.

		5S			Throughput				Turnaround		Standard work					Waste elimination/process analysis					
Strategic goals	Importance	5S OR	5S Pathology lab	5S Nursing stations	Improve emergency department throughput	Reduce bottlenecks in the orthopedic institute	Improve women's center value stream throughput	Streamline endoscopy throughput	Improve OR turnaround	Improve lab turnaround	Improve pre-op patient readiness	Standardize medical practice coding process	Standardize emergency department sepsis protocols	Standardize congestion heart failure treatment protocols	Standardize stroke treatment protocols	Streamline outpatient rehab processes	Streamline OR instrument tracking	Improve women's center intake processes	Improve inpatient bed assignment	Reduce outpatient wait time	Improve outpatient imaging diagnostic time
Reduce OR costs	6	9							9		9						9		3		3
Increase OR capacity	7	5							9		9						9		3		3
Increase emergency department capacity	3				9								9	9	9						9
Improve customer satisfaction by reducing wait times	9	3				9	9	9	9	3	9					9	3	9	9	9	9
Streamline processes to improve productivity	8	1	1	1	9	9	9	9	9	9	9	3	3	3	3	9	9	9	9	9	9
Improve patient safety	10	3	3	3	9	3	3	3	9	3	3	9	9	9	9	3	3	3	3	3	3
Improve quality of processes	9	3	3	3	1	1	1	1	1	1		1	9	9	9	1	9	9	9	3	3
Reduce operational costs	4	1	1	1	9	9	9	9	9	9	9					9	3	9	9	9	9
Absolute priority		213	69	69	234	228	228	228	405	174	336	123	222	222	222	228	339	300	339	246	312
Relative priority		16	19	19	8	9	9	9	1	17	4	18	13	13	13	9	2	6	2	7	5

FIGURE 23.12
Strategic business process architecture: goals and initiatives prioritization.

Potential Lean initiatives	Priorities for priority impact effort	Priorities for Pugh initiative selection matrix	Difference impact effort—Pugh (number levels)		Priorities for QFD house of quality	Priorities for SBPA	Difference (QFD – BA)
5S							
5S OR	Med	High	1		6	16	−10
5S Pathology lab	Low	Med	1		13	19	−6
5S Nursing stations	Med	Med	0		14	19	−5
Throughput							
Improve emergency department throughput	Low	High	2		1	8	−7
Reduce bottlenecks in orthopedic institute	Med	Med	0		14	9	5
Improve women's center value stream throughput	Med	Low	1		4	9	−5
Streamline endoscopy throughput	Low	Med	1		17	9	8
Turnaround							
Improve OR turnaround	Low	High	2		1	1	0
Improve lab turnaround	Low	Med	1		18	17	1
Standard work							
Improve pre-op patient readiness	Low	Med	1		4	4	0
Standardize medical practice coding process	High	Low	2		7	18	−11
Standardize emergency department sepsis protocols	High	High	0		7	13	−6
Standardize congestive heart failure treatment protocols	Med	High	1		7	13	−6
Standardize stroke treatment protocols	Med	High	1		7	13	−6
Waste elimination/process analysis							
Streamline outpatient rehab processes	Low	Low	0		19	9	10
Streamline OR instrument tracking	Med	High	1		3	2	1
Improve women's center intake processes	Med	Low	1		14	6	8
Improve inpatient bed assignment	Med	Low	1		11	2	9
Reduce outpatient registration wait time	Med	Low	1		20	7	13
Improve outpatient imaging diagnostic time	Med	Low	1		11	5	6

FIGURE 23.13
Comparison of the examples for the four prioritization techniques.

SBPA helps the organization focus on important Lean improvement projects and the processes that truly matter to the business and its customers and enables us to better meet our customers' expectations through streamlined and optimized processes.

Conclusions

Figure 23.13 shows the priorities of Lean initiatives based on the different prioritization techniques. We calculated the difference in prioritization levels (low, medium, and high) for the impact effort and Pugh approaches and the difference in relative priority number between the QFD and business architecture approaches. All four approaches provide different priorities. Returning again to the advantages and disadvantages listed at the beginning of this chapter, the technique that the organization uses depends on the time it has to prioritize and whether it wants to prioritize based on criteria important to the organization or align to the business strategies. We presented four different techniques, and we suggest that you try each one and apply the one that best fits the needs of your organization's Lean program.

References

Dale, B. (1994). A Framework for Quality Improvement in Public Sector Organizations: A Study in Hong Kong. *Public Money & Management* 14 (2), pp. 31–36.

Evans, J.R., and Lindsay, W.M. *Managing for Quality and Performance Excellence*, 8th edition. South-Western Cengage Learning, Independence, KY, 2007.

Furterer, S., *Blazing the Trail to Operational Excellence: Leveraging Information Systems Business Architecture Methods to Enable Operational Excellence*, IIE Magazine, Norcross, GA, 2009a.

Furterer, S., *Lean Six Sigma in Service: Applications and Case Studies*, CRC Press, Boca Raton, FL, 2009b.

Furterer, S., *Lean Six Sigma Program Success Factors in a Retail Application*. International Conference on Industry, Engineering, and Management Systems, Cocoa Beach, FL. March 2008.

Hoffman, J., and Mehra, S. (1999). Management Leadership and Productivity Improvement Programs. *International Journal of Applied Quality Management* 2 (2), pp. 221–232.

IBM Corporation, *Component Business Models: Making Specialization Real*, IBM Business Consulting Services, IBM Institute for Business Value, Somers, NY, 2005.

Newman, J. (April–June 1994). Beyond the Vision: Cultural Change in the Public Sector. *Public Money & Management*.

Porter, M., *Competitive Advantage: Creating and Sustaining Superior Performance*, Free Press, New York, 1985.

Sureshchandar, G., Chandrasekharan, R., and Anantharaman, R. (2001). A Holistic Model for Total Quality Service. *International Journal of Service Industry Management* 12 (4), pp. 378–412.

24

Emerging Technologies Influencing Lean

David M. Dietrich, Elizabeth A. Cudney, and Sandra L. Furterer

CONTENTS

Introduction

As technology grows at an unprecedented rate, it is difficult to capture emerging technology trends into Lean practitioner's best practices. Emerging technologies are considered innovative pieces of modern-day technology that manifests in hardware, software, or other improvement venues. Often, these technologies gain a foothold in commercial industries, government labs, or academic settings and provide a disruptive alternative to a current service or practice.

By nature, Lean provides an opportunity to identify waste, streamline process efficiency, and provide solutions through iterative improvement that benefits the customer and/or society. From a high-level perspective, many emerging technologies share a similar focus. The challenge is to leverage the full potential of Lean by understanding how emerging technology attributes align to Lean initiatives. This process of mapping emerging technology attributes to Lean focus areas may be decomposed into a series of steps that assist the Lean practitioner in emerging technology deployment.

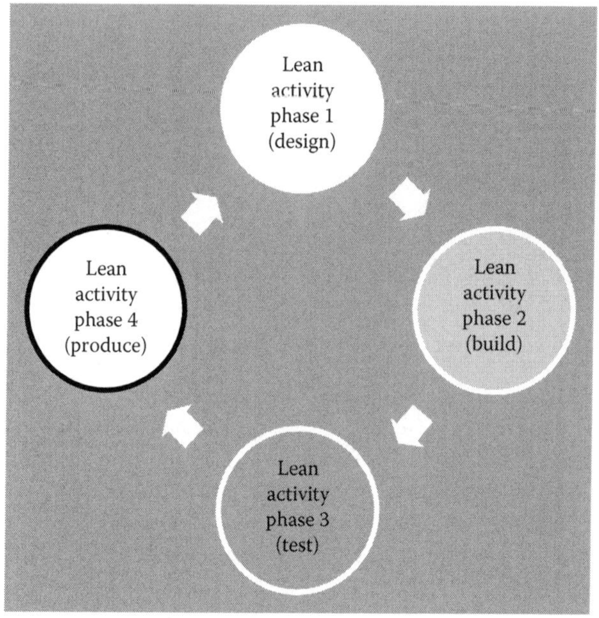

FIGURE 24.1
Design, build, test, and produce Lean improvement cycle.

Consider an example in which a company is currently engaged in a Lean initiative surrounding new product development aimed toward process improvement efficiency of an entire design (phase 1), build (phase 2), test (phase 3), and production release (phase 4) sequence. As illustrated in Figure 24.1, after phase 4 completes the cycle returns to phase 1 again for a new project selection.

It could be argued that with the help of concurrent engineering best practices all the phases may be conducted at the same time. However, for the sake of simplicity consider the simple sequence discussed in the following subsections.

Phase 1: The Design Phase

The company determines that the design phase could use a Lean office exercise to streamline the design cycle. Therefore, within this phase traditional Lean office techniques are implemented that include 5S data management systems, colocating printers closer to employees, optimizing the use of time efficiency of design engineering, and so on. This effort aligns with the Lean theme of design cycle time reduction.

Phase 2: The Prototype Build Phase

Next, as design engineers are streamlining the Lean office practices in phase 1, the prototype lab workers are conducting their own 5S initiative, cleaning, organizing, and standardizing the lab area to be able to efficiently construct the computer-aided designs (CADs) produced in phase 1. This effort aligns with the Lean theme of throughput efficiency.

Phase 3: The Testing Phase

Next, in the testing, testing engineers are busy conducting single-minute exchange of dies (SMED) activities on mechanical test equipment with the intention of speeding up the mechanical testing of the prototypes physically produced in phase 2. Again, this effort aligns with the Lean theme of throughput efficiency.

Phase 4: Low-Rate Initial Production

Finally, industrial and manufacturing engineers are conducting a value stream mapping exercise of a future state cell layout system for the production of the product that has now moved through the first three phases. They have identified four areas of focus: fabrication, assembly, quality inspection, and packaging. This value stream mapping effort focuses on the elimination of wasted motion and the standardization of work centers.

Emerging Technology Identification

Now, let us consider how emerging technologies may benefit this simple case study. By identifying the specific Lean objective of each activity into simple themes, it becomes clear that the survey of emerging technologies must exhibit the attributes of throughput efficiency, design cycle reduction, wasted motion, and standardized work centers. As a response to this exercise, quick research is conducted through a web search tool to identify recent emerging technologies that are commercialized. Because the aforementioned case example manufactures a product design, a search filter is placed by adding the key word "manufacturing" in the web search tool to obtain the search phrase "emerging technology manufacturing." Throughout the search, the key terms "additive manufacturing," or "3D printing," or "rapid prototyping," or "solid free form fabrication" appears frequently.

Additive Manufacturing

Additive manufacturing (AM) is an emerging manufacturing technology capable of tremendous process efficiency gains. The process works by additively constructing objects, layer by layer, directly from three-dimensional

(3D) digital definition as an input. A user may design a concept in CAD and "grow" a part out of plastic or metal, often in less than a day. Being able to offer this type of quick response provides a Lean practitioner with the ability to positively affect Lean improvement activities through the following:

- Fabricate new concept designs within a day or less.
- Fabricate fixtures within a day or less.
- Integrate assemblies of multiple components into a complex, single-piece design.
- Generate standardized work instructions directly into parts, removing the need for labels or stickers.
- Eliminate stored tooling inventory because no tooling is needed to produce the part.
- Without manufacturing constraints, poka-yoke design is enabled through increases in design complexity.

As shown in Figure 24.2, AM is added as an emerging technology candidate into this case scenario.

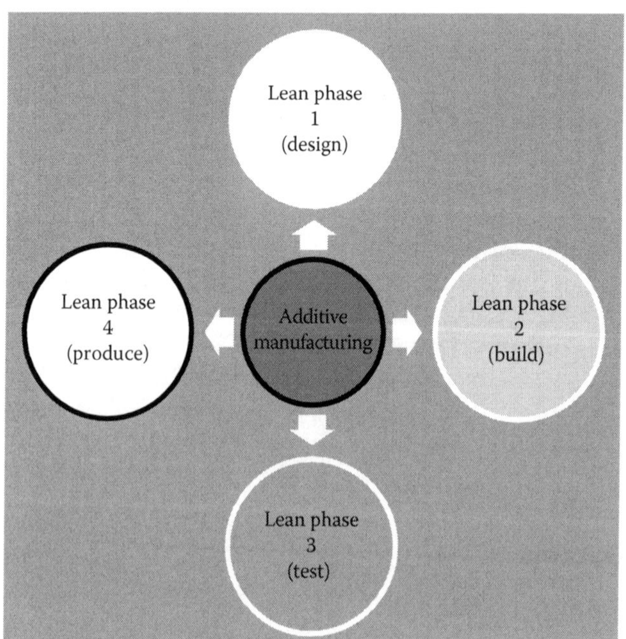

FIGURE 24.2
Additive manufacturing, an emerging technology candidate.

Phase 1: The Design Phase

In addition to using the traditional Lean office techniques to reduce time in the design engineering cycle, now design engineers are challenged to see how AM may also play a role in the Lean initiative. Design engineers determine this by using AM equipment; they are able to release designs to the prototyping lab to inexpensively produce physical prototypes within hours or days of finalizing designs.

By physically being able to handle and assemble designs, designers are able to poka-yoke their design prior to releasing it to the prototype build phase. Since parts may be fabricated directly from digital definition, there is no need to produce blueprint paper drawings anymore.

In addition, designers realize that by using AM they are able to produce incredibly complex designs that are not easily manufactured using traditional processes. This allows the designers to take a multi-piece assembly and redesign it as a single piece, thereby reducing the need for workers to assemble the components together.

Phase 2: The Prototype Build Phase

The prototyping lab workers are able to produce prototypes much faster using AM machines than before. Being able to produce physical hardware faster provides the time for workers to sustain their recently implemented 5S initiatives. In addition, this time savings allows the workers to concentrate on more value-added activities.

Phase 3: The Testing Phase

In the testing lab, testing engineers are busy conducting SMED activities on mechanical test equipment with the intention of speeding up the mechanical testing of the prototypes physically produced in phase 2. The test engineers are also challenged to see how AM may fit into their Lean improvement activity. Through discussions with the design engineers in phase 1, they determine that they will be able to create rapid tooling and fixtures to assist in converting internal steps to external steps through SMED. By being able to create these fixture concepts quickly, they are able to add significant time savings to the testing process.

Phase 4: Low-Rate Initial Production

Industrial and manufacturing engineers have finished their future state value stream map of a future state cell layout system for the production of the product. They too have been asked to identify where AM technology fits into their low-rate initial production waste elimination. After talking with the engineers and lab workers, they understand that AM technology

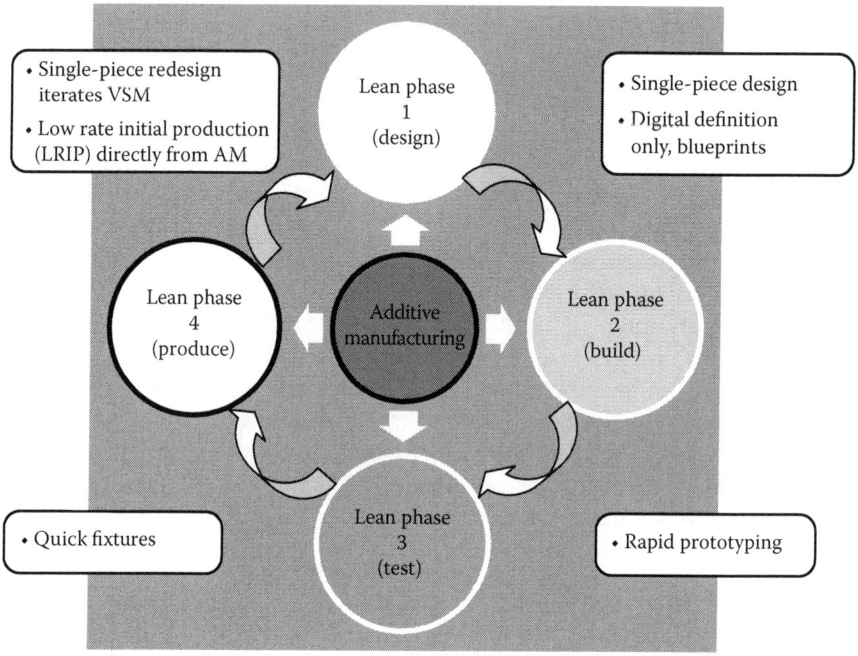

FIGURE 24.3
Complete integration of additive manufacturing into emerging technology deployment: VSM, value state map.

allows the designers to group multi-piece assemblies into a single part. This attribute of AM allows manufacturing engineers to conduct an iterative value stream mapping exercise to eliminate several assembly work areas and fastener installation work areas. By eliminating these work stations, significant inventory reduction and, subsequently, motion waste reduction is achieved. This concept is shown in Figure 24.3.

Examples of Additive Manufacturing Applied to Lean

Over the years, Lean practitioners have developed a comprehensive list of wastes associated with production. Notably, as time progressed, the original 7 types of waste, or *muda,* were expanded to 10 wastes commonly associated with Lean production. These types of wastes come in the following forms:

- Complexity: Reducing process, product, or system complexity.
- Labor: Reducing the amount of labor required on a systematic basis.
- Overproduction: Producing more than the customer requires.

- Space: Too much wasted space in assembly, digital storage, and so on.
- Energy: Too much raw power used in processing; may come in the form of motion or electrical.
- Defects: Producing products or systems that do not satisfy customer requirements.
- Materials: Shelf life spoilage and waste.
- Idle materials: Bulk inventory waiting for movement.
- Time: Throughput increases.
- Transportation: The physical time in distance between workstations.

However, using the design flexibility of AM, designs may be changed from traditional multi-piece designs and integrated into a single monolithic structure, as shown in Figure 24.4.

After understanding the benefits of an emerging technology, the next step is identifying how exactly it will reduce waste. Using the same aforementioned wastes, responses are developed with respect to AM to gain a clear picture of the benefits of using the emerging technology. These responses include the following:

- Complexity: Single-piece design reduces part complexity.
- Labor: Assembly labor is saved using single-piece designs.
- Overproduction: AM produces parts on demand, thus reducing the propensity to overproduce stock inventory.
- Space: No tooling storage is needed to produce additively manufactured parts. Conventional monument machinery may be eliminated, thereby freeing up space.
- Energy: Compare a single-piece assembly to a multi-piece assembly; producing monolithic designs will reduce the amount of energy required to produce components.
- Defects: The probability for assembly defects decreases with single-piece monolithic designs.
- Materials: Shelf life spoilage and waste is reduced.

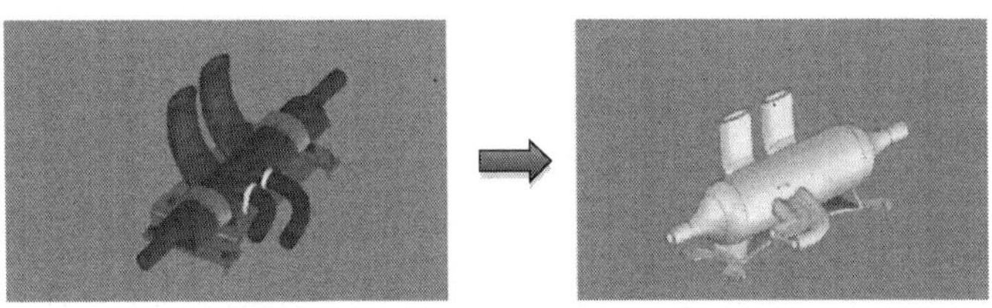

FIGURE 24.4
Example of design integration utilizing additive manufacturing.

- Idle materials: Bulk inventory waiting for movement is reduced as standard fasteners are eliminated in favor of a single-piece monolithic design.
- Time: Throughput is increased through workstation consolidation based on single-piece monolithic design.
- Transportation: the physical time in distance between workstations is lowered due to a reduction in the number of workstations.

Conclusion

Many other types of emerging technologies exist besides AM that may be applicable to company sponsored Lean initiatives. Some other types of emerging technologies include, but are not limited to, genetic bioengineering, nanomaterials, alternative energy technologies, electrical storage, artificial robotic intelligence, adaptive information technology networking, cloud-based computing, 3D digital display, biofuel, and so on. By using AM as an example, this chapter merely illustrates how to integrate emerging technologies into not only a single Lean activity insertion but also an entire Lean initiative within a company.

25

Future and Challenges of Lean: Engagement and Success Factors

Elizabeth A. Cudney and Sandra L. Furterer

CONTENTS

Introduction

This book provides an overview of Lean and the various Lean techniques such as value stream mapping, 5S, and single-minute exchange of dies. In addition, it provides real-world manufacturing, healthcare, and service-oriented case studies applying these methods and tools. This chapter describes a view into the future with the attempt to project where Lean will evolve over the next decade.

Applying Lean Methods and Tools to Streamline the Design and Development Life Cycle

Lean is a compilation of several diverse methods and tools that provide a holistic and integrated toolkit for process improvement. These same tools can be applied for designing and developing products and services. As new products and services become increasingly complex and multifaceted, it is necessary to more tightly couple and integrate the entire service and product development process with Lean.

A comprehensive methodology is necessary utilizing Lean techniques to design a product (application), process, or service right the first time. Design typically accounts for 70% of the cost of the product, and 80% of quality problems are unwittingly designed into the product (application), process, or service. Therefore, one-third of the budget must be devoted to correcting the problems that are created with the first two-thirds of the budget.

The three aspects of improving the product development process are the following:

1. Maximize profitability.
2. Minimize time.
3. Minimize cost.

These must be balanced without compromising value from the customer's perspective.

Lean design/product development aids in identifying and reducing or eliminating waste in the product development process. Lean design focuses on removing waste from all aspects of the product and associated development process before the start of manufacture. Lean design addresses the entire life cycle of a product. More specifically, Lean design targets cutting manufacturing costs during the design cycle and accelerating the time-to-market. Kearney (2003) identified the most common forms of waste in product design, as shown in Figure 25.1.

Mascitelli (2004) developed five principles of Lean design, which are as follows:

Principle 1: Precisely define the customer's problem and identify the specific functions that must be performed to solve that problem.

<table>
<tr><th>Area of waste reduction</th><th>Percentage of design waste</th><th></th></tr>
<tr><td>Designs never used, completed, or delivered</td><td>Unknown</td><td></td></tr>
<tr><td>Downtime while finding information, waiting for test results, etc.</td><td rowspan="2">33–50</td><td></td></tr>
<tr><td>Unnecessary documents and prototypes</td><td></td></tr>
<tr><td>Underutilization of design knowledge, for example in costly parts</td><td>18</td><td rowspan="4">58%</td></tr>
<tr><td>Overdesign, such as features customers don't need</td><td>8</td></tr>
<tr><td>Validating manufacturing errors early in the design process</td><td>17</td></tr>
<tr><td>Poor designs producing product defects</td><td>15</td></tr>
</table>

FIGURE 25.1
Waste in design.

Principle 2: Identify the fastest process by which the identified functions can be integrated into a high-quality, low-cost product.

Principle 3: Strip away any unnecessary or redundant cost items to reveal the optimal product solution.

Principle 4: Listen to the voice of the customer frequently and iteratively throughout the development process.

Principle 5: Embed cost reduction tools and methods into both your business practices and your culture to enable cost reduction.

Key areas of waste in product design stem from process delays, design reuse, defects, and process efficiency. Process delays are caused by time lost in looking for information, waiting for test results, and waiting for feedback. Waste from design reuse is the result of not learning from past design experiences; not reducing unnecessary features; and not reducing designs that are never used, completed, or delivered. Defect wastes stem from poor designs and warranty issues. Finally, process efficiency waste is caused by the underutilization of design knowledge and not validating manufacturing errors early.

There are several ways to decrease costs in the design cycle by reducing direct material cost, direct labor cost, operational overhead, nonrecurring design cost, and product-specific capital investments. Direct material costs can be reduced by using common parts, design simplification, defect reduction, and parts count reduction. Direct labor costs can be reduced through design simplification, design for manufacture and assembly, and standardizing processes. Operational overhead can be reduced by increasing the utilization of shared capital equipment and modular design. Nonrecurring design costs can be decreased by standardization, value engineering, and platform design strategies. Product-specific capital investments can be minimized by using value engineering, part standardization, and one-piece flow.

Huthwaite (2004) developed five laws of Lean design, which are as follows:

1. Law of strategic value: Ensure you are delivering value to all stakeholders during the product's life cycle.
2. Law of waste prevention: Prevent waste in all aspects of the product's life.
3. Law of marketplace pull: Anticipate change to deliver the right products at the right time.
4. Law of innovation flow: Create new ideas to delight customers and differentiate your product.
5. Law of last feedback: Use predictive feedback to forecast cause and effect relationships.

By incorporating Lean principles into product and process design, further improvements can be made to the design of a product or service. The product development process can be shortened, bringing the product to market faster while still ensuring value to the customer.

Adapting Enterprise Business Architecture Modeling to Product, Service, and Process Design

With the complexity of products, processes, and technologies that impact design decisions, enterprise business architecture modeling can be applied in the future to product and process design as it has been applied to the information system development and design life cycle. Enterprise architecture (EA), also referred to as business architecture (BA), is a relatively recent body of knowledge that comes from the information systems realm (Bieberstein, Laird, Jones, and Mitra, 2008). EA can provide an enterprise-wide understanding of a business. It attempts to connect business strategies to planned change initiatives focusing on information technology (IT) projects that can provide tactics to meet business strategies. In many organizations, BA is documented and developed by the IT organization as a way to understand business processes. The business processes enable the extraction of key business elements that support required capabilities of the business to meet customers' needs. This provides traceability from the business strategies to the business requirements through to the implemented IT. Demonstrating the alignment between IT initiatives and the business strategies helps to ensure that resources of people, time, and money are applied appropriately.

BA modeling techniques and methods can be used to also provide prioritized alignment with the key strategic initiatives related to design of new products, services, and processes in an enterprise. This can provide alignment between the business strategies and goals and the organization's new product and service development initiatives and goals.

BA helps us to understand the 3–5 year strategies of businesses. BA also provides models that describe the business entities (business processes and relevant business information) and the relationships, dynamics, and rules that govern their interaction to achieve enterprise-wide objectives. These same modeling techniques can be used to understand the key strategic initiatives, products, services, and processes that can be designed to meet the organization's strategic plans and identify the customer requirements that are met through the product and service designs. The modeling can identify the business capabilities and processes that provide the products and services, linking manufacturing and service processes to key design decisions.

The elements of BA describe the business enterprise, and are shown in Figure 25.2 (Furterer, 2011). BA first includes understanding the customers and their needs and expectations. From a product or service design, we can capture the customer needs for a new product or service. Next, we capture and document the business strategies and goals, as well as the external and internal influencers on the business. The relationship between the business goals and the business capabilities that support the goals should be understood. We use the value chain showing a chain of activities that provide value to the customer (Porter, 1985) to understand and decompose the processes that help us to meet the new product or service designs. The functional decomposition provides a hierarchical organization of functions and the processes that they include. Each value chain and the subsequent business functions will be used to further decompose the processes. This ensures traceability from the value chains to business processes that provide customer value.

The functional decomposition could be used in product design to decompose and relate the conceptual design of customers' requirements to the technical product requirements that meet the customers' needs.

Business capabilities enable business functions. Although the way in which a business implements its processes is likely to change frequently, the basic capabilities of the business tend to remain constant. Business processes and their activities describe the sequence of activities that enable the business to meet the customer's expectations and provide value through value chains.

Business components are identified to optimize the activities that support the business. These components consist of the activities that require similar people, processes, and technologies. They allow the standardization of

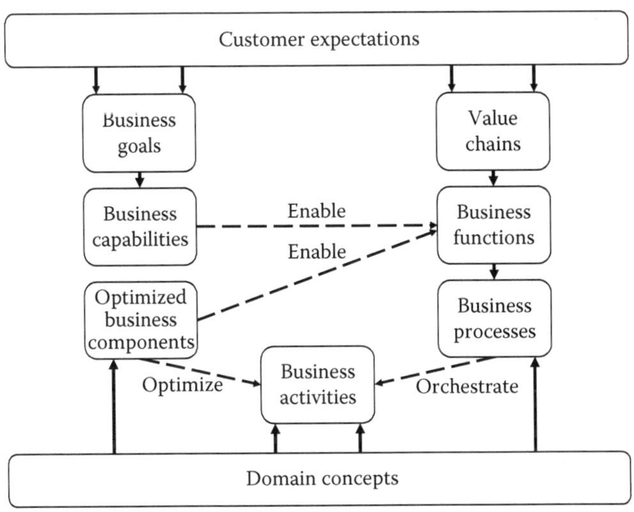

FIGURE 25.2
Business architecture elements.

the business processes by componentizing the activities that can be used in multiple areas of the business, across many business units and markets (IBM Corporation, 2005).

In product design, the business components could become product components or even bills of materials that meet the design.

Domain concepts describe the information and roles that exist in the business, which are part of the business processes. The domain conceptual model in product design describes the product specifications and identifies the materials, components, and specifications required in the design of a product.

Incorporating knowledge, methods, and tools from the Lean body of knowledge and enterprise business architecture can help us to simplify and deal with the growing complexities of product, service, and process design in the future. You can be part of the journey to enhance product, service, and process design through the application of these bodies of knowledge.

Engagement and Success Factors

There are several essential elements to becoming Lean and sustaining the Lean journey. There must be a clear, compelling, and urgent reason to change. Cross-functional leadership must proactively and visibly lead the organization through the change process. This means getting the right leaders to work together to develop a vision of what the organization needs to become and a strategy for getting there. The right leaders are those with enough power to lead the change throughout the organization. Leadership must continually communicate and role model the new vision and strategies. Leadership must break down barriers to make the necessary improvements.

In addition, leadership must engage the people closest to the top-priority problems, or the opportunities, to identify, design, develop, plan, and implement the improvements. It must leverage the successes and best practices for making improvements by eliminating waste in other areas. Leadership must help everyone in the organization understand the connection between the improvement activities, results, and the vision of the organization so that the new behaviors become part of the "way we engage our people and run our business."

Successful Lean implementation requires the total immersion of top management. Leaders must create an environment that allows members to participate in making the decisions that affect their work, voice honest opinions, and constructively criticize and challenge tradition. This may involve reorienting the organization and changing entrenched behavior. It is critical that the leadership communicates a clear vision, creates a sense of urgency emphasizes continual training, and stimulates workers and managers alike to engage in the kind of cooperative experimentation that is the cornerstone of a vital, learning organization.

Leadership must identify a vision, align employees at all levels of the organization to that vision, and motivate the employees to achieve that vision.

Change requires leadership. Key aspects of providing leadership include the following:

- Clear direction
- Focused goals and rationale
- Willingness to let people make mistakes
- Open communication
- Allow honest mistakes without blame or reprisals
- Give credit for their ideas, work, and successes

The elements required for change include the following:

- Change objectives
- Change vision
- Champion, sponsors, facilitators, and teams
- Training
- Systems that enable Lean
- Process performance metrics

As described in Chapter 22, Lean implementation requires careful planning with a structured approach. There are several implementation success factors. First, leadership must prepare and motivate people through widespread orientation to continuous improvement, quality, training, and recruit employees with the appropriate skills. A key element of this orientation is to create a common understanding of the need to change to Lean.

Leadership must also structure their Lean implementation with a focus on employee involvement. This is enabled by leadership pushing decision making and system development down to the lowest levels. It is also important to share information; manage expectations; and identify and empower champions, particularly operations managers. Another key aspect is to create an atmosphere of experimentation by tolerating mistakes and being patient. Leadership should install enlightened and realistic performance measures, evaluation, and reward systems and do away with rigid performance goals during implementation. The need to execute pilot projects prior to changing the culture across the organization is essential.

Leadership should also focus on providing clear and consistent channels of communication. The fact that management would like employees to know how the day ended induces employees to get involved in the big picture. This can be accomplished through a centrally located board with key metrics and three light poles with green, yellow, and red in production cells. Visual performance monitors are an important part of clear communication. Metrics should be published and displayed to be used as daily performance metrics through team or departmental Lean boards that allow the posting of metrics and enhance communication within subgroups. The chief executive officer and upper management should be actively and visibly involved in monitoring and updating the metrics.

Leadership should also address the issue of "What is in it for me?" Leadership should tackle the issues of compensation discrepancies. This involves cleaning out the skeletons in the corporate cupboard through clear and transparent communication. High priorities for leadership to address are: broken promises, unfair incentive plans to select groups, and lack of fairness, among others. Therefore, a clear definition of what's in it for me should emerge before beginning the Lean implementation since the way people are compensated and motivated directly affects how successful the implementation will be.

Part of a successful Lean implementation relies on the organizational framework. Identify champions from top management who will actively support the Lean implementers. Champions need to commit time to attend all Lean presentations and be ready to make quick decisions based on their understanding of the benefits that will accrue from implementation.

References

Bieberstein, N., Laird, R.G., Jones, K., and Mitra, T., *Executing SOA: A Practical Guide for the Service-oriented Architect*, IBM/Pearson, Upper Saddle River, NJ, 2008.

Furterer, S., *Systems Engineering Focus on Business Architecture: Models, Methods and Applications*, CRC Press, Boca Raton, FL, 2011.

Huthwaite, B., *The Lean Design Solution*, Institute for Lean Design, Mackinac Island, MI, 2004.

IBM Corporation, *Component Business Models: Making Specialization Real*, IBM Business Consulting Services, IBM Institute for Business Value, Somers, NY, 2005.

Kearney, A.T., The Line on Design: How to Reduce Material Cost By Eliminating Design Waste, AT Kearney Inc., Chicago, IL, 2003.

Mascitelli, R., *The Lean Design Guidebook*, Technology Perspectives, Northridge, CA, 2004.

Porter, M.E., *Competitive Advantage: Creating and Sustaining Superior Performance*. Free, New York, 1985.

Glossary

Additive Manufacturing: An emerging technology that "grows" files directly from CAD, thus eliminating the waste of tooling or extra material usage from machining. Additive manufacturing may be used to augment Lean production and prototyping efforts. Also known as rapid prototyping, solid freeform fabrication, additive fabrication, direct digital manufacturing, or direct manufacturing.

Andon: A line indicator light or board hung above the production line to act as a visual control. Andons are used to visually signal an abnormal condition.

Autonomation: Automation with a human touch or transferring human intelligence to a machine. This allows the machine to detect abnormalities or defects and stop the process when they are detected. Also known as Jidoka.

Bar Chart: A graphical method that depicts how data falls into different categories.

Batch-and-Queue: A mass production practice of producing lots and sending the batch to wait in a queue before the next operation.

Benchmarking: An activity to establish internal expectations for excellence, based on direct comparison to "best." In some cases, the best is not a direct competitor in your industry.

Black Belt: A person trained to execute critical projects for breakthrough improvements to enhance the bottom line.

Brainstorming: A technique to generate a large number of ideas in a short period of time.

Breakthrough Objectives: In policy deployment, those objectives characterized by multifunctional teamwork, significant change in the organization, significant competitive advantage, and a major stretch for the organization.

Cell: A logical and efficient grouping of machines or processes that enables one-piece flow.

Cellularization: Grouping machines or processes that are connected by work sequence in a pattern that supports flow production.

Cellular Manufacturing: Manufacturing with the use of cells. See Cell.

Chaku-Chaku: Japanese term for "Load-Load." It refers to a production line raised to a level of efficiency that allows the operator to simply load the part and move on to the next operation. No effort is expended on unloading. See Hanedashi.

Champion: An individual who acts as the sponsor or owner of a project and has the authority and responsibility to inform, support, and direct a team. Typically, the individual is a Director or VP-level manager. Also known as a mentor or sponsor.

Changeover: Altering a process to manufacture a different product.

Changeover time: As used in manufacturing, the time from when the last "good" piece comes off of a machine until the first "good" piece of the next product is made on that machine. Includes warm-up, first-piece inspection, and adjustments. Changeover times can be reduced through the use of single-minute exchange of dies (SMED).

Cost of Poor Quality (COPQ): Costs associated with not doing things right the first time. Examples of COPQ include scrap, rework, and waste.

Countermeasures: Immediate actions taken to bring performance that is tracking below expectations into the proper trend. Requires root cause analysis.

Curtain Effect: A method that permits the uninterrupted flow of production regardless of external process location or cycle time. Normally used when product must leave the cell for processing through equipment that cannot be put into the cell (i.e., heat treat, painting). Curtain quantities are calculated using the following formula:

$$\text{Curtain quantity} = \frac{\text{Per unit cycle time of curtain process}}{\text{Takt time}}$$

Customer: Anyone who uses or consumes a product or service. A customer can be internal or external to the provider.

Cycle Time: The time from the beginning of one operation in a process until it is complete.

Defect: A nonconformance in a product or service.

Design for Manufacture and Assembly (DFMA): A philosophy that strives to improve costs and employee safety by simplifying the manufacturing and assembly process through product design.

Deviation: The difference between an observed value and the mean of all observed values.

Downtime: Lost manufacturing time due to equipment, material, information, or manpower.

Economic Value Added (EVA): A residual income measure that subtracts the cost of capital from the net operating profits after taxes (NOPAT). It is the financial performance measure most closely linked to shareholder value and the cornerstone for a financial management and incentive compensation system that makes managers think and act like owners.

Emerging Technologies: Technologies on the forefront of development that offer the ability to significantly disrupt current supply chains and/or ways to produce goods or services.

Failure Mode and Effects Analysis (FMEA): A structured approach to assess the magnitude of potential failures and identify the sources of each potential failure. Corrective actions are then identified and implemented to prevent failure occurrence.

5S: A method of creating a self-sustaining culture that perpetuates an organized, clean, and efficient work place. Also referred to as the five pillars of the visual workplace.

Five Whys: A simple problem-solving method of analyzing a problem or issue by asking "Why?" five times. The root cause should become evident by continuing to ask why a situation exists.

Fixed Costs: Costs of production that does not change when the rate of output is altered.

Flexibility: The ability to respond to changes in demand, customer requirements, and so on.

Flow Chart: A pictorial representation of a process that illustrates the inputs, main steps, branches, and outcomes of a process. A problem-solving tool that illustrates a process. It can show the "as is" process or "should be" process for comparison and should make waste evident.

Flow Production: A philosophy that rejects batch, lot, or mass processing as wasteful. Product should move (flow) from operation to operation in the smallest increment, one piece being ideal. Product should be pulled from the preceding operation only as it is needed. Often referred to as "one-piece flow," only quality parts are allowed to move to the next operation.

Gage Capability Study: A method of collecting data to assess the variation in the measurement system and compare it to the total process variation.

Green Belt: An individual trained to assist a Black Belt. This individual may also undertake projects of a lesser scope than Black Belt projects.

Hanedashi: Device or means of automatic unload of the work piece from one operation or process, providing the proper state for the next work piece to be loaded. Automatic unloading and orientation for the next process is essential for a Chaku-Chaku line.

Heijunka: Production leveling process. This process attempts to minimize the impact of peaks and valleys in customer demand. It includes level production volume and level production variety.

Hoshin Action Plan: Form used by the team to detail specific activities required for success, milestones, responsibilities, and due dates.

Hoshin Implementation Plan: A form used to track performance (Plan vs. Actual) on policy deployment objectives. Usually reviewed with top management on a monthly basis, but reviewed by the policy deployment team more frequently.

Hoshin Kanri: A strategic decision-making tool that focuses resources on the critical initiatives to accomplish organizational objectives. This process links major objectives with specific support plans throughout the organization.

Hoshin Strategic Plan Summary: Form used to show relationships between 3- and 5-year objectives, improvement priorities, targets, resources required, and benefits to the organization.

Input: A resource that is consumed, utilized, or added during a process.

Jidoka: Automation with a human touch or transferring human intelligence to a machine. This allows the machine to detect abnormalities or defects and stop the process when they are detected. Also known as Autonomation.

Just-in-Time (JIT): A strategy that concentrates on delivering the right products in the right time at the right place. This strategy exposes waste and makes continuous improvement possible.

Kaikaku: Radical improvement to eliminate waste.

Kaizen: Japanese for continuous improvement. The term is composed of *kai,* meaning "to take apart," and *zen,* meaning "to make good." Based on the philosophy that what we do today should be better than yesterday and what we do tomorrow should be better than today, never resting or accepting status quo.

Kaizen Event: A planned and structured event to improve an aspect of a business.

Kanban: Japanese term meaning "signboard" or "signal." It is a means of communicating a need for products or services. It is generally used to trigger the movement of material where one-piece flow cannot be achieved, but is also used to signal upstream processes to produce product for downstream processes.

Keiretsu: A grouping of Japanese companies that allows each to maintain operational independence but establishes a permanent relationship with other members in the group.

Key Performance Indicator (KPI): A method for tracking or monitoring the progress of existing daily management systems.

Lead Time: The total time required to deliver an order to the customer.

Machine Cycle Time: The time from when the "start" button is pressed until the machine returns to the original starting position.

Manufacturing Lead Time: The total manufacturing time beginning from raw material to creating the final, saleable product.

Material Requirements Planning (MRP): A computerized system to determine the quantity and timing of material supplies based on a master production schedule, a bill of materials, and current inventories.

Metric: A performance measure that is linked to the goals and objectives of an organization.

Milk Run: A supply and/or delivery vehicle that is routed to various locations to pick up or deliver products and supplies. A milk run can be external to customers or suppliers. A milk run can also be internal to a factory with a material handler.

Mission: A statement of an organization's purpose.

Mixed Model: A value stream that accommodates multiple product models.

***Muda*:** Japanese for waste.

Multiskilled Worker: Associates at any level of the organization who are diverse in skills and training. They provide the organization with flexibility and grow in value over time. Essential for achieving maximum efficiencies for JIT.

***Mura*:** Japanese for unevenness.

***Muri*:** Japanese for unreasonableness.

Noise: Unexplained variability in a response.

Non-Value-Added (NVA): Those process steps that take time, resources, or space, but do not transform or shape the product or service toward that which is sold to a customer. These are activities that the customer would not be willing to pay for.

One-Piece Flow: A manufacturing process in which product moves one piece at a time through all necessary operations.

Operator Cycle Time: The time for an operator to complete one cycle of an operation. The total operator cycle time includes walking, loading, unloading, and inspection.

Output: A product or service delivered by a process.

Pacemaker: The point in the process where the customer's order enters the process.

Pareto Chart: A vertical bar graph for attribute or categorical data that shows the bars in descending order of significance, ordered from left to right. Helps to focus on the vital few problems rather than the trivial many. An extension of the Pareto principle suggests that the significant items in a given group normally constitute a relatively small portion of the items in the total group. Conversely, a majority of the items will be relatively minor in significance (i.e., the 80/20 rule).

PDCA Cycle: Plan–Do–Check–Act cycle. PDCA is a repeatable four-phase implementation strategy for process improvement. PDCA is an important item for control in policy deployment. Referred to as the Deming or Deming cycle.

Pilot Cell: An experimental exercise in a cell to determine the viability of a concept.

***Poka-Yoke*:** A Japanese expression meaning mistake proof. A method of designing production or administrative processes which will by their nature prevent errors. This may involve designing fixtures that will not accept an improperly loaded part.

Policy Deployment: See *Hoshin Kanri.*

Process: An activity that blends inputs to produce a product, provide a service, or perform a task.

Process Map: A visual representation of the sequential flow of a process. Used as a tool in problem solving, this technique makes opportunities for improvement apparent.

Production Smoothing: See Heijunka.

Productivity: Output per unit of input, for example, output per labor hour.

Pull: A system in which replenishment does not occur until a signal is received from a downstream customer.

Push: Conventional production where product is pushed through operations based on sales projections or material availability.

Quality Characteristic: An aspect of a product that is vital to its ability to perform its intended function.

Queue: Inventory authorized by a push signal.

Rework: An activity to correct defects produced by a process.

Root Cause: The ultimate reason for an event or condition.

***Sensei*:** A teacher with a mastery of a body of knowledge.

Setup Time: One of the three elements of changeover time. The other two elements are clean-up and start-up. Setup Time is the time required to convert and/or adjust the equipment for the next product.

***Shusa*:** A strong team leader in the Toyota product development system.

Sigma (σ): Standard deviation. A measure of the variation or dispersion from the average or expected value.

Sigma Capability: A measure of process capability that represents the number of standard deviations between the center of a process and the closest specification limit. See Sigma Level.

Sigma Level: A measure of process capability that represents the number of standard deviations between the center of a process and the closest specification limit. See Sigma Capability.

Signboard: English for the Japanese term kanban. See Kanban.

Single-Minute Exchange of Dies (SMED): Method of increasing the amount of productive time available for a piece of machinery by minimizing the time needed to change from one product to another. This greatly increases the flexibility of the operation and allows it to respond more quickly to the changes in demand. It also has the benefit of allowing an organization to greatly reduce the amount of inventory that it must carry because of the improved response time, while maximizing ROI and EVA.

Six Sigma: A quality improvement and business strategy that emphasizes impacting the bottom line by reducing defects, reducing cycle time, and reducing costs. Six Sigma began in the 1980s at Motorola.

Spaghetti Chart: A map that illustrates the path of a product as it travels through the value stream or the path a person must travel to complete their tasks.

Standard: A prescribed documented method or process that is sustainable, repeatable, and predictable.

Standard Deviation: A measure of variability in a data set. It is the square root of the variance. See Sigma.

Standardization: The system of documenting and updating procedures to make sure everyone knows clearly and simply what is expected of them. Essential for the application of the PDCA cycle.

Standard Work: A tool that defines the interaction of people and their environment when processing a product or service. It details the motion of the operator and the sequence of action. It provides a routine for consistency of an operation and a basis for improvement. Standard work has three central elements: takt time, standard work sequence, and standard work-in-process.

Standard Work-in-Process: The minimum amount of material for a given product which must be in process at any time to ensure proper flow of production.

Stretch Goal: A goal designed to create out-of-the box thinking for breakthrough improvement.

Supermarket: An inventory storage location authorized by a kanban pull system.

Supplier Partnership: An approach to business that involves close cooperation between the supplier and the customer. It provides benefits and responsibilities that each party must recognize and work together to realize.

Takt Time: The frequency with which the customer wants a product. How frequently a sold unit must be produced. The number is derived by dividing the available production time in a shift by the customer demand for the shift. Takt time is usually expressed in seconds.

Target Cost: The cost a product cannot exceed for the customer to be satisfied with the value of the product and for the manufacturer to obtain an acceptable return on investment.

Throughput Time: The total time for a product from concept to launch, order to delivery, or raw material to customer delivery.

Total Cost: The market value of all resources used to produce a good or service.

Total Productive Maintenance (TPM): Productive maintenance carried out by all employees. It is based on the principle that equipment improvement must involve everyone in the organization, from line operators to top management.

Total Revenue: The price of a product multiplied by the quantity sold in a given time period.

Total Utility: The amount of satisfaction obtained from the entire consumption of a product.

Toyota Production System (TPS): A manufacturing model built on reducing lot sizes to allow for flexibility, control of production parts, and the logical arrangement of production equipment.

U-Shaped Cell Layout: An optimized floor layout that efficiently combines worker flexibility and training with workstation positional layouts.

Utility: The pleasure of satisfaction obtained from a good or service.

Value: A capability provided to a customer for an appropriate price.

Value-Added: Any process or operation that shapes or transforms the product or service into a final form that the customer will purchase.

Value Stream: All activities required to design and produce a product from conception to launch, order to delivery, and raw materials to the customer.

Value Stream Mapping: A process for identifying all activities required to produce a product or product family. This is usually represented pictorially in a value stream map.

Variance: A measure of variability in a data set or population. Variance is equal to the squared value of standard deviation.

Variation: A process is said to exhibit variation or variability if there are changes or differences in the process.

Visual Control: Visual regulation of operations, tool placement, and so on, which provides a method for understanding a process at a glance.

Visual Management: Systems that enable anyone to immediately assess the current status of an operation or given process at a glance, regardless of their knowledge of the process.

Voice of the Customer (VOC): Desires and requirements of the customer at all levels, translated into real terms for consideration in the development of new products, services, and daily business conduct.

Waste: Also known as *Muda*. Any process or operation that adds cost or time and does not add value. Eight types of waste have been identified:

1. Waste from overproduction
2. Waste from waiting or idle time
3. Waste from unnecessary transportation
4. Waste from inefficient processes
5. Waste from unnecessary stock on hand
6. Waste of motion and efforts
7. Waste from producing defective goods
8. Waste from unused creativity

Work-in-Process: Material in the process of being converted into saleable goods.

Work Sequence: The specific order in which an operator performs the manual steps of the process.

Index

Q

R